14

MATHEMATICS WITH LOVE

MATHEMATICS WITH LOVE

The Courtship Correspondence of Barnes Wallis, Inventor
of the Bouncing Bomb

Mary Stopes-Roe

Macmillan

First published 2005 by
Macmillan
Houndmills, Basingstoke, Hampshire RG21 6XS and
175 Fifth Avenue, New York, N. Y. 10010
Companies and representatives throughout the world

Macmillan® is a registered trademark in the United States, United Kingdom and other countries.

ISBN–13: 978–1–4039–4498–6
ISBN–10: 1–4039–4498–9

This book is printed on paper suitable for recycling and made from fully managed and sustained forest sources.

A catalogue record for this book is available from the British Library.

A catalog record for this book is available from the Library of Congress.

10   9   8   7   6   5   4   3   2   1
14   13   12   11   10   09   08   07   06   05

Printed and bound in China

This story celebrates the delights of sending and receiving letters. It is dedicated to the memory of my parents, Barnes and Molly, who wrote the letters printed here; and to all those over the centuries who have left us their stories in this way.

"And oft the pangs of absence to remove
By letters, soft interpreters of love"
Matthew Prior 1684–1721 *Henry and Emma*

"She'll vish there wos more, and that's the great art o' letter writin'"
Charles Dickens 1812–1870. Sam Weller, *Pickwick Papers*

CONTENTS

ACKNOWLEDGEMENTS

My first debt is to my siblings Barnes, Elisabeth and Christopher, to whom, with myself, the letters belong.

I am grateful to the many friends who have patiently encouraged me with comment, suggestion and discussion. They have supported me in the moral dilemmas confronting those who read private material and make it public. In particular I thank Ray Brown who first introduced 'Barnes and Molly' to others in his radio play of that name.

I owe a debt to the Textbooks Colloquium who first helped me to view the mathematics lessons as an unusual text; and especially to the late John Fauvel of the Mathematics Department of the Open University, who assured me they were a correspondence course of serious teaching material.

My thanks to Professor Twyman of the Department of Typology and Graphic Communication at Reading University; and to Kevin Moran, a final year student, who set out my first draft in printed form.

My husband Harry has saved me with patience, kindness and experience from despair and defeat in the battle with the computer. My thanks to him are heartfelt. And I warmly appreciate and am grateful for the sensitive, friendly guidance and cooperation of my Editor Sara Abdulla, without whom I might well have fallen at the final hurdle.

NOTES ON THE LETTERS

There are more than 250 letters in this correspondence, without including the 152 pages of mathematics lessons. Some of the letters are mere notes, while some are 12 pages long. All are written in beautifully clear writing: Barnes's an elegant italic hand, Molly's a rounded and still rather girlish script. Each letter is still in its own separate envelope; therefore the whereabouts of sender and receiver, and dates of writing and posting, are recorded for the personal letters. The mathematics lessons were despatched under separate covers which have not survived, and the exact dating of these is certain for only a few. Handling these documents is a privilege, a step into the lives, situations, pressures of time and emotion, of two private people. Each letter, after 80 years carefully folded, refuses to give up its creases.

As children we knew the story of the mathematics lessons, and how they were sent by our father to our mother as he was courting her; but we had no idea that they still existed. In 1979, as I cleared out our family home, I found them tightly packed in an old-fashioned dress-box at the top of a cupboard.

There are vagaries of spelling and punctuation, particularly on Barnes's part, which I have carefully kept. The volume of text is considerable and I have omitted much of the more descriptive and conversational exchanges, retaining the passages that more clearly show the personalities and emotions of the two. I have had to be brief with explanations of the characters in their family and social lives, trusting that sufficient details can be derived from what they themselves say in their letters.

All the manuscripts used here belong to Barnes and Molly's four children. My siblings have generously and kindly allowed me to delve into the letters, private though they are. Some of the characteristics which the writers describe in themselves are different from those appearing later in the well-known public figure and his supportive wife; I sincerely hope that I have done them justice. I have certainly written with increasing love and respect.

There is a considerable archive of Barnes Wallis material which is not held by the family. It is lodged in the Science Museum; Churchill College, Cambridge; the Barnes Wallis Memorial Trust at Elvington, York; the R.A.F. Museum, Hendon; and the Imperial War Museum. The family archive will finally be deposited in the British Library.

PROLOGUE

In the dog days of summer, 1962, a 74-year-old man lay in a London hospital bed. He was bored and in some discomfort after a prostate operation; but instead of trying to sleep, read or use his earphones, he amused himself by writing to his daughter:

Great-grandfather   died aged 72 years in about 1860–70
Grandfather   d.   "   81   "   "   "   1906
Father   d.   "   86   "   "   "   1945

Now it is not really possible to apply mathematical laws to an <u>individual</u>, but only to very large numbers of individuals as is done by insurance companies, etc., but it is amusing, if no more, to take advantage of the singular regularity of advance shown above, to see how old I should be, assuming that mathematical laws <u>did</u> apply to an individual. If we neglect end effects, as we must do, since historically they are buried in prehistoric times, and for the future are unpredictable for a variety of reasons, then we can treat the above sequence as a very small part of a curve that extends, virtually, to infinity in both directions.

The simplest form is parabolic, and the curve that fits the 3 known points reasonably well is $N = 72x^{.166}$ where N is age at death, and x is the number of the generation.... On this basis I should have lived to just over 90, but now I have had an op. that I am assured will add 10 years to my natural term!! Estimated age at death = 100 years. What a dreadful thought. Of course its all nonsense anyway, because one cannot just deal in generations – but it has passed a tedious hour in hospital..."

Seventeen years later, at a memorial gathering soon after the patient's death at 92 (not far off his original estimate), his eminent surgeon was one of the speakers. Very many patients, a large number of them notable people, had passed through his clinic in the meantime; but the description of symptoms by this particular one had stuck in the surgeon's mind. The patient described his complaint in unusual terms: he could not pass water in the customary "beautiful parabolic arc". Such ways of thinking, however, came naturally to this man.

The patient was Barnes Wallis, an engineer, designer and inventor of considerable note. He is most widely remembered as the inventor of the bomb which breached the dams of the Ruhr valley in Germany in the Second World War; but this was only one of his achievements and interests. These ranged over airships, heavier-than-air-craft (most notably the Wellington and the trophy-winning Wellesley), variable geometry (or swing-wing) aircraft, submarines, bridge design, telescope construction, medical callipers, racing skiffs, school buildings and furniture, household gadgets, and a technique for accurate wood carving. He was creative for three quarters of a century, and was always imparting knowledge and arousing appreciation for mathematics and good design. Whenever an interested listener was available he launched into explanations and illustrations, using paper napkins, menus, programmes, envelopes – whatever came to hand. He was a teacher who followed his own methods, and paid no heed to the rules. He held just one brief teaching post in his long life; but throughout this life he inspired insight and understanding in many people by the dedication, the enthusiasm, the delight which he brought to his explanations.

Chapter 1
BEGINNINGS

In the autumn of 1922 Barnes had been teaching for just two months when he wrote his first lesson in mathematics to his cousin Molly Bloxam. The two of them met for the first time in April of that year, when Molly was 17 and in her last year at school, and Barnes was 35. He had not only lived twice as long as Molly, but his years had been full of experience, more hard and turbulent than happy.

Barnes, having gained a Foundation Scholarship, boarded from the age of 12 at the ancient public school Christ's Hospital; but by his own wish and determination and in spite of discouragement from his family and teachers, in 1904 he left at the age of 16. From his childhood he had been inventive, and interested in mechanics and machines; but the public schools of the time did not prepare their boys for a career in engineering. Barnes was convinced that the most appropriate way to learn was on the job, and apprenticed himself to a shipbuilding firm on the Isle of Wight. He learnt the hard way, living in cheap lodgings and working his way up from the shop floor to the drawing board. In 1913 he was appointed one of the chief designers in the airship development programme of the great engineering company Vickers.

Barnes came from a close and loving family in which he was the second child, with an older and a younger brother and one sister. His father was a doctor in East London, dedicated to his work, loved and valued by his patients; but he had been crippled by poliomyelitis caught from one of them, dragging round a lame leg for the rest of his life. He found it hard to maintain his practice. The family was thus permanently impecunious. Barnes saw his mother struggle with ill health to cope with the heavy domestic chores of the time, maintaining the respectability necessary for a doctor's household. He understood her daily, heartfelt prayers and her dutiful shouldering of the burdens laid upon her. He was her favourite child, her close companion and emotional support. During his school years, and the following apprenticeship away from home, the two exchanged more than a thousand letters. His mother shared with him every detail of her daily life. The adolescent Barnes was less communicative and her many questions were not always answered, but he never failed to write. He planned to relieve her burdens by earning a living as soon as possible, choosing a hard course out in the world; but her early death in 1911 ended his hope, and the failure haunted him. From now onwards he must go on alone; the constant, loving, anxious, intruding, advice was ended, and he withdrew into himself. His mother stood on a pedestal, and for years no other woman entered his heart. His emotional distress was eased only by the comradeship of colleagues at work, and in his sporadic experiences of military life. After eleven years he was pulled, without any expectation that this might happen, from such desolation. A second great love took over his life.

Barnes's father Charles was also bereft by his wife's death. Five years later, in 1916, he married one of his late wife's girlhood friends, Fanny Bloxam. In 1922, with the war safely in the past, Fanny's brother Arthur Bloxam brought his family from the country back to London. His two eldest daughters Barbara – 'Baba' – and Molly were about to enter University College London. The two girls were glad to visit their Aunt Fanny, but apprehensive about meeting an unknown elderly male cousin. Would he be superior, fashionable, condescending? Neither had any experience whatever of male company other than their father's and the occasional visits of cousins. Like Barnes, Molly was the second child in a warm and loving family. Below her came Betty, Nancy, Pam and, nearly ten years younger than herself, George. Her family, so far, had been all to her, but it was also somewhat exclusive and smothering. The hub of this cheerful activity, security and comfort was her father to whom, as his favourite child, she was very close. Their mother Winifred Bloxam loved her children warmly. She was caring and dutiful, a gifted musician with no time, opportunity or encouragement to express her talents; but she was also eccentric and neurotic. A beloved nanny, Nan to two generations of the family, nurtured them with daily care and affection, and a structured nursery rule. Far from venturing out on to the troubled and uncharted waters of the world as Barnes had done, Molly had scarcely been allowed to dabble her fingers.

Thus in April 1922 the two met: Barnes with years of hard experience, and highly respected in his profession, but without any close female contact since his mother died; Molly half his age, and without any knowledge of work or male company or indeed of anything beyond the family fence, and more naive than was usual for a well-brought up girl. Her innocence and straightforwardness appealed to Barnes; and the attention of one so mature and clever intrigued Molly.

When they met, Barnes was in fact out of work. The closing of airship production had made him redundant. He had turned this setback into advantage, and began working for a London External B.Sc. Molly was busy with the final school exams which would gain her entrance to University College London. In the six months since meeting in April, a sporadic correspondence had grown between the two. Then Barnes's circumstances suddenly changed. In some anxiety about work, he had applied to an agency for a teaching job, and was surprised to be offered one in a young gentlemen's academy in Chillon, Switzerland. He, who had left school 18 years earlier, and whose subsequent learning had all been self-motivated and by correspondence, took on the post and left for Chillon in September of 1922. He was one of the few of his generation still alive who had never been abroad, since his war years had been spent in a reserved occupation. In early August 1914, full of the zeal that swept the country, he had enlisted but was immediately called back to the drawing board on the airship development programme. Twice more he enlisted, and to his disappointment, was recalled. He was intensely patriotic, believing sincerely throughout his life in the rightness of the allied cause and in particular the preservation of the British ethos and way of life. His feelings as he crossed the Channel were a mixture of anticipation and apprehension coloured by homesickness and a deep regret that his burgeoning friendship with Molly would be disrupted.

Being brought up in a religious family still attached to Victorian manners and ideals, he was not affected by the loosening of codes of conduct of the 1920s. He had thus intended to get Arthur Bloxam's permission to continue to write to his daughter Molly. His stepmother Fanny had arranged that he and she would visit the Hampstead family for Barnes to say his farewells, but for some reason Fanny could not go. Barnes, longing to say a personal farewell to Molly, battled with his besetting diffidence and went on his own.

Barnes: 26th Sept. 1922. In the Dover train.
I did call, but rather late – my business took me rather longer than I had expected. I easily identified the house, tho' I walked past it the first time thinking it must be the Vicarage for St Lukes Church – "Vicars Moor" – the place where they moored the Vicar, a little unusual, but still quite explicit. Then I saw 25, followed by 23, so by mathematical induction

concluded that "Vicars Moor" was 27. So I went in a rang and knocked – and waited. Then my courage failed, no one came to the door, I had time to think that perhaps you were all busy and I should only be in the way so I simply bolted, and didnt stop till I got to Charing Cross, where I treated myself to a lonely and miserable cup of tea. I'm awfully sorry Molly. I <u>did</u> want to see you all so very much, but having to wait gave me complete cold feet. I dont suppose you will understand, but that is really and truly what happened.

♦ The haste of his unexpected departure left him no time to visit again, and he had to leave Fanny to intercede for him with her brother. Mr Bloxam conceded that Barnes could write as often as he wished, but Fanny's remarks raised some doubts in his mind. There was to be no 'nonsense', only the chat and news which might safely be exchanged between two pen-friends, nothing that might pose a threat to Molly's life at College and her open-minded contact with men her own age. Two weeks later having pondered the outcome, Fanny wrote to the hesitant suitor, whom she well knew to be diffident and lacking in self-confidence. There was no reason why he should not write in friendship to Molly, so that they could get better acquainted. Barnes was a lonely man, and in so far as one with a warm and dependent heart can endure loneliness, he was accustomed to endure it. But putting many miles of land and water between himself and Molly sharpened the fear of loss. Letters could blunt this and, encouraged by Fanny's report, to letters he turned.

Barnes to Molly: 1st Oct. 1922. Chillon.
The Swiss train left Gare de Lyons at 9 pm Tuesday and I got here 9.10 am Wednesday. I spent the most appalling night Ive ever had in my life. By the way I had a jolly good crossing and made a hearty lunch on board, tho' it was pretty rolly. I had a carriage with 2 fat French men and 2 fat French ladies, and they hermetically sealed the carriage and turned on the heat! I had no sleep whatever, and felt simply done up next day. I'm feeling fairly happy now, but was very miserable the first 3 days. Ive never been home-sick since I first went away to school, when I was 12, and thought I was beyond such weakness, but I <u>have</u> been so wretched...

This is a most gorgeous place. There are no words. I am now (Sunday afternoon) sitting out on the balcony of our common room in blazing sun-shine, dressed only in tennis flannels and perspiring freely, while towering all round are mountains covered with snow and above that the most priceless blue sky.... My bedroom window looks out over Chillon Castle, about which Byron wrote a poem (which I have never read).

Right under my window is a vineyard! The whole mountain sides are stiff with vineyards. They have scare crows with old bits of tin plate hung on them and when the wind he blow, tin plate he go cling-cling. This goes on all night. Then every cow has a huge leather collar and a large and very mellow toned bell. So that:-

1) Tin plate he go cling cling
2) Cow he go clong clong (by the hundred)
3) Lake he go lap-lap
4) Trees they go rustle rustle
5) Mountain stream he go tinkle tinkle

All together, some cacophony. Still I manage to sleep

I like the Head _immensely_. He's the sort of man you fall down and worship at once. He was head of Dover Coll. and chucked it to go as Chaplain to the Guards at the beginning of the war, being subsequently invalided out.

I suppose by the time you get this you will be just starting at London University. Do tell me all about it. Are you taking Science? and if so what subjects? I'm most awfully interested.

Molly: 8th Oct. 1922. Hampstead.

The first two or three days were horrid because I was continually getting lost. Its such an enormous place; but I like it awfully now. There was a "freshers" social on Friday, which was great fun. I actually had the temerity to try to dance a fox-trot, and as my partner was very good, we got on fairly well, but I always have to be warned if anything unusual is about to happen.

There are some awfully nice men and women in our classes. There are two funny little Japanese men, who always go about together; and there are three black men with very tight tiny curls. There's one most learned youth, who knows all about everything, with great big spectacles, and there's one who holds his pen in such a queer way that it quite fascinates me; and there's one who limps and is awfully decent. There is the prettiest girl I have ever seen in all my life. Don't you love looking at pretty people? I wish it wasn't rude to stare. In fact heaps of them are so pretty that I just love sitting and watching them.

I am taking science – Botany, Chemistry, Zoology, and Physics. I can manage the first three fairly well, but physics is positively hopeless. The old boy dashes along at such a rate that I can't possibly keep up with him, and the terms he uses are all so much Greek to me. I've never done any before, and when I read by myself, its dreadfully muddling, but I suppose it'll get better soon.

Barnes: 11th Oct. 1922. Chillon.
Thank you very much indeed for your perfectly ripping letter. You must have spent hours of your all too scarce leisure, and I have loved reading it, and have re-read it about 20 times. I will take you at your word, and go on writing, for the next best thing to getting a letter from you, is to write one to you. For somehow, out here, where one is sometimes somewhat lonely the mere fact of sitting down to write gives a feeling of company. It was very nice of you to be so understanding about that Monday [his abortive farewell visit]. Of course the family were in no way responsible, it was all my own fault for being so wretchedly shy. Do you remember how I funked you and Baba when you first came to New Cross?

I am most awfully interested in your science.... I cannot see how you can do physics and not maths. You cannot get really far in physics without a good groundwork of maths to help you.... I suppose it is silly of me to think that I might be of any help by writing to explain things to you? Would you let me try? You see I've had so much difficulty with things myself, as I have had to learn all I know without any tutor, that I can often see the hard parts better than other people.

I wonder what your lecturer has started on. I have had to start a physics class here from the beginning, and after a general explanation of the ground covered by physics, I had [the boys] up to units, fundamental and derived, and so to the idea of dimensional equations, which I think helps one to realise what one is measuring and observing. I find dimensional equations a terrific help myself. I mean this sort of thing:

$$V = [L^3][T^0][M^0]$$

where V stands for Volume, L for unit of length, T unit of time and M unit of mass. Then when you come to deal with forces and accelerations you can set them down like that, and see just what you are doing. Please let me try?

Chemistry I dont know much about, having never done any since I left school.

If you <u>do</u> think I could help just send a postcard and I will write by return.

I've been getting so cross with some of my people – I thumped a desk today. People seem so <u>stupid</u> over maths. I dont mind how much explaining

I do, or what pains I take to make them understand, but inattention and wilful stupidity I cannot tolerate.

Dont forget about the physics. If you would only send a <u>card</u> to say what you have difficulty on, I should so love to try to help.

Molly: 18th Oct. 1922. Hampstead. On a crowded postcard.
Thank you most awfully, Barnes, for offering to help me. I'm afraid the matter is that I am so terribly stupid. Our lecturer goes so fast that I simply can't follow him. He started by telling us what displacements, all the velocities and acceleration etc are. He uses so many letters and there are so many equations that I get hopelessly muddled. I am reading very slowly straight through our mechanics text book, at the rate of about 2 pages a day; and at present it is quite understandable, if I can only get it to stick in my head. I've never heard of fundamental and derived units, or of dimensional equations. If there is anything which is hopelessly muddling, I will let you know. Thank you very, very much.

I'm afraid you'd thump the desk pretty hard if I were in your class.

Molly: 22nd Oct. 1922. Hampstead.
Once again thank you ever so much for your offer of help. As I told you, I think at present anyhow, it is merely a matter of reading the book. The lectures are scarcely any use at all, except when he does experiments, because he goes so fast. I suppose when I can remember all the different formulae and how they are used, it will be all right. I wish they had taught us physics at school.

Yes I do remember how you funked me and Baba, and how you didn't come in till very late – after we'd gone to bed on Saturday. What did you think we'd be like? I know I funked you too – I was dreadfully afraid you'd be horrid and superior, and altogether hopelessly lazy and fashionable. Instead of which you turned out to be as decent as possible.

Barnes: 25th Oct. 1922. Chillon
About the physics – dont bother about fundamental and derived units and dimensional equations. Sorry I mentioned them if your prof. is not doing them. They help me personally – but it is not necessary to know them.

You will often find that in any one group of things there is a key formula from which all the others may be readily derived.

If only you knew the elements of Calculus, all these things become so simple. Take however as an example:

$$s = ut + \tfrac{1}{2}at^2.$$

where s = displacement or distance
u = initial velocity
a = acceleration
t = time

Now from this one "fundamental formula" you can get all the others connected with velocities, accelerations, and time, so you only have to remember the one.

See: if u is 0 (i.e. the body has no initial velocity) then $s = \tfrac{1}{2}at^2$.

This is the equation for all things connected with free falls under gravity when we write g for a (g you know is 32.2 ft/sec² at London).

Again, given u, s, and a you can write $\tfrac{1}{2}at^2 + ut - s = 0$, whence $t = -u/s \pm u^2 + 2as/a$ from the theory of quadratic equations.

Then suppose you want the final velocity you simply differentiate a with regard to t so:-

$$ds \,/\, dt = u + at$$

and again for acceleration differentiate again and

$$d^2s \,/\, dt^2 = a$$

Again for uniform acceleration you know that final velocity as above = $u + at$ = initial velocity + (acc × time)

so that mean vel^y = $\tfrac{1}{2}$\{initial + max\}
$= \tfrac{1}{2}\{u + (u + at)\}$
$= \tfrac{1}{2}\{2u + at\}$
$= u + \tfrac{1}{2}at$

Now $s = t$ × mean vel
$= t \times \{u + \tfrac{1}{2}at\}$
$= ut + \tfrac{1}{2}at^2$ as before.

If you haven't done this, don't bother to read it, and don't worry to understand it. One thing I do advise, and that is to learn the solution of a quadratic equation by heart – not understanding it if necessary – you can often apply correctly things you do not understand fully. The solution is this

if say $ax^2 + bx + c = 0$
then $x = -b/2a \pm \sqrt{(b^2 - 4ac)}/2a$

You see you can write down at once the solutions of an eqn such as $\frac{1}{2}at^2$ + $ut - s = 0$ simply by saying to yourself $a = \frac{1}{2}a$, $b = u$ and $c = -s$

$$\text{whence } t = -u/2 \times \tfrac{1}{2}a \pm \sqrt{(u^2 - 4 \cdot \tfrac{1}{2}a \times -s)/2 \times \tfrac{1}{2}a}$$

$$= u/a \pm \sqrt{(u^2 + 2as)/a} \text{ as before.}$$

Then if the problem states values for u, a and s you can immediately find t. This is awfully useful in exams.

Oh dear, am I getting to be too like a schoolmaster Molly? I dont mean it, indeed I dont. Only I am so keen on maths and on helping you if I can but I daresay I am only making things worse. Anyhow I'm not one of those awful people who always expect you to take their advice. Please dont think that. What suits me may not help you in the least. All I do is to try to lay my experience at your feet, for you to take or leave as you please.

Another fundamental formula is $f = ma$

where f = force in poundals
$\quad m$ = mass in lbs
$\quad a$ = acceleration

according to whichever set of symbols you use. Suppose from this you want to obtain the formula for Kinetic energy. Taking $S = ut + \frac{1}{2}at^2$ and starting from rest,

so that $S = \frac{1}{2}at^2$

then work done on body = force × distance

$$= f \times S$$

$$= \text{kinetic energy}$$

but $f = ma$

therefore $f \times S = ma \times S$, but $S = \frac{1}{2}at^2$
therefore $f \times S = ma \times \frac{1}{2}at^2 = m\frac{1}{2}.a^2t^2$

But $at = v$ (final vel^y for body starting from rest = acc × time)

therefore $a^2t^2 = v^2$

therefore $f \times S = \frac{1}{2}mv^2$

All done you see by remembering only two formulae – and there are several more wh. could be got.

What did I expect you to be like? Honestly I hadnt formed any mental picture at all. The mere fact that you were a girl was enough to frighten me when I heard you were coming. And how could I ever dream you could be as you are?

I love your expression "as decent as possible". Poor fellow, he has no doubt made the best of a bad job!

Molly: 29th Oct. 1922. Hampstead
How absolutely topping it is of you to trouble to help me with the physics. It is a relief to know that one needn't remember all the formulae, and to know the really important ones. We had the solution for a quadratic equation drummed into us at school, and I don't think I can possibly forget that. The only thing I didn't understand was about the differentiating. How can you say that $ds/dt = u + at$, and why for acceleration do you say d^2s/dt^2 instead of squaring both d's or s and t? I've never heard the expression "differentiate" before, and I don't know what it means.

There was a Science Society Social last Friday. It was great fun. We had dancing afterwards and it was simply too funny for words. First of all I danced with a youth who was even worse than I, so after about two rounds we stopped and I made him dance with a very tall girl who was quite capable of pulling him round. Afterwards I got on fairly well, but I always had to tell every partner to warn me when he was going to do a different sort of step, and after that we got on very well.

It has been raining all afternoon so there was no netball, and I came home early. George and I had tea up in Pam's room. We had an awfully cosy time, and toasted bread and had hot buttered toast. Don't you love it? Then we had tea by the light of the fire alone. It was very nice except that George emerged into the light of day (or rather electric light) simply smothered in jam. He also put his foot into the butter because there wasn't room for it on the table and it was on the floor. However there was very little of it, and he ate it so it was alright.

Barnes, you know really what I meant by "as decent as possible". I didn't mean anything about making the best of a bad job. I meant as decent as it was possible for any human being to be.

Barnes: 6th Nov. 1922. Chillon
Yes, I did know, or at least guessed, which meaning was to be attached to your phrase "as decent as possible". I dont know what to answer Molly – I've been stuck here for over a quarter of an hour, from which you will conclude that there isnt any answer.

[Next Day]. The only answer I can give, Molly, is this, that what you say is very far from being the truth, but if you chose to believe it of a man, he would make the most tremendous efforts to live up to your ideal, for <u>your</u> sake.

My lecture on Airships came off last Thursday. I spoke for nearly $2\frac{1}{2}$ hours and everyone was kind enough to say it was a great success. We had to hire a lantern, and some of the boys were so keen, that they came to me yesterday to ask whether I would give a further lecture if they raised a sub-scription amongst themselves to pay for the lantern again! I dont think they could have paid me a better compliment do you? Personally I hope the Head wont be able to find a vacant date, as one could hardly repeat such a success, and I should hate it to fall flat!

I had another curious compliment too, Mrs Lushington [Headmaster's wife] came up directly after and said "you've got such a wonderful voice". I cant say I was aware of it, but all she meant I think was that I could go on and on and on without the slightest falter or sign of fatigue....

The Head has asked me to stay on here as a master as long as I possibly could. I told him that I was really an engineer, and that when engineering revived in England, especially Airships I must go back and I had only come as a schoolmaster here to see something of the world and to learn French, and to amuse myself generally. But I said I would [stay] for a bit, Molly. What do <u>you</u> think? He said some very nice things. Altogether it pleased me very much as at the moment I was feeling a bit downhearted, and it is some consolation to realise that one has "made good" at a totally new profession. .... Dont you think that is the best? You see all my heart is in Airships, and I <u>have</u> worked so hard.

What a task you have set me – to explain the Calculus by post. But how very gladly I will do it if only you can put up with me! Its most awfully simple really – for some things. And fortunately it is only the really easy things that you will want.

Now here begins lecture one, from me Barnes to you Molly, on the very delightful subject of the calculus. In the first place, perhaps it is best to say that I do not propose to give the full mathematical treatment of the sub-ject – I do not see that that is at all necessary (also it is rather difficult). The calculus is a very beautiful and simple means of performing calcula-tions, which either cannot be done at all in any other way, or else can only be performed by very clumsy, roundabout and approximate methods. No one would suggest that you must be able personally to manufacture a needle, before you are to be allowed to sew, or that I must be able to make a watch, before I am allowed to tell the time, – these things are tools, or instruments, put at our disposal by the accumulated experience of our forefathers, and we are quite justified in making such use of them as our skill and ingenuity can contrive. So with the calculus – it is a mental tool left at your disposal by the great mathematicians of the past. (I mean that you need not worry because you may not fully grasp the fundamental principles on which it is based, any more than you will worry because you cannot make a needle!).

The calculus was first discovered – I think one can almost say discovered – perhaps formulated would be a juster term, by Newton in 1667 I think. Simultaneously by Leibnitz in Germany. On looking up my notes here, I see that Leibnitz published his first statement in 1684, while Newton had the idea long before that. Each accused the other of plagiarism, but historians now think that there is little doubt they may be regarded as simultaneous discoverers. Simultaneous discoveries in science are not rare things – they are usually the result of the accumulation of perhaps hundreds of years of thought, and two great investigators arrive at the same result at the same time.

♦ In her next letter on the 16th November, Molly quite unintentionally poured cold water over Barnes's efforts. He had already started on the next chapter explaining the orders of magnitude, but school affairs interrupted him, and the letters crossed in the post. Receiving Molly's before he had sent his own letter, Barnes concluded that his explanations were not appreciated. He dispiritedly crossed out the two pages; but he had no time to write another letter and the lesson has survived.

Molly: 16th Nov. 1922. Hampstead.
Of course I don't want you to explain the Calculus. I didn't know that differentiate had anything to do with the differential Calculus.

We had an awfully interesting debate at Coll. yesterday. The motion was that "Women should not continue their professional careers after marriage". I was for it. There were some awfully good speakers on both sides, and it was only carried by a very small majority. One American girl, who was very much against it, said that she knew a mother with a baby boy of two, and she taught all day at school, and saw her child only on Saturdays and a little while before he went to bed. The speaker seemed to think that that was quite enough. I can't understand a mother who would <u>want</u> to go out and leave her children all day, when she didn't have to. Are you in favour of the motion? But perhaps you never thought anything about it either way.

Barnes: 16th Nov. 1922. Posted 25th. Chillon.
I am sorry this letter has been so long delayed, it will not get to you till next week now. Shall I go on with the Calculus? Poor Molly you haven't a chance to say 'no'. Well, first I must talk a bit about "small quantities". Mathematicians have a queer way of speaking about small quantities. They say 'quantities of the 1st, 2nd, 3rd &c "order of magnitude", when they really mean quantities of the 1st, 2nd, 3rd order of smallness. Thus a star of the 2nd order of magnitude is a size smaller than a star of the 1st order. Funny people, aren't they?

Now I dont know how they define the orders of magnitude in astronomy, but in mathematics it is done like this:- to take an example, in the time I think of Elizabeth, people began to find that the subdivision of the hour into quarters of an hour, was not quite minute enough to suit their purposes. People began to get busier, and the time of appointments for merchants had to be made with more exactitude. Clocks in those days had I believe only the one hand – the hour hand, and each of the 12 divisions marking the hours was divided into 4 parts, not 5 as we have them now. Then the hour hand, as it moved from one hour to the next, moved over these 4 subdivisions, thus telling the quarter and half hours, and this was as near as people could get. So they went to their mathematicians and said they wanted something better, and the mathematicians said "very good, we will divide the hour very minutely. We will divide it into sixty minute (with the accent on the -ute) parts; and further we will give you another hand, geared so that it rotates once every hour, and we will call it the minute hand. Then you only have to divide the hour spaces into 5 instead of 4 parts, and the whole thing will be complete". So they did, and it was so; and because the English always have a tendency to throw the accent in a polysyllabic word as far forward as possible, it was not long before the new divisions came to be called "minutes".

This contented them for quite a long time, until – again I am uncertain, but I think Newton, found that in his scientific experiments even the minute was not minute enough; so he established a subdivision of the "second order of magnitude", and following the usual mathematical procedure in such cases, since the minute was one sixtieth of an hour, he made the new division one sixtieth of a minute. And since it was of the second order, it very soon became generally known as the "second" for short. So you see a "second" is a 1/60th of 1/60; while a "3rd" would be 1/60 of 1/60 of 1/60; and so on. The fraction need not be 1/60 – it could be 1/100 or 1/1000 or anything – the point is that the second order becomes

1/100 of 1/100 or 1/1000 of 1/1000, and the 3rd order becomes 1/100 of 1/ 100 of 1/100 or 1/1000 of 1/1000 of 1/1000, etc, etc, so that you see that the orders of magnitude in mathematics diminish very quickly....

You must have had a most interesting debate, tho' personally I cannot see what valid arguments could possibly be put forward by the opposition. Marriage is a career, the greatest and most wonderful in the world. Fancy watching your sons ripening into manhood, moulding their characters, forming their views; your sons, your very very own. I dont mean you personally of course, I mean any Mother's. Surely the career of careers; and what can be happier than a large and well managed family, with a devoted mother. No, I fear I am not in sympathy with the opposition.

Oh Molly this is such a miserable letter, and was going to be such a nice one. I've crossed out the bits about the calculus, it was written before I got your letter, and I haven't heart to rewrite it all. Of course I must bore you to tears with my silly old maths, when you told me you weren't interested in them.

Molly: 2nd Dec. 1922. Hampstead.
Barnes, why did you say that about the maths? They are not old or silly, and I am most certainly not the least little bit bored by them – would you be if somebody were explaining them to you as rippingly as you are explaining them to me? It isn't that I'm not interested in maths, but the point is that I'm so stupid that unless everything is put most awfully clearly, I can't understand it. Our Maths mistress at school was very clever, but she used to go so fast that I could never keep up with her, and I used to have to go over everything again and again at home. That's what I try to do now, but there is so very little time because we have work set for us which has to be given in, and that takes me nearly all my time. Absolutely the only reason why I asked you not to explain the calculus was because I thought that you hadn't time. You see you had said how dreadfully busy you were, and I thought that you must be tired of maths and physics by the end of the day; it most certainly wasn't because I didn't want it. Do you remember when you told me about $s = ut + \frac{1}{2}at^2$? That was most awfully useful, and has helped me a lot. So please don't think I don't like the maths and don't want them explained; it's only that I'm afraid it must be so dreadful for you to have to bother to explain them to me. Thank you very much indeed for what you have done, and I'd love you to go on if you can possibly spare the time....

Of course I did read all that about the calculus. Did you really think I wouldn't? I had no idea that that was the way they got seconds and minutes. I thought that the minutes and seconds came all at once when they first had clocks. I can just imagine those merchants and people coming to the decision that they must divide the clock into smaller divisions than quarters of an hour. What funny slow old times those must have been.

Barnes: 6th Dec. 1922. Chillon.
I think you are very clever, Molly, and I am sure I am right. One doesn't see maths straight off. It is very easy to appear to be clever at maths, when one knows them – I can impress my boys quite simply, by doing something rather quickly which they do not understand. But that is not being clever, its simply showing off and I try to avoid it. If I could solve quickly and easily something which I had never seen or done before – I mean something very tricky in advanced calculus, or something of that sort – then I might think I was clever. But I know I jolly well couldn't, so I dont pretend I am. I think in one way I am like you – we are both "hard witted". And it is because I myself have had such a struggle, and found things so very hard that I am able to teach as well as I do – (or so they are kind enough to say). But dont make the mistake of accusing yourself of being stupid, because you then start by thinking things are difficult, which I find makes them hard, even if they are quite simple.

I've often been astonished to find that really I understood a thing long ago, only I had been told it was hard, and immediately concluded that I couldn't really understand it, and kept on thinking there must be something much more abstruse in it than what I could see.

♦ In two more weeks term would end for both Barnes and Molly. Barnes looked forward eagerly to seeing Molly: Molly only said she looked forward to Christmas, but she did ask whether the Wallis family would visit for the day. In the meantime there was much activity. Barnes's letters were filled with winter sports, and his extra-curricula activities. He described his impressive lecture to the assembled College on airships, with diagrams of the illustrative models, the 'taking gadgets' which he had devised. The success of these prompted the Headmaster to ask for some more 'taking gadgets' for his own lecture on astronomy. From Barnes's descriptions and diagrams, these were indeed clever, and led to a request to manufacture sunsets and firelight for the end-of-term dramatics. He responded, as he did throughout life, with versatility and ingeniousness. Molly responded with happy stories of family goings-on and College Socials and dances. Worries about lectures and learning were put aside.

Barnes arrived home exhausted after a long journey and a rough Channel crossing, but he lost no time in writing to Molly, asking if he might see her alone before he and his father and step-mother joined the Bloxam family for Christmas Day. On 21st of December, they managed to escape the curious and prying eyes and ears of the sisters by taking a walk in the pouring rain of the Hampstead streets. Barnes described his feelings for Molly with a suggestion of his hopes for the future. The rain dripping off the brim of his hat onto her face dampened demonstrative ardour, but in any case he was much too gentlemanly to press her in any way; but he now had some idea of her feelings towards him. Armed with this, his intention was to speak formally to her father on Christmas Day, asking permission to make suit to his daughter. Molly unwittingly thwarted this plan. Young and naive, soaked and excited, she went home and broke the news to her parents. Barnes returned home dissatisfied with what he had said to Molly, and the manner in which he had said it. To put matters straight, he wrote a letter of explanation which he intended to hand to her on Christmas day. That evening, on an odd sheet of notepaper, Molly recorded what Barnes had said to her without making any comment as to her own feelings or responses.

On Christmas day Mr Bloxam was angry with Barnes who, in his opinion, had not kept to the restrictions imposed on him. He required stricter limitations on letters, no more than once a fortnight for both of them, absolutely no love letters, and no more visits. Barnes was desolated, and although he honourably said nothing at the time, he was aggrieved. He explained later to Molly that he had kept his word since the embargo on sentiment had been laid only on letters, and this he had certainly observed. No word had escaped him until they were together in the rain; he had no intention whatever of interrupting her academic career, and his chivalrous mind rejected any idea of asking for any commitment. In this situation Barnes was unable to give his letter of explanation to Molly, and it remained burning a hole in his pocket. She had no sight of it nor knowledge of what its contents revealed until a whole year later. Barnes left London on Boxing Day and exhausted the rest of the holiday in visiting friends until the time came to return to Chillon. He did, however, venture to visit Hampstead before he again left England.

Reprimanded and rebuffed he may have been, but he was not defeated. The mathematics came to his aid. Molly had to pass her exams. She was certainly in need of help from somewhere, and why not from himself? Over the next fifteen months he posted the mathematics lessons in separate envelopes. Mr Bloxam was himself too honourable a man to pry into the fortnightly personal letters, although for the next year he would have found nothing to disturb him. Barnes's first letter of the New Year was impeccably unemotional, and Molly answered likewise, with long descriptions of theatre visits, birthdays, a family play and concert; but she could not entirely overlook what had happened on Christmas Day.

Barnes: 9th Jan. '23 c/o Col. Thompson, Grange-over-Sands, Lancs
A very happy New Year to you. I wonder what sort of a holiday you are having? I've been wandering about all over the place – on Boxing Day I partly trained, and then walked with a rucksac down to Sussex, and from there on to Hampshire, and then came North, to stay with friends in various places – this is my third visit up here. It funny how soon one gets used to sleeping in a different bed every other night or so – counting home as one it comes to 6 different beds in 18 days, or an average of 3 nights per bed!

Life's rather muddly just at the moment, the directors of Vickers up here [in Barrow-in-Furness] seem very anxious for me to rejoin the Company, but have no work whatever for me to do.... It rather looks as if they might offer me a post in London – if so it is almost certain to be "come at once or not at all", and here am I engaged to conduct about 18 young gentlemen out to Switzerland on Tuesday next!....

Molly: 14th Jan. 1923. Hampstead
I was most awfully glad to get your letter; thank you most awfully for it. We've been having jolly decent holidays. I don't know if it will interest you, but I'll tell you everything we have been doing....

Barnes, will you read what I am going to say, and then put it right away out of your mind, and take no more notice of it, please.

I told Daddy and Mum what you said to me on Thursday evening, [Dec. 21st] directly after supper. But please don't think it was because I hadn't anything to talk about and wanted something to say, or anything awful like that. I told them simply because it was so lovely and I was so glad, that I had to tell someone; and you have to tell your father and mother when anything great and exciting happens to you.

And there's one other thing. Daddy was angry when I first told him, because he thought you'd broken your word or something. But I knew all the time that you hadn't, and that it was all right; I knew that what you did must be all right simply because you <u>are</u> Barnes, but he didn't realise that that would make it any better than if you were somebody else. Oh, I'm hopelessly muddled, so I'll stop. Goodbye, Molly.

Chapter 2

A MODERN FAIRYTALE

Life back in Chillon was a source of anecdote and amusement, and winter sports – skating, skiing, enjoying the snowy mountains – provided health and pleasure. Molly responded cheerfully with tales of her family's post-Christmas entertainments including a family dramatic production and Betty's birthday treat. Mr Bloxam took his three oldest children and his wife to the cinema, the first such visit they had ever made. They saw 'The Four Horseman of the Apocalypse', taken from a Spanish novel of the Great War. An exciting event in which Rudolf Valentino leapt to stardom, but a strange choice, particularly for the neurotic Mrs Bloxam and the over-sensitive Betty.

The debacle of Christmas Day still overshadowed Barnes. Molly loved and honoured her father and heeded his bidding; the second term in College would be even more enjoyable than the first, with more friends and more familiarity. Fortnightly letters, however cheerful, could become less absorbing. Barnes cast the hook of mathematical assistance again, and Molly took the bait readily.

Barnes: 22nd Jan. 1923. Chillon.
It was very nice to be able to see you all to say goodbye, and to tell you about rejoining Vickers. I happened to mention to the Boyds [friends in the North] that I was going back to call on Craven [a director of Vickers] sometime during the next week, in order to ask his advice about applying for the post of manager of a motor works near there, as of course I didn't want to get anything permanent to do if there was a chance of airships starting again soon. When he heard that I was thinking of going for another post, he said at once that he thought they wanted me at the Firm's London headquarters, and they offered me the post. Its going to be very interesting work, and one may have to do a lot of travelling. If any client in a foreign country wants Vickers to build them some electric generating machinery, or sink an oil well, or things of that sort which require consultation and advice, I may have to go out to meet their engineers on the spot to talk over things.

My American mother – I mean the mother of an American boy here, whom I am coaching for Cambridge in maths, – says she is coming to England too, as "dear Herbert" has never been able to learn maths from anyone in all America and Europe so well as he learns from me!!! She thinks I might keep an eye on the dear boy while he is working London. I told her she had <u>much</u> better send him to a special coach in Cambridge itself. "Dear Herbert" by the way is 6'5" high and weighs 13 stone! Truly I find much humour in schoolmastering.

I had a terrible time at Victoria last Tuesday when about 20 mothers, assembled to see the last of their various dear Herberts, examined me with critical and merciless eyes, as tho' they really doubted whether I were a fit person for the job.

I wonder how the physics are going this term. I did not have time to ask you in the holidays. Next time I will write another chapter on the calculus if you like.

Molly: 28th Jan. 1923. Hampstead.
I'm afraid I didn't seem half sympathetic enough about Vickers when you told us on Sunday. Don't you hate it when you are frightfully excited about something, and you tell somebody, and they don't care a bit about it? Anyway I did care about it; I am most awfully glad and I think it is simply topping.

Our Inter science is in June, worse luck. We are at present doing electricity in Physics, which is one degree better than Mechanics and Hydrostatics. I'd love you to write me another chapter on the calculus, if you don't mind. But there is something else I want to ask you. What does Lt mean? The first time I saw it was in this formula:- Instantaneous velocity

at $t = \text{Lt.} \dfrac{s_2 - s_1}{t_2 - t_1} \quad t_1 \rightarrow t_2,$

and I haven't an idea what $t_1 \rightarrow t_2$ is for. Another time we had it was in

$\text{Heat} - H = KA \text{ Lt.} \dfrac{t_1 - t_2}{d} T \quad d \rightarrow 0$

Supposing you had to work out an example and you wanted one of these formulae, what would you do with the Lt, or would you never use a formula like that? I know that it has something to do with fractions getting smaller and smaller, but that is all.

It's breakfast time now, and if I don't stop this letter, you won't get it for ages.

Barnes: 7th Feb. 1923. Chillon

I am very sorry this letter is so late, but I did want to finish the explanation of your formulae, and I could not do that without writing much more about the calculus than I had meant to at one go, although I had started it last week. So with a brief explanation in this letter I think you will be able to understand them – only you had better read the enclosed chapter III first. [sent under separate cover] Its rather hurried at the end, as I haven't had as much time for it as I should have liked....

What you suggest about Vickers that Sunday never so much as entered my head. I thought you were all awfully nice and kind about it. Of course it means a terrific lot to me, but I never expect anyone else to get excited over my affairs. I suppose I've got so used to the dear old Pater, who rarely does more than grunt whatever one tells him, tho' I must do him justice to say that he did say "Congratulations" when I told him this time.

Now for the formulae. "Lt" simply stands for "limit" or if one reads it in full, "the limiting value of". The little arrow → is read "tends to", so that $t_1 \to t_2$ is read "t_1 tends to t_2" that is t_1 approaches the same value as t_2.

The whole thing

$$\text{Instanteous velocity at } t = \text{Lt.} \frac{s_2 - s_1}{t_2 - t_1} \quad t_1 \to t_2$$

would therefore be read in full like this:-

"the instantaneous velocity at time "t" is equal to the limiting value of the fraction $s_2 - s_1/t_2 - t_1$ when t_1 tends to (or approaches and finally reaches) the value of t_2"".

The other formula I recognise as a modified form of the general expression for the flow of heat thro' a plate, where H = the no. of heat units, k is the coefficient of conductivity, A the area of plate, t_1 and t_2 the temps on the 2 sides, T the time and d the thickness. "$d \to 0$" would mean 'as "d" tends to the value "0"', tho' I confess I am puzzled as to why he should want to consider the flow of heat through a plate of zero thickness! But perhaps it is not the formula of which I am thinking after all.

Poor Molly, I do not wonder that you are puzzled when they give you things like that. It would be so much simpler if they would just teach you a little elementary calculus, instead of adopting these useless and puzzling evasions of what is so beautifully simple. That is why I started talking about it. I simply <u>must</u> post this letter now, or you will not get it till next week, and yet really I do want to write even a little more on the calculus, introducing you to differential coefficients. But my beanstalk [referring to

Calculus Chapter III] will enable you to understand what your first formula means very clearly I hope.

If you like to think of the "rate of growth" of the bean as its "velocity"
which it really is, or rather "velocity" is simply "rate of growth of distance
"s" with regard to time "t", then your t_2 and t_1 correspond to Jacks Q'' and
P'', s_2 is the height of the bean at Q'' and time t_2 (i.e say 2.35 p.m.) and s_1
is the height of the bean at P'' and time t_1 (i.e. 2.25 p.m.). So that $s_2 - s_1$ is
the height grown, or distance covered, between 2.25 and 2.35. Now here
comes Jacks discovery, that he could make P'' and Q'' move towards each
other as close as he wished (or $Q'' \to P''$) and <u>in the limit</u> when Q'' coincides with P'' he gets the <u>instantaneous velocity</u> or rate of growth of the
bean, as represented by the slope of the tangent at the coincident points.

No, you cannot make any use of a formula like that. [Molly's query of
28th Jan]. You see at once that in the limit when t_1 does reach the value t_2
that s_1 must also have reached the value s_2. Look at this little diagram [in
the letter] and imagine t_1 to be balancing s_1 on the end of its nose. Then
as t_1 moves outwards towards t_2, so s_1 gets longer and approaches s_2, and
finally equals s_2 when $t_1 = t_2$. So that in the limit $s_2 - s_1 = 0$ and $t_2 - t_1 = 0$
and the fraction, if you attempt to assign any definite values to t_1, t_2, s_1, s_2,
becomes useless.

I have talked to you about small quantities but for a very different
reason, which we shall see later on. Personally I find it is much better
boldly to explain what I mean by instantaneous values than to puzzle
people with these terrible limiting values and vanishing fractions.

But how sick you must be of maths. If anything is not clear, do tell me,
and I will put it another way.

Calculus: Chapter III
We have seen in the previous chapters how if a small quantity only be
taken small enough in the first case, then the small quantities of the
second and subsequent orders of smallness are really quite negligibly
small.

Suppose we leave this question of small quantities for a moment, and
turn on to something at first sight quite different. You must be quite familiar with the idea of a rate; for instance one speaks of walking at "a rate of"
say 4 miles an hour, or "he was living at the rate of £1000 a year" and so
on. In nearly every case where the word is employed in the popular way,
we refer to time, – walking at 4 miles an hour really means that distance
increases with regard to time at that rate. But there are heaps of other

"rates", we can make anything we like vary with regard to anything else – the rate at which pressure varies with regard to volume (Boyles Law) – the rate at which volume varies with regard to temperature (Charles Law) – the rate at which the length of a steel rod varies with regard to tempera-ture – the rate at which a piano wire vibrates with regard to its tension – in practically everything we do we get this underlying idea of the rate at which some one thing is happening, or growing, with regard to some thing else.

By vary I may mean, increase or decrease, "grow" or "shrink" in some definite way connected with the increase, decrease, growth or shrinkage of some other thing or quantity. For instance in walking at say 4 miles per hour, distance (s in your symbols) may be said to "grow" 4 miles in every hour, i.e. while time (t) "grows" one hour, distance "grows" 4 miles.

Put mathematically we say "The rate of growth of distance with regard to time is 4".

Here 4 is simply a ratio or number – just as one might say "This river flows 4 times as fast as that" – it doesnt matter whether 4 miles or 4 inches; the units in which we happen to measure the stream speed do not affect the <u>ratio</u> of the speeds.

♦ [The next paragraph Barnes considered so significant that he wrote it in red ink].

But you may argue "Since "time" is growing "hours" how can distance grow 4 times as fast – distance grows "<u>miles</u>" not <u>hours</u>, therefore since the dimensions are not the same (i.e. one is time, the other space) you cannot compare them – you might as well tell me to divide 2 cows by 4 black-birds!!!"

Yes but I can readily say "Very easily done – there are twice as many blackbirds as there are cows. It doesnt affect the ratio of their <u>numbers</u> that they are of different kinds. So that I can equally say that space grows 4 units (of space) while time grows one unit (of time). We must of course always remember the particular units (i.e. miles, hours) in which we made the original comparison. Does this make it clear? The statement about 4 to 1 would no longer be true if I suddenly turned to <u>minutes</u> and said, "Oh then I am walking 4 miles in <u>one minute</u>". Our statement <u>only</u> applies <u>to the units in which we first made the comparison</u>.

We can even fall back on dear old x and y, well tried friends, and we can make y grow with regard to x.

I have even amused my classes in calculus with Jack and the beanstalk.

The modern Jack had had a London County Council education, and when he noticed the phenomenal growth of the Beanstalk, he promptly took readings of the time, and the height of the beanstalk, every half hour.

What he got was something like this – we imagine that the time that he heaved the beans out of the window in disgust was 12 noon, and he measured the height of the beanstalk in yards with his mothers tape measure!

Time	noon 12	p.m. 12.30	1.0	1.30	2.0	2.30	3.0	3.30	4
Height of Beanstalk in yards	0	1	8	27	64	125	216	343	512

Being a very bright lad indeed he conceived the idea of representing the way the bean was growing on paper, and he got this pretty result.

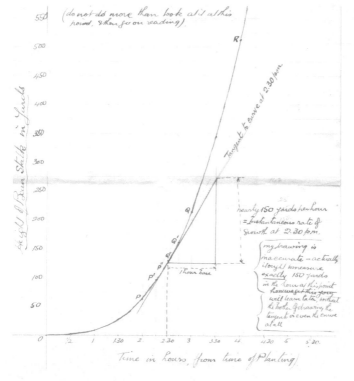

Do not do more than look at this point, and then go on reading.

His mother came along about 3 o'clock, and saw what he was doing. "Wonderful" says she – like most mothers she was very easily impressed where her own children were concerned, – "wonderful, – at whatever rate is the bean growing now, Jack"? "Rate!" says Jack, scratching his head, "thats just whats puzzling me. If only I could tell the rate at which it was growing I could calculate when it will have reached the top of the cliff, and I could make my plans accordingly. But the wretched thing wont continue to grow at the same rate for two moments together!

Look here, Mother – between 12 and 1 o'clock, it only grew 8 yards". "Only!!" put in his mother. "Yes" said Jack, "then between 1 and 2 it grew 64 less 8 yards, that makes 56 yards; and then between 2 and 3 it has grown 216 less 64 yards that makes 152 yards in one hour".

And he drew this little diagram to show his mother how the bean had grown between 2 and 3 oclock.

Later on after his mother had left, he drew another, to show how the bean had grown from 3 to 4 oclock. Here it is:-

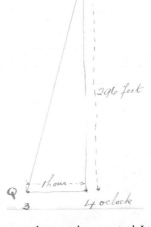

Poor old Jack, he got awfully fed up when he drew this one. With a thing that grows faster and faster like this he thought – why it will soon be shooting up like a rocket! And even for one hour it wont grow at a uniform rate for the line from P to Q is curved, so that it was growing much faster at say 5 minutes to 3 than it was at say 5 minutes past 2. Oh dear!

Then he suddenly had a bright Idea. I have spelt it with a capital I because it really was a very bright idea. "Supposing" thought Jack "I take a smaller interval than one hour, I shall get a more nearly accurate idea of its rate of growth". So he took the points P' and Q' (read this pee-dash and q-dash), at 2.15 and 2.45 respectively (see large diagram [p. 23]). Even this did not satisfy him for he could still see that the line was curved between P′ and Q′ showing that the bean had been growing faster at 2.45 than it had been at 2.15. So he took two more points P″ and Q″ (pee two dash and q two dash) at 2.25 and 2.35 – very close to 2.30, – so close that he had to make a 3rd little diagram, because when he tried to join P″Q″ he found the line practically coincided with his curve, and he couldn't see what he was doing. Here it is:-

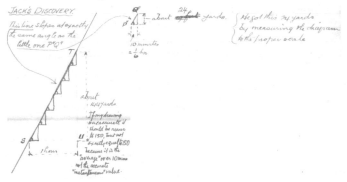

All of a sudden he sprang from his seat with a YELL! which brought his mother running from the kitchen. But all he would do was to run round and round murmuring tangents, tangents, over and over again. When he was sufficiently calm to explain, what he told her was this:- He found that if he prolonged the little line P″Q″ until it was as long as ST, that is until its projection on a horizontal base (i.e. SU) was just one hour long then the vertical projection UT was equal to the rate of growth of the bean in yards per hour.

AND MOREOVER, there was nothing to prevent his taking the two points P″ and Q″ as close together as ever he liked, he would still, by producing the line to cover a base of 1 hour be able to see the rate of growth per hour.

AND AGAIN, the closer he took P″ and Q″ the less variation there could possibly be between the rate of growth at P″ and Q″, and therefore the nearer together they were, the more nearly was he getting the actual value of the rate of growth.

AND FINALLY, if he chose to take P″ and Q″ so close together that they actually coincided, THEN his little sloping line would become the tangent to the curve at the point of coincidence, and by producing it as before to cover a 1 hour base, he would actually obtain [*next sentence in red*] the instantaneous value for the rate of growth of the bean at the point

P″Q″ (P″ coinciding with Q″). This is what he has drawn [*in graph above, p. 23*] where he has drawn a tangent to the curve at 2.30 p.m. thus finding the rate of growth of the bean in yards per hour <u>at that</u> <u>particular instant in the growth of time</u>.

A fraction of a fraction of a second <u>before</u> 2.30 p.m. the bean was growing a little slower, – a fraction of a fraction of a second <u>after</u> 2.30 p.m. the bean was growing a little faster, than the value shown.

FOR NO SINGLE PERIOD OR <u>DURATION</u> OF TIME DID THE BEAN <u>CONTINUE</u> TO GROW AT THAT PARTICULAR RATE.

2.30 P.M. is NOT AN INTERVAL OF TIME any more than a milestone is distance on the road. 2.30 p.m. and a milestone are merely names given to "points" one in time, the other in space.

Like Euclidean or geometrical "points" they possess position, but no magnitude.

You may if you like think of ourselves as sitting still while these artificial points in time flash past us, just as we flash past the mile posts when going in a fast train.

At the <u>instant</u> that 2.30 p.m. flashed past the bean its rate of growth was 150 yards per hour.

Another way of putting it – If the bean had continued to grow steadily (or better, "Uniformly") at its instantaneous rate at 2.30 p.m. then it would have grown 150 yards in the next hour and the continuation of the graph showing its height would have would have been the tangent which I have drawn [*in graph*] instead of the red curve.

Be very certain that you realise clearly that the red curve [*in graph*] does not show the <u>rate of growth</u> of the bean. It only shows the height of the bean at any given time. The <u>slope of the tangent</u> at any point on the curve gives the instantaneous rate of growth at that point.

We can quite easily plot a second curve, or "derived curve" showing the rate of growth. We only have to draw a number of tangents, and measure from them and plot the rates obtained. I have done this on [p. 27]

Since ac, df, gj are all one hour bases therefore bc, ef, hj represent to scale the rate of growth per hour at the points a, d and g where the tangents are drawn.

I have plotted these at 1, 2, and 3 p.m. and drawn a fair curve in red on [the second figure]. You will see it is a some-what similar curve to the other but rather flatter. Now the ordinates of this red curve give the instantaneous rate of growth at the corresponding time.

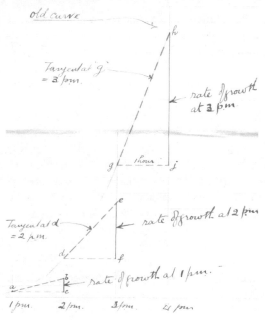

old curve

Tangent at g
= 3 p.m.

rate of growth
at 3 p.m.

h

g ⌞ 1 hour ⌟ j

e

Tangent at d
= 2 p.m.

rate of growth at 2 p.m.

d f

a b

c rate of growth at 1 p.m.

1 p.m.   2 p.m.   3 p.m.   4 p.m.

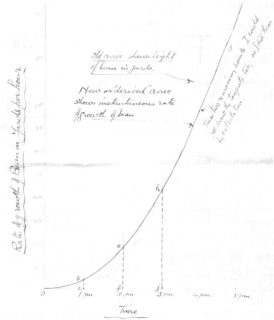

Old curve shows height
of bean in yards.

New or "derived" curve
shows instantaneous rate
of growth of bean

These two remaining points I could
not draw the tangents too, so I did them
by calculation.

Rate of growth of Bean in yards per hour.

h

e

b
c

O   1 p.m.   2 p.m.   3 p.m.   4 p.m.   5 p.m.

Time

Please do not scoff at my little fairy tale, it gets heaps more technical in the next instalment.

To be continued – si vous le voulez!

♦ One of Molly's sisters had caught German Measles and Molly, along with all the others, was quarantined within the house. Infectious diseases were taken seriously, and Molly would have to miss College until the last sufferer was cleared. If the children caught it sequentially, she would miss many weeks of lectures and classes. But she cheerfully expected it would be 'alright', and chattered on about more pleasant matters. February 1923 was warm, and Molly talked of the early flowers she loved, the servant problems suffered by her eccentric mother, and a visit to Auntie Fanny in Barnes's home. Servants were very much a part of life even in less wealthy households; the kitchen was largely out of bounds for the children and daily household chores were not included as part of the family's routine. Hence when they occurred, it was like playing at housekeeping. Harriet was a mainstay of the Wallis household, the faithful cook-general who had served for many years. Auntie Fanny had no servant problem as did poor Winifred Bloxam, endlessly upset and angry with her staff.

Molly: 14th Feb. 1923. Hampstead
I'm not the least little bit surprised at what Herbert's mother said; of course he couldn't learn maths from anyone in all America and Europe as well as he could from you. I can't think how you manage to do it so beautifully, it's all as clear as daylight. I love Jack and his mother and the beanstalk, and I really enjoyed reading it. Please go on; je le veux tres beaucoup (that's supposed to mean very much, I don't know if it is right). Thankyou very much indeed, Barnes, it is topping of you, specially as you must be tired after teaching boys all day.

That second formula [from her letter of 28th Jan.] is meant to be an expression for the flow of heat through a plate; it certainly does seem funny about the plate having no thickness, but it seems to me more sensible than a good many things he talks about.

There's just one thing you mention which I am not quite sure about – are the ordinates of a curve the points on a curve, or are they the points you'd get if you were to drop perpendiculars from the points on the curve to the base?

Baba and I went to tea to Auntie Fanny last Sunday. Harriet was out, so we helped Auntie Fanny get tea. It was a pity, because we were told that we must come home directly after tea, so we were as long as we could possibly eating it, but we couldn't stay to help wash up, and that of course is half the fun.

Barnes: 22nd Feb. 1923. Chillon
About the ordinates, if we have any curve the ordinate is the "Y" height always. [And he drew one of his neat free-hand diagrams]. Considering the point P, its "ordinate" I have marked "y", and the horizontal measurement, called the "abscissa", I have marked "x". They are just two names to enable us to distinguish between the vertical and horizontal distances from the axes.

I've spent so long on Chapter IV that I fear I must hurry this letter, or you will not get it this week. The last hope of reaching London leaves here this (Thursday) afternoon, and the letter gets to you on Saturday some time, probably Saturday evening. Still I thought the calculus might amuse you while you had to stop away from coll. I will not make other chapters so long.

I've done heaps of things since last I wrote. Last Saturday week, M.Bernheim and I climbed up to 4000 ft in the mountains just behind the College. We never really intended to go so high – both rather tired at the weekend, we meant to have a stroll in the woods above Chillon Castle, when we came upon a notice board nailed to a tree, saying "Chalet Souchaux, The, Cafe, Chocolat". On my map, Souchaux was about 2500 feet above us, so we thought we would go up, and as it was then 2 p.m. we would get there in nice time for tea. I had no winter sports clothes on, just knickers, [plus fours or knee breeches] shoes and stockings, and Bernheim – or the Baron, as we always call him, had an overcoat, hat, stick, gloves and trousers. We never can cure him of his continental habit of putting on an overcoat and hat whenever he goes out. When we got near the top I was awfully tired, and simply longing for tea, and the going was rather hard, as I had no nails in my shoes, and the track was all frozen snow.

On the way up I got the most glorious glimpse of Chillon Castle thro' the trees, about 800 feet below, with the sun shining on the lake. I took a photo, for a "moonlight" one.

At the top, we fell into quite a respectable snowdrift. Lots of the snow had already gone in avalanches, as you can see by the bare patches in the photos. That was due to the sudden thaw two weeks ago. When we got out of the deeper snow I took a photo of the Baron, because he looked <u>so</u> funny, with his long hair all ruffled (he's very musical) and hat, stick and gloves, on top of a mountain.

When we got <u>to</u> the top we found the chalet fast closed and shut for the winter, as no one stays there in winter!

♦ Enclosed in the letter he sent four photographs, three of them very amateur snaps of the mountains and the lake and one postcard size enlargement which had all the details mentioned clearly shown. There was also a little dried flower. When Molly saw it, the gentian was blue; time has taken its colour, but after 80 years its shape remains as Barnes pressed it.

Chapter 3
COUNT VON INTHELIMIT AND THE DUKE OF DELTA EKS

Molly: 28th Jan. 1923. Hampstead
Baba and I went to a lecture by Walter de la Mare on "Wuthering Heights" last Friday. It was very interesting and he lectures beautifully. I love "Wuthering Heights" in spite of, or perhaps because of its queerness. What sort of books do you like?

Barnes: 7th Feb. 1923. Chillon
I am afraid I am rather old fashioned in my tastes for books. Very few of the modern authors appeal to me much. Really I never get any time for reading and scarcely read one novel a year now. Its awfully difficult to say who is my favourite – one likes one sometimes, and then another, but perhaps R.L.Stevenson would come first at all times.... After him, it all depends on how you feel, and I vary between Thackeray, Dickens, Bronte, George Eliot on the one hand, and people like Kipling and Ian Hay on the other. Tho' last August on my holiday I had a spasm of Jane Austen and thoroughly enjoyed her. I <u>cannot</u> stand the Tarzan man. I cannot get thro' even one book. I have the (to me) rather pleasant gift of completely forgetting what a book is about soon after I have read it, so that I read the same ones over and over and over again. I suppose some I have read even 6 or 10 times. Ive quite a little collection of true sea stories – I mean old sailing ship ones, which really fascinate me. And yet mathematical formulae and telephone numbers (but <u>not</u> dates in history) seem to stick in my head for ever! Memory is a queer thing.

I have been playing fives, and I have given up smoking again. Really and truly, when I am very fully occupied I prefer not smoking to smoking – one feels so <u>free</u> somehow. It is only on holidays when one has to sit about and talk to people that I ever feel tempted to start again, and once started, although it is a nuisance, it is hard to give up.

Skiing is <u>much</u> more exciting than skating, one goes such fearful speeds, and come such awful croppers.

Molly: 14th Feb. 1923. Hampstead

You are the very first person I ever met who is glad he forgets a book soon after he had read it. It never occurred to me before that it could be considered a pleasant gift, but of course it must be topping to be able to read the books you love over and over again without remembering exactly what is going to happen next. I don't forget books very soon, or remember them for very long; I'm kind of middling, which is rather feeble. I'm so glad you like R.L.Stevenson and Dickens and Thackeray and Jane Austen and all those lovely people, because so many people seem to like only futile Tarzan-y sort of books.

This is about the warmest February I can remember....

When I went into the garden a little while ago, I was most awfully surprised to see that the Christmas Roses, which we brought with us from Wycombe [their previous house], were in flower; I never expected them to bloom this year. They are the loveliest kind I have ever seen – three petals speckly and three plain, and very big and creamy-coloured... lilac – especially purple lilac – is almost my favourite flower.

Barnes: 22nd Feb. 1923. Chillon.

What is "quite" your favourite flower, if lilac is "almost"?

I will not make other chapters so long. Perhaps you will think I make too much of a very simple matter – of course it is very difficult to tell, as I cannot know whether you have grasped a point or not, as I can when I am lecturing; but I find that most people find the preliminary ideas of the Calculus rather hard to understand straight off, so I reiterate freely. Once you have got hold of the fundamental ideas, you will find the remainder really quite simple, and most useful in physics.

Calculus: Chapter IV

We must part company with Jack. He poor fellow got very little further, and indeed he had no need to bother further with mathematics after he had climbed the beanstalk, killed the giant, and rescued and married the beautiful Princess. (I must confess I dont remember if that is really how it ends up, but it sounds very suitable).

So far we have considered two apparently unrelated ideas; the idea of small quantities, and the idea of instantaneous values.

1) <u>Small quantities</u>. We saw that if only a small quantity were in the 1st place made small enough, then the corresponding small quantity of the 2nd order of smallness was for all practical purposes negligible. (e.g. small

quantity $= \frac{1}{1,000,000}$ then small quantity of 2nd order $=$
$(\frac{1}{1,000,000})^2 = \frac{1}{1,000,000,000,000}$

There is no reason in maths. why we should not choose our small quantity
in the 1st place as small as we please, say $\frac{1}{10,000,000,000,000,000,000}$

then the second order is $\frac{1}{100,000,000,000,000,000,000,000,000,000,000,000,000}$
which is really a very small, small quantity indeed.

Finally, there is no reason why we cannot say "Go to! we will choose
our 1st small quantity smaller than any quantity which we can imagine or
conceive". What then becomes of our small quantity of the 2nd order?
Poor little fellow, he simply doesnt exist.

We may speak of this as making our first quantity "indefinitely small" –
that doesn't mean more or less small, it means that it is reduced in magnitude beyond any finite or conceivable limit. And since our first quantity is
so small that no smaller quantity can be conceived, it follows that the
smaller quantities of the second and higher orders simply dont exist at all.

Our friend Count von "InTheLimit" crops up here. I've made him a
Count because he's a wandering cosmopolitan sort of fellow that is always
turning up in unexpected places. And so we say that if we continue our
process of choosing our first quantity smaller and smaller then "In the
Limit" (Hullo, Count, you here!), when our first quantity is indefinitely
small then the quantities of the 2nd and subsequent orders are in fact
zero.

2) Instantaneous values. We have seen how by drawing a tangent, to a
curve which is plotted to represent the size of one quantity we can obtain
the rate of growth of the first quantity with regard to the growth of the
second, at any point on the curve.

Instead of a "bean" and "time" let us consider the area of a square and
its length of side.

If we say let $y =$ area   (in square inches or square feet or square
                             "units" of any kind)

and if we say let $x =$ length of side   (in inches or feet or length "units"
                             of any corresponding kind)

Then we may write:-

$y = x^2$

Let us plot the function. (If you have any equation connecting two
variables, we always talk about them in the following terms.

a) y (i.e. in this particular case; it may be h, or s, or p or anything) – y is called the "dependent variable".

b) x is called the "independent variable" or "argument".

c) y is said to be "a function of x" and the equation is some times written for short y = f(x) which is read "y equals a function of x" or "y equals the function of x" if you have previously quoted some particular function.

d) In general any expression containing a variable (e.g. $3x^3 + 2x^2 + 6x + 15$) is called a function of that variable; or if you equate y to $3x^3 + 2x^2 + 6x + 15$ then y becomes a function of x, because if you assign any series of values to x, and work out the value of the expression, poor old y has to follow suit, in other words its value depends on that of x.

Similarly when $S = ut + \frac{1}{2}at^2$ we say s is a function of t, or S = f(t) (read "s equals a function of t"). In this case t is the argument and s the dependent variable.

For you can readily see that given the value of u and a, s must depend entirely on t. u and a are called constants, i.e. once you have given them a value, that value does not change. I expect you know all this.

Let us then plot $y = x^2$.

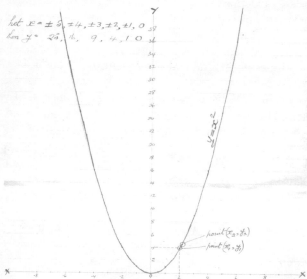

The curve shows the relationship between the area y and the length of side x. It is not necessary that x and y should be to the same scale – indeed how can they be, seeing that y is "square" units, while x is length units. I have plotted both +ve and -ve (positive and negative) values of x; really of course one cannot have a square with a minus length of side, but I did it without thinking, and have left it, as it shows you how the curve becomes a parabola; and also because of course the -ve values have a mathematical interpretation, wh. we need not go into now.

Now looking at the curve we see it is very like the beanstalk curve. And it resembles it in this, that the area is obviously growing faster than the length of side, i.e. the "ordinates" grow faster than the "abscissae". Consider for instance the area for sides 2 and 4 long.

side 2, area = 4

side 4, area = 16

So that <u>doubling</u> the side has resulted in the area growing <u>4 times</u> as great as it was before.

Jack would find the rate of growth of area with regard to that of the side for any length of side by drawing a tangent at the required point.

Can we do better than this?, for after all Jacks method is neither convenient nor accurate. Let us try.

We saw that Jack in leading up to his great discovery passed thro' a stage where he took two points on the curve very close together, and this gave him very nearly the actual value he was seeking. He obtained his actual instantaneous value only when his two points coincided.

We will follow suit, only with numbers instead of point on the curve. At least the numbers will of course be represented by points on the curve, but we shall deal with the actual numerical values instead of graphically represented values.

Consider then a square of side length 2. Then from $y = x^2$ we have $y_1 = 2^2 = 4$. Now take a point very near, say $x_2 = 2 + .001$. If you like to think of it in inches then the new length of side is 2 inches + $\frac{1}{100}$ of an inch.

and $y_2 = (2 + .001)^2 = 2^2 + 2 \times .001 \times 2 + (.001)^2$ C

{on the analogy of $(a + b)^2 = a^2 + 2ab + b^2$}

$\therefore y_2 = 4 + .004 + .00001$

(read these little suffixes y one, y two, etc. They are used merely to distinguish different <u>quantities</u> of the <u>same kind</u>).

Subtract from this equation our original $y_1 = 2^2 = 4$ and we get

$y_2 - y_1 = .004 + .00001$ C

or, the increase in area for an increase in length of side of .001 is .004 + .00001 or $\dfrac{\text{Increase in Area}}{\text{Increase in Side}} = \dfrac{.004 + .00001}{.001}$

or $\dfrac{y_2 - y_1}{x_2 - x_1} = 4 + .01 = \underline{4.01}$ A

Note that the ratio $\dfrac{\text{Increase in Area}}{\text{Increase in Side}}$ is the rate of growth of area with regard to growth of side just as

$\dfrac{\text{Increase in Distance}}{\text{Increase in Time}} = \dfrac{s_2 - s_1}{t_2 - t_1} = \text{velocity}$

or in other words = rate of growth of distance with regard to time.

Let us take a point nearer still. Let the side grow from 2 to 2 + .0000001

then $y_2 = (2 + .0000001)^2$

$\qquad = 2^2 + 2 \times .0000001 \times 2 + (.0000001)^2$ D

$\qquad = 4 + .0000004 + .0000000000001$

Subtracting our original $y_1 = 4$ we get

$y_2 - y_1 = .0000004 + .0000000000001$ D

and $\dfrac{y_2 - y_1}{x_2 - x_1} = \dfrac{.00000040000001}{.0000001}$ B

$\qquad = \underline{4.0000001}$

Look at the two results that I have labelled A and B. When the growth in x was 1/1000 we get the Average rate of growth of y between the points $x_1 = 2$ and $x_2 = 2.001$ with regard to the growth of x was 4.01. When the growth in x was 1/10,000,000 we get the Average rate of growth of y between the points $x_1 = 2$ and $x_2 = 2.0000001$ with regard to the growth of x was 4.0000001.

This seems to indicate to us that as we reduce the growth of x to smaller and smaller dimensions, i.e. that as we bring the point x_2 closer and closer to the point x_1 at value 2, so the rate of growth of y with regard to x at that point in the curve, approaches nearer and nearer to the value 4.

Notice also the quantities I have marked C and D. These are clearly quantities of the <u>second order of smallness,</u> so that we know that <u>In the Limit</u> (My <u>dear</u> Count) when the growth of x is reduced indefinitely (i.e. when $x_2 \to x_1$) the quantities C and D vanish entirely and we are left with the instantaneous rate of growth of y with regard to x, at the point where x

= 2, and this instantaneous ratio appears to us as if it will have the value of exactly 4.

You see at last I have joined together the two separate ideas which I laid before you, the small quantity and the instantaneous value.

But how can I really and truly get at the instantaneous value, i.e. can I possibly show this value to be what I suspect it to be namely 4?

Yes I can but I must introduce to you a new character, the Lord Chamberlain of the Lilliputians, the Duke of Delta Eks. "Molly, the Duke of Delta Eks" – "Duke, ~~Molly~~ [the name Molly is crossed out but is quite visible] Miss Bloxam" (Hanged if I am going to allow the fellow to get as familiar as that straight off, insignificant tho' he is).

I didn't introduce the Count to you on purpose, he gives one the impression of being rather a shady sort of fellow, dont you think, always "fading away" into the illimitable background. Besides who could possibly know a person who lives in the place where parallel straight lines are popularly supposed to meet!

Well, the Duke of Delta Eks is a very wee and obliging sort of fellow, who has the peculiar property that he too will fade away into the background at the merest suggestion on our part that he should become "indefinitely reduced". Poor little fellow, it must really be rather hard lines to start life quite a decent, measurable finite size, even tho' only a very small size, and then just as he is beginning to look around and enjoy life to be told to fade away until he is smaller than any imaginable quantity. It's not even a decent death, its annihilation! Here he is:- "δx" [in red], the Greek letter for 'd' and our familiar 'x'. They are inseparable. Like the Caterpillar (wasn't it?) in 'Alice', we pay old Delta Eks double and make him mean just what we like. So I hereby pronounce his doom, – that from now onwards he shall be taken to mean "A very little bit of x" so that if x happens to be a length, then δx is a very very short length. Similarly, if we are dealing with time, and t is time, then δt means a very very short interval of time. Again, δs means a very short distance; δh generally means a very little bit of height; δv means a very small increase (or decrease) of velocity. "δ" is not to be regarded in the algebraical sense as a coefficient reading as $\delta \times x, \delta \times t, \delta \times s,$ or $\delta \times v,$ – it is simply a sign tacked on to the quantity with which we happen to be dealing at the moment, meaning "a very little bit of that particular thing", whether length, area, volume, height, distance, velocity, time, or any one of the hundreds of things with which we may want to deal. You see we have indeed paid "δ" double and given him a funny meaning!.

So that instead of on [p. 36], saying that the length of side should grow from 2 to 2 + .0000001, I could put this into our new symbols, and say "let the length of side grow by a very small amount, from the value x, to the value $(x + \delta x)$.

Notice that δx begins life as a finite measurable quantity. We afterwards ask him to oblige by fading away, when Count von InTheLimit comes on the scene. But while we are making <u>use</u> of the poor little Duke, he is real and commensurable, and can be multiplied, divided and squared, or subtracted or added, just like any other finite, real, algebraic quantity.

Let us take then our general equation for the curve:- $y = x^2$, where x is not a particular value, but entirely general. Notice that when I want to indicate some definite point on the curve, I call it x_1 or x_2 or perhaps x' or x'' (x dash; x two dash). When x (or y or t or d or s) may have <u>any</u> value, we just use the simple letter.

Good; then let x grow by a very small amount to $x + \delta x$.

Then y must also increase very slightly; let us indicate the corresponding growth in y by δy; so that while x has grown from x to $x + \delta x$, y will have grown from y to $y + \delta y$.

We can equate the new values of x and y by substituting them in the old general equation thus

$$(y + \delta y) = (x + \delta x)^2$$
$$\text{or } (y + \delta y) = x^2 + 2x.\delta x + (\delta x)^2$$

Notice that δx behaves like one solid quantity thus I write $2x\delta x$, just as I might write say $2xz$ – I <u>cannot</u> say $2x\delta x = 2\delta x^2$ because the only connection between x and δx is that δx is a very little bit of the same kind of thing, but otherwise it is a completely separate entity. Notice also that I write $\delta x \times \delta x = (\delta x)^2$. When you get used to regarding δx as one symbol, I shall not bother to put the brackets, but shall write δx^2. Supposing $\delta x = 1/1000$, then $\delta x^2 = (1/1000)^2 = 1/1,000,000$

Again suppose $x = 2$ and $\delta x = 1/1000$ then $2x\delta x = 2 \times 2 \times 1/1000$. Does this help to make the little Duke's individuality clear?

Very good then. Now to find out the relationship between δy and δx. Note that the ratio $\delta y / \delta x$ = Increase of y/Increase of x = rate of growth of y with regard to x. We are therefore hunting the ratio $\delta y / \delta x$.

Take our 2nd equation, and from it subtract the 1st thus:-

$$y + \delta y = x^2 + 2x.\delta x + (\delta x)^2 \qquad 2$$
$$y \quad\ = x^2 \qquad\qquad\qquad\qquad\quad 1$$

$$\delta y \quad = 2x\delta x + (\delta x)^2$$

Before I go any further I must explain another peculiarity of the little Duke's. When I tell him to fade away, he does so, but his spirit hovers about in a sort of disembodied state. Getting quite creepy! The disembodied spirit of the little Duke is called "dx" (dee-eks), and it suddenly makes its appearance with a click, as Count von InTheLimit springs up out of a trap door and the poor little mortal remains of the unhappy Duke vanish into the regions of the "indefinitely small".

Look out, Molly! Strong smell of sulphur, blue flames, waily noises in the orchestra, Bang! Thump! Hey Presto, Abracadabra, "Count come forth", And

$dy/dx = 2x$ (read this dee-y by dee-x equals two x)

Allow me to present to you the instantaneous rate of growth of y with regard to x, for any value of x in the equation $y = x^2$, or in Calculus language, "the differential coefficient of y with regard to x".

That is, at any point in the curve the rate of growth of y with regard to x, is just twice the <u>value of</u> x at that point. Hence if x is 2, then $dy/dx = 2 \times 2 = 4$, <u>or y is growing four times as fast as x at the instant when x has the value 2</u>. Turning back you will see that we expected this result on [p. 37].

How has the trick been done. Quite simply thus:-

Turn [back] and look at our relation

$$\delta y = 2x\delta x + (\delta x)^2$$

Now $(\delta x)^2$ is a small quantity of the second order of smallness, and therefore when δx itself is indefinitely reduced $(\delta x)^2$ vanishes and in the limit is in fact zero.

That enables us to dispose of $(\delta x)^2$.

So I give the little Duke instructions to fade away. Thinking back you will see that when I tell the little Duke to fade, I am really beginning to bring the point x_2 up to point x_1 ($x_2 \rightarrow x_1$) so that $(x_2 - x_1)$ or δx, – {for $(x_2 - x_1)$ = increase in $x = \delta x$} begins to vanish. At the same time of course δy begins to vanish.

When x_2 coincides with x_1 (and y_2 with y_1) nothing of the little Duke remains but his disembodied spirit – which is in fact the <u>instantaneous value</u>, or slope of the tangent to the curve at the given point x_1, and so we use our spirit symbols dy and dx to indicate that the little Duke has ceased to exist as a finite quantity, he has passed to the regions of the indefinitely small, and only his spirit remains with us.

So, put more mathematically, we say:- As long as δx has a finite, commensurable value (although a very small quantity)

$\delta y = 2x.\delta x$ <u>approximately</u>

because until δx is indefinitely reduced $(\delta x)^2$ is not actually zero.
or dividing by δx we get

$\delta y / \delta x = 2x$ approximately

BUT <u>in the Limit</u> when δx (and consequently δy) is indefinitely reduced (i.e. smaller than any quantity we can imagine or conceive) then this statement is in fact <u>absolutely, rigidly, mathematically true</u>,

or $dy/dx = 2x'$

What I have written is in no sense a proof of this statement. I only give you a demonstration. The rigid proof is absurdly difficult and you have no need to worry about it.

I'm sorry this is such an awful chapter. I feel quite ashamed of it, but this is the end.

What we have found dy/dx is the "rate of growth of y with regard to x.

We can illustrate in this particular case of an area, our method graphically.

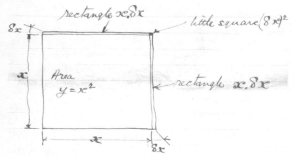

If I make a square, side x, then Area $= y = x^2$. Increase side to $(x + \delta x)$ then area $(y + \delta y) = x^2 + 2x.\delta x + (\delta x)^2$. You will se that with quite a big finite value of δx, how very little $(\delta x)^2$ contributes to the total increase.

Hence <u>rate</u> of Increase $=$ Increase in y/increase in x

$$= \delta y / \delta x \approx 2x \text{ and in the limit } dy/dx = 2x.$$

<u>Note</u>: $\approx$ means "is approximately equal to"

♦ With a mixture of elation and regret, Barnes was facing problems. Vickers Engineering, who had been paying him a retainer since making him redundant, had now recalled him to work again on the airship development programme. He had been profoundly hoping for this, but he insisted on fulfilling his term's commitment to Chillon College until early April.

Molly: 26th Feb. 1923. Hampstead
Thank you very much indeed for your letter and for Chapter IV, it is all absolutely understandable. I was sorry not to hear any more about Jack, but I have grown quite fond of the poor little Duke; I'm sure he wears a black coat which is rather shiny about the seams, and a shabby top hat. I don't like the Count so much, because he is so much fatter and better fed. You certainly don't make too much of a very simple matter as far as I am concerned, because I'm sure I shouldn't understand it if you didn't do exactly as you are. I didn't know all that part about functions; I'd never heard about them or arguments or variables before. I'm quite clear about the ordinates now; I had no idea before that they were anything else but points. In the middle of the chapter, where the Duke's spirit comes in, I was beginning to get rather puzzled, because $dy/dx = 2x$ certainly did seem to me like a trick; and very stupidly, I went back to see if I had missed out anything; but when I had the sense to go on, of course it became quite clear. I am going to read the next chapter straight thro' once, and then read it very slowly and carefully all the other times.

Sad to relate, Betty and George have succumbed; George got them last Sunday week, and Betty the day after... so we cannot go back till a fortnight from to-day. We are missing all sorts of jolly things at Coll.- lectures and dances and socials and matches. Neither George nor Betty have been at all bad; in fact George had been enjoying himself immensely. Nothing can keep him in his room, and he wanders about the passage upstairs shouting "unclean! unclean!" At first he had a bell which he rang at the same time (you can imagine the noise) but Nan soon took that away.

What a gorgeous walk you and the Baron must have had; but how disappointing to find the chalet closed after all your climbing. I guess you were very hungry when you got home.

Thankyou so much for the dear little gentian. I should think he <u>was</u> a wonderful little fellow, growing all up there amongst the snow, and what a lovely place to live in.

It's practically impossible to say what your favourite flower is, because they are all so lovely; but I think roses, specially red and white one, come first.... Then of course I like purple lilac and lavender and carnations and flowering currant and pheasant's eyes and love-in-the-mist and daffodils and wild thyme and speedwell and meadowsweet and peonies and sunflowers and heaps of others. What is your favourite?

Barnes: 8th 1923. March.
I am glad you like roses, for I think they are my favourite too. Tho' I am an awfully bad hand at flowers, and dont know the names of any. More particularly, what is a Christmas rose? I can follow you as far as a daffodil, but wild thyme, flowering currant, speedwell, meadowsweet, love in the mist and peonies have "got me guessing" as "dear Herbert" says.

I am beginning to feel quite sad at the thought of leaving this beautiful place. I do wish one didn't get so very fond of places. This day four weeks I shall be in London again! I am coming home by the night train on Thursday April 5th, getting to London 4.p.m. I always get attached to a place where I have been happy, and I get quite miserable when I have to go. Anyhow, I think all very beautiful places make one just a bit sad, tho' I dont know why they should exactly. I dont feel it with beautiful things, or pictures, or anything else but places.

I lunched with my American mother today. She asked me to spend a week with her and dear Herbert in Paris at the end of the term. Alas, I <u>must</u> get to Vickers as soon as possible, and I have already had to say I cannot go to Milan with the Head Master, for he and Mrs Lushington have invited me there!

♦ In spite of his life-long devotion to forms of transport, their efficiency and speed, Barnes himself surprisingly was no traveller and never really enjoyed visiting foreign places. The strength of the roots which he put down, even if temporarily, and the difficulty of cutting them, contributed to this unwillingness. Molly, on the other hand, loved travelling and was fascinated with exploring and adventuring; she found the restrictions laid on a young lady's activities frustrating.

Barnes: Thursday evening [posted 8th March1923]
I'm most awfully sorry, but I have been so interrupted that I haven't even got my letter started, but I will send it tomorrow. I hear that for 15c extra one can send by aeroplane, so I will try that for fun, – do let me know if it reaches London any sooner. This should get to you last post Saturday – my last chance for reaching England this week, but perhaps that plane will catch up.

I have bought an aeroplane stamp, and they tell me that if I catch the 7p.m. poste at Montreux Gare, my letter will reach Londres at 2p.m. tomorrow (Saturday) so you will get this about the same time as the other. Do you collect stamps?

Molly: 14th March 1923. Hampstead
Have you really never seen a Christmas rose? Here is one,.... It is the last rose of Winter and it is practically dead....

I expect you've often seen and smelt wild thyme and meadow sweet, and all those other wild flowers, though you don't know their names.

I am most awfully proud of having a letter by aeroplane; thank you so much for it. Do you know, it never arrived here till Tuesday morning; [13th March]. I suppose there was a storm or something which delayed it. I don't collect stamps; Barbara does; but I am most certainly not going to give her my aeroplane stamp, considering it is the first one I have ever had... all the family has seen it and admired it and envied me because of it. I suppose one day it will be an ordinary, everyday thing to have an airpost, but any way I shall treasure this first one of mine.

I got the ordinary land-and-water letter on Saturday morning at breakfast time.

I should think you would feel sad at the thought of leaving Switzerland. I can quite understand how you feel about beautiful places. I haven't been to many myself, but I remember once when we were among the mountains in North Wales one day, and there was nothing but heathery hills and a big lake, it felt just like that – sad and solemn and very big and lonely.

Barnes: 21st March 1923. Chillon
I'm most awfully sorry about the aeroplane letter – how perfectly stupid, because if I had posted it in the ordinary way you would at least have got it on Monday. They assured me it would get to London by 2 p.m. on Saturday, and I dashed in to Montreux to catch the fast train at 7 p.m.! Going in by tram I showed the conductor the stamps and had a most exciting conversation. We both waved our arms and I jabbered about "avions" and "hydravions" until my head whirled. But he beat me in the end and I was forced to lapse into silence. I was much relieved when more passengers got in, because I feared every moment he would fall on me again.

Calculus: Chapter V
We have thus discovered two different ways of finding the rate of growth of y with regard to x, or of <u>any</u> one thing compared with <u>any</u> other thing,

provided that we are given some connection between the two things. This 'connection' we call "expressing one thing as a 'function' of another. If the function we happen to be handling does not refer to anything in particular we generally use y and x and we say $y = f(x)$ (read y = a function of (x)).

Several other things that we know about we can express as functions, thus

$$s = f(t) \qquad \text{(distance equals a function of time)}$$

$$v = f(t) \qquad \text{(velocity \quad " \qquad " \qquad ")}$$

also $\quad v = f(a)f(t) \qquad$ (" \qquad " \qquad " \qquad " of acceleration and of time – useful if <u>both</u> are varying)

$$p = f(v) \qquad \text{(pressure is a function of volume)}$$

$$v = f(p) \qquad \text{(volume " " \quad " \quad " pressure according}$$
to <u>which</u> we are making the <u>argument</u>)

$$a = f(f) \qquad \text{(acceleration is a function of the force } (f))$$

(perhaps you use $p = ma$ or $f = ma$ or $p = mf$ as your dynamical equation?)

All these things do not tell us what the function is, they are merely a conventional sort of mathematical shorthand, so do not be terrified of them.

1) Our first way was to plot the function, and then draw a tangent at the point at which we required to find the rate of growth (Jack's method).

2) Our 2nd was was to employ the little Duke and then make him fade away, leaving us with his spirit, which we called dy/dx, to distinguish it from his poor little body $\delta y/\delta x$.

Now note that these two methods are not really dissimilar, for in drawing the tangent we first of all took points a little way apart on the curve and then made them approach closer and closer together until they coincided.

Thus in fig I δx is the finite little body of the Duke and the ratio $\delta y/\delta x$ represents the <u>average</u> rate of growth over the finite distance δx. In fig II we have instructed the Duke to fade away, and δx and δy have both vanished <u>BUT the slope of the line remains</u>, and this is the spirit of the Duke or dy/dx.

So you see we may say either:-

1) that our method no. (2) is an arithmetical dodge to save us the trouble of drawing curves and tangents; or:-

2) our method no. (1) is merely a <u>Geometrical illustration</u> of what looks at first sight rather a mysterious trick, when performed arithmetically.

Whichever view you adopt does not matter, it is only important to remember that there is nothing uncanny about dy/dx; it is only the measure of the slope of the tangent to the graph of the function with which we are dealing.

Perhaps here I had better explain one further very useful term, adopted from trigonometry.

If we have a right angled triangle ABC then we speak of the <u>slope</u> of the

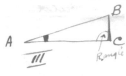

line AB as the "tangent" of the angle A. This slope is <u>measured</u> by means of the ratio BC/AC i.e. length of line BC/length of line AC. This idea of <u>slope</u> is not a new one to you – you would understand me quite well if I said "the slope of this hill is one in eight". Here I should mean that for every 8 feet measured <u>horizontally</u>, we went 1 foot up <u>vertically</u>, thus

This is merely a popular way of expressing the tangent of the angle α. (alpha – greek "a")

If I said to you "oh, look at this hill, tan α equals .125" (i.e. $\frac{1}{8}$ of course), you would probably regard me with curiosity tinged with suspicion, but really it is an identical statement. Or if I said, panting for breath, "the differential coefficient of the hill in Kidderpore Avenue is one-eighth" you would undoubtedly send for the nearest policeman!

Yet all I should have intended to convey would have been that the growth of the vertical height was $\frac{1}{8}$ of the growth of horizontal distance, or "the <u>rate</u> of growth of height with regard to distance $= \frac{1}{8}$". (or in symbols dh/ds $= \frac{1}{8}$).

Or we write tan A = BC/AC where "tan" is short for "tangent".

So you see we can have a tangent to a curve, and also a tangent (or slope) of an angle. It is perhaps a little unfortunate that the name for two things is the same, and apt to be confusing, but the name of the one is taken from the other, for in the old days mathematicians used to define the trigonometrical tangent as the length of the line PQ. They made OP <u>unit</u> length and then the ratio PQ/OP was of course = to PQ, since OP=1 and since PQ is the <u>tangent</u> to the circle, they called the length PQ the "tangent of the angle POQ".

Of course you we
know that angle
OPQ = a rt angle, from
geometry,

So you must try not to get muddled as to which kind of tangent I am speaking of (what awful English!). The connection between the two does not exist in the case of any curve except the circle, so that when I speak of the slope of a tangent to a <u>curve</u>, and wish to express this slope as a tangent of an angle, I must mentally construct a little right angled triangle in order to get the ratio BC/AC (fig III). In fig V PQR is any curve, and I have drawn a tangent to it. Then it is always the slope of this tangent with regard to the XX axis in which we are interested. I have marked the angle which the red tangent makes with XX as θ (Greek letter for th, called "theta"). We can measure the value of tan θ by taking the ratio BC/AC (fig V). As in the case of a hill this measures the slope.

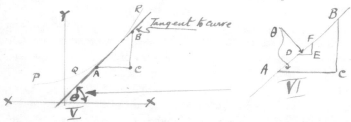

Notice that angle BAC = θ (from parallel lines) hence tan BAC = tan θ = BC/AC. Note also that it didn't matter what length I made AC above, since in similar triangles the ratios of the sides are the same – thus if I make all the sides parallel, then the two triangles are similar and BC/AC = FE/DE = tan θ.

So that, as I said, dy/dx, although obtained so mysteriously, is merely the measure of the slope of the tangent to the curve. That is, put in other words, dy/dx is the tangent (i.e. the trigonometrical tangent) of the angle θ.

Having wearied you to the limit of your patience with the meaning of dy/dx, let us go on again with the more interesting process of finding dy/dx for a variety of functions.

Let us take $y = x^3$.

(Enter the Duke, black coat, shiny seams, top hat and all.) You have only left out one thing – he has a copy of "The Times" tucked into his coat, because only prosperous and solid people take "The Times". Like us! And whoever was it used to do that, and one day it fell out, and someone picked it up and handed it back, and in doing so noticed that it was dated about 3 years previous! Its a character in some book, and with my usual happy (?) (I fear you have placed a doubt in my mind as to whether it is really as happy as I thought) forgetfulness, I have forgotten everything except the fact, but it sounds like Thackeray, and yet ("Gummy, what a sentence!" – who does that quotation come from?) I cannot recall any character of his like that, so perhaps it is Dickens, but in any case it is completely irrelevant, I've just eaten two oranges running, I mean not <u>me</u> running, but consecutively, one-two, one-two, and think they must have monté à la tête; there's the gong for 1st period, so I simply must put a reluctant full stop, tho' I'm not nearly out of breath yet!

<u>Much later (and saner)</u> I do occasionally go mad. I made one of my VI form boys quite cross the other day by setting in one of my mental alertness tests the sum – one of ten to be done in 10 minutes, entirely in the head – "If 3 claps of thunder wake 5 people, how many people will 6 claps wake?" He solemnly put down 10 as the answer, and when I hooted at him he persisted that one could do it by proportion!!!

Poor little Duke, what a time I have kept him waiting – but then he's just the sort of person who <u>would</u> always be waiting, is he not?

Let x grow from x to $(x + \delta x)$ and let y correspondingly grow to $(y + \delta y)$

then $y + \delta y = (x + \delta x)^3$

and multiplying out

$$y + \delta y = x^3 + 3x^2.\delta x + 3x(\delta x)^2 + (\delta x)^3$$

subtract $y \qquad = x^3$

$$\delta y \quad = 3x^2.\delta x + 3x(\delta x)^2 + (\delta x)^3$$

As before $(\delta x)^2$ and $(\delta x)^3$ become in the limit zero,

and $\qquad \delta y = 3x^2.\delta x$

or $\qquad \delta y/\delta x = 3x^2$

(Approx. because we have not reached the limit, – the little duke is still fading!)

And in the limit

$$\frac{dy}{dx} = 3x^2 \text{ really and truly.}$$

Or, the rate of growth of y with regard to x is $3x^2$.

Thus, at the point where $x = 4$, y is growing $3 \times 4^2 = 48$ times as fast as x!

or, where x is 4 the slope of the hill is <u>48 in 1</u>.

48 up   Some Hill!

θ

1 along

or, $\tan\theta = dy/dx = 48$.

But do not lose sight of the fact that this is an "instantaneous" value only. Just beyond $x = 4$, y will be growing faster still, just before $x = 4$ y was not growing quite so fast.

Let us try again, say $y = x^4$

then as before $y + \delta y = (x + \delta x)^4$

then multiplying out

$$y + \delta y = x^4 + 4x^3.\delta x + 6x^2\delta x^2 + 4x\delta x^3 + \delta x^4$$

Subtract $y \qquad = x^4$

$$\delta y = 4x^3.\delta x + 6x^2\delta x^2 + ...$$

and $\quad \delta y/\delta x = 4x^3$ approx

and $\quad dy/dx = 4x^3$ really and truly.

Let us try again, say $y = x^5$

then $y + \delta y = (x + \delta x)^5$

Little Duke's getting a little flustered!

and $\quad y + \delta y = x^5 + 5x^4.\delta x + 10x^3\delta x^2 + 10x^2\delta x^3 + 5x\delta x^4 + \delta x^5$

and $\qquad \delta y = 5x^4\delta x + $ etc

and $\quad \delta y/\delta x = 5x^4$ approx.

and $\quad dy/dx = 5x^4$ really and truly.

When $x = 4$ in this case y is growing $(5 \times 4 \times 4 \times 4 \times 4) = 1280$ times as fast as x. I'm not going to draw this hill, it makes me feel giddy – regular cliff in fact.

Let us have a look at our results, while the Duke brushes his hat, and rubs the seams of his coat with turps. (I too have been poor and shiny).

Function	$\dfrac{dy}{dx}$
$y = x^2$	$= 2x^1$
$y = x^3$	$= 3x^2$
$y = x^4$	$= 4x^3$
$y = x^5$	$= 5x^4$

$= 2x^1$ only we don't bother to write the 1

Note the coefficients & indices of $\dfrac{dy}{dx}$

Could you, do you think guess the answer of $y = x^6$, or $y = x^7$ when I ask for dy/dx? Try to work it out too, as I have done.

♦ The question Barnes asked Molly to solve was the first he had put to her, and she rose to the challenge. He was doubtful about the coherence of his chapters since, having constructed his own original syllabus, he had no routine to which he could refer. There is no evidence as to whether he made out a plan before he started, and no hint that he kept any sort of copy for his own record. But in fact he did refer back not infrequently to points and examples (and indeed mistakes) from previous chapters, indicating that he maintained an overview of his course. He was approaching the point in his argument at which he could properly answer Molly's question of the previous October, still hanging in the air, on the meaning of d^2s/dt^2.

Molly: 14th March 1923. Hampstead
I guess that dy/dx = $6x^5$ when $y = x^6$, and $7x^6$ when $y = x^7$. I worked it out without looking at the book (I call it the book because it is growing so nice and big and fat, just like a book. I have copied out chapters I and II because I couldn't very well put the letters they were in with the other chapters, and anyway it would look rather queer to have bits of letter mixed up with the other chapters!) and if the little Duke was flustered with the lengthiness of $(x + \delta x)^5$ (I can't make proper δs; they look just like 8s), what do you think he was like with $(x + \delta x)^7$?

Barnes: 21st March 1923. Chillon
Thank you very much for the Christmas rose. No, I have never heard of one, certainly never seen one before. I think he's perfectly ripping. How on earth does it know which petals to make speckly. Can roses count do you think?

Are you keen on wild flowers? You would simply love this place. I never knew Switzerland was a great place for flowers before, but apparently in the Spring its just a mass of daffodils and narcissi and already it swarms with prims and violets, scented ones too. The Baron and I went the most gorgeous walk last Thursday, straight up the mountain at the back of the College, and then along a little telegraphists track above the surface of the Lake. Every step one comes on masses of flowers very like anemones, only beautiful shades of red- and blue-purple, heaps of rich moss, and lizards basking in the sun. And then, where the sun hasn't got at it, a great mass of snow. The little track goes round quite a precipice at one part – a great rock face, covered with trees, moss and flowers, on one hand, and an almost vertical drop, thickly wooded, on the other. And you can see for miles and miles over the lake - and far away a dog barks, and then an old bee comes droning by. It fascinated me so, I went again by myself on Sunday and lay for nearly an hour in the sun.

I eat two oranges ~~runn~~ [this half-word is crossed out] consecutively (thankyou) and started a letter to you, but a bee came and sat on my pen, which was all sticky from a lump of nougat I had put in my pocket by way of tea, and most of it had no paper on, so I had to eat it long before tea time. So that your letter did not prosper, and the address was "Up a hill"; such an obviously orangey beginning that in the end I gave it up.

Oh, I have persuaded the Head to let me go on Maundy Thursday, by the night train. I leave Montreux about 7 pm and reach Victoria about 7 pm (Good Friday). So I shall get a weeks holiday, as I do not start at Vickers until the 9th. This is my last letter from Switzerland, Molly.

You are quite right in your answers, jolly good, I am most awfully pleased. This instalment gives you the key to differentiating the functions which puzzled you in October, tho' I haven't answered your question about d^2s/dt^2 yet – that is coming next. It is jolly decent of you to have bothered to keep the first chapters, and to copy them. I fear it must all be awfully disjointed, because its awkward not having anything to refer to. Your δs look as though you started them from the top so – δ^{start} – you would find it easier, if you are doing so, to start from here $\rightarrow \delta$ and go round the other way.

I'm feeling jolly sad at leaving – I <u>hate</u> leaving. I <u>must</u> finish this to catch the post.

Calculus: Chapter VI
I wonder if you have done the Binomial theorem? If you haven't it doesn't much matter, because you can easily learn to apply its results, without bothering to wade thro' the processes and reasoning by which the result is obtained. The Binomial was another great discovery of Newton's. The French refer to it as "Le Binome de Newton", just as we would speak of Boyle's Law. I think its rather decent of them to give old Newton the credit for it don't you? So different from the Germans, who always try to rob him of the Calculus and give it to Leibnitz.

The Binomial, tho' not such a revolutionary step in mathematics as the Calculus, is yet a very important theorem. It is more in the nature of a labour saving device. A Binomial expression is simply an expression with two unlike terms, e.g. $(a + x)$, $(x + y)$, $(3p + 2q)$ are all examples of binomials, also $x + \delta x$. Now, as you have found, when we have a binomial raised to a high power the process of multiplying out is very laborious and cumbersome. When the index of the power is fractional or negative (e.g. $(x + a)^{1/7}$, $(x + a)^{-15}$) the process by ordinary algebraical methods would be impossible.

So Newton just gives us a very nice formula which enables us to write down <u>at sight</u> the 'expansion' as it is called, of <u>any</u> binomial expression raised to <u>any</u> power, whether positive, negative, or fractional.

Never mind how he got it, you are not to worry over that. The strict proof that it is true for <u>all</u> indices is very difficult – I never understand it properly myself but I have no hesitation in making use of the powerful weapon which it places in my hands – or head rather!

It is just this, – if we have any binomial – {let us take $(x + a)$} and we wish to raise it to any power; (to be perfectly general let us take 'n') then

$$(x+a)^n = x^n + n.x^{n-1}a + \frac{n.\overline{n-1}}{1.2}x^{n-2}a^2 + \text{(read eks}+a\text{ to the nth)}$$

$$+ \frac{n.\overline{n-1}.\overline{n-2}}{1.2.3}.x^{n-3}a^3 +$$

$$+ \frac{n.\overline{n-1}.\overline{n-2}.\overline{n-3}}{1.2.3.4}.x^{n-4}a^4 + \ldots + \ldots + a^n$$

That is all. (Don't stop to puzzle over it, but read on).

Let us examine each term, and see how they are constructed.

The 1st term you see is nothing but x raised to the nth power $= x^n$. The last term is nothing but 'a' raised to the nth power $= a^n$.

$$\text{2nd term} = n.x^{n-1}a$$

$$\text{3rd term} = \frac{n.\overline{n-1}}{1.2}.x^{n-2}a^2$$

$$\text{4th term} = \frac{n.\overline{n-1}.\overline{n-2}}{1.2.3}.x^{n-3}a^3$$

and so on.

Note how the <u>coefficient</u> of x is made up. In the 1st term it is 1; in the 2nd term it is n, in the 3rd term it is $\frac{n \times (n-1)}{1 \times 2}$; in the 4th term it is

$\dfrac{n\times(n-1)\times(n-2)}{1\times2\times3}$; in the 5th term $\dfrac{n\times(n-1)\times(n-2)\times(n-3)}{1\times2\times3\times4}$ and so on.

Note how the <u>indices</u> of x and a go.

In the 1st term x has it all its own way x^n

In the 2nd term, 'a' creeps in and bags one from x thus $x^{n-1}a$.

In the 3rd term, 'a' bags another one, so, $x^{n-2}a^2$

You see 'a' wins one more every time, and the sum of the indices of x and a always makes n.

Let us try our hand at expanding $(x + a)^5$

By formula

$$(x+a)^5 = x^5 + 5^{x-1}a + \frac{5\times(5-1)}{1\times2}x^{5-2}a^2$$

$$+ \frac{5\times(5-1)\times(5-2)}{1\times2\times3}x^{5-3}a^9$$

$$+ \frac{5\times(5-1)\times(5-2)\times(5-3)}{1\times2\times3\times4}x^{5-4}a^4$$

$$+ \frac{5\times(5-1)\times(5-2)\times(5-3)\times(5-4)}{1\times2\times3\times4\times5}x^{5-5}a^5$$

$$= x^5 + 5x^4a + 10x^3a^2 + 10x^2a^3 + 5xa^4 + a^5$$

($x^{5-5} = x^0 = 1$ you know – if not tell me and I will explain.)

In practice one makes it a little shorter thus, taking $(x + a)^7$

$$(x+a)^7 = x^7 + 7x^6a + \frac{7.6}{1.2}x^5a^2 + \frac{7.6.5}{1.2.3}x^4a^3 + \frac{7.6.5.4}{1.2.3.4}x^3a^4$$

$$+ \frac{7.6.5.4.3}{1.2.3.4.5}x^2a^5 + \frac{7.6.5.4.3.2}{1.2.3.4.5.6}xa^6 + \frac{7.6.5.4.3.2.1}{1.2.3.4.5.6.7}a^7$$

$$= x^7 + 7x^6a + 21x^5a^2 + 35x^4a^3 + 35x^3a^4 + 21x^2a^5$$

$$+ 7xa^6 + a^7$$

I always remember the formula as I have written it first above. You do not need to remember it however, as you can always refer to it when you need it. When we have other letters or numbers we can simply substitute them in the general formula.

Writing δx instead of a, we get

$$(x+\delta x)^5 = x^5 + 5x^4\delta x + 10x^3 8x^2 + 10x^2 8x^3 + 5x\delta x^4 + \delta x^5$$

Do not let yourself get confused if you have to expand say $(1 + x)n$. Notice here, that all you have to say to yourself is, "Aha! I must write 1 wherever the formula has an x, and I must write x wherever a formula has an a," thus

$$(x+a)^n = x^n + nx^{n-1}a + \frac{n.\overline{n-1}}{1.2}x^{n-2}a^2 + \text{etc.}$$

Therefore

$$(1+x)^n = 1^n + n.1^{n-1}x + \frac{n.\overline{n-1}}{1.2}1^{n-2}x^2 + \text{etc.}$$

Now "one" raised to any power is of course 1, so we write

$$(1+x)^n = 1 + nx + \frac{n.\overline{n-1}}{1.2}x^2 + \frac{n.\overline{n-1}.\overline{n-2}}{1.2.3}x^3 + ...$$

If we make n represent a negative number, and write $-n$ we get

$$(1+x)^{-n} = 1 + (-n)x + \frac{-n \times (-n-1)}{1.2}x^2$$
$$+ \frac{-n \times (-n-1) \times (-n-2)}{1.2.3}x^3 + ...$$

Collecting all the minuses we get

$$(1+x)^{-n} = 1 - nx + \frac{n.\overline{n+1}}{1.2}x^2 - \frac{n.\overline{n+1}.\overline{n+2}}{1.2.3}x^3 + ...$$

You see the terms have become alternatively positive and negative, and by taking the minus signs outside the terms such as $(-n-1)$ we have got $(n + 1)$ etc.

If n is positive and x negative we get a similar result thus,

$$(1-x)^n = 1 - nx + \frac{n.\overline{n-1}}{1.2}x^2 - \frac{n.\overline{n-1}.\overline{n-2}}{1.2.3}x^3 + ...$$

only the $(n - 1)$ etc remain $(n - 1)$ as before. The terms simply become alternately + and – because $(-x)$ raised to an <u>odd</u> power remains -ve and $(-x)$ raised to an <u>even</u> power becomes +ve.

One warning here, which will not however affect the use you will require to make of the theorem:-

The Binomial theorem is only true for negative and fractional values of n when the 'a' term is <u>numerically less than one</u>. If 'n' is positive and a whole no. then it is universally true for any values of the 'a' term.

I have had to bring this in, in order that we may obtain a "general" method of differentiating (i.e. finding rate of growth i.e. finding dy/dx when we are dealing with y's and x's).

We have up till now, done several <u>special</u> cases, and overworked the little Duke considerably in doing it; and we have a very shrewd notion as to what the rule will be.

We can guess for instance that if $y = x^{15}$, then $dy/dx = 15x^{14}$. We did this by noticing that in every case we have tried, we get $dy/dx = x$ multiplied by its <u>original index</u>, and with its <u>new index</u> one less.

Or to put our rule into words we might say, "to differentiate y with regard to x, reduce index of x by one, and multiply it by its original index".

Shall we see if this is generally true? Let us see if we can differentiate y $= x^n$.

As before we must call up the little Duke. "Morning, Duke, here's a tough job for you today" – Duke removes top hat and carefully places copy of Times inside it – he daren't remove his coat, because that would disclose the fact that he had no shirt; poor little fellow, he has to make do with one of those false shirt fronts, and dummy cuffs. He has a hard time, for this is almost the last time we shall employ him, and goodness only knows what other work he gets. Count von IntheLimit bagged all the little fellows estates years ago, but he used to be quite important.

Come along then little man!

$$y + \delta y = (x + \delta x)^n$$

Expanding this by the Binomial we get

$$y + \delta y = x^n + n. x^{n-1}\delta x + \frac{n. n - 1}{1.2}. x^{n-2}\delta x^2$$
$$+ \frac{n. n - 1. n - 2}{1.2.3}. x^{n-3}\delta x^3 + \ldots$$

Subtract $y = x^n$

$$\therefore \ \delta y = n. x^{n-1}\delta x + \frac{n. n - 1}{1.2}. x^{n-2}\delta x^2 + \ldots$$

Notice that the 2nd and subsequent terms of this result contain δx^2 and higher and ever higher powers of δx, therefore as before we may neglect them when the Duke fades. "Thanks awfully Duke – do you mind?"

So $\delta y = n.x^{n-1}\delta x$ approx

$$\delta y/\delta x = n.\, x^{n-1} \qquad "$$

and In the Limit

$$dy/dx = n.\, x^{n-1}$$

Poor Duke, this accursed Binomial cuts down the time his clients want him so enormously.

Well, look at our result; as we thought all we have to do is to subtract one from the index of x and to multiply by the original index.

Let us, as a grand reminder, write down all the complex names which are given to this simple result.

1) We have found the instantaneous rate of growth of y with regard to x, for any value of x we may choose to insert.

2) We have differentiated 'y' with regard to 'x'.

3) We have found the "differential coefficient" with regard to x (i.e. diff. coeff. is nx^{n-1}).

4) We have found the "first derived function" or "derivative" of y with regard to x.

They all mean the same thing.

There are other ways of expressing dy/dx, all useful in their respective places.

Thus, we also write Dy for dy/dx, the 'argument' with respect to which y has been differentiated being either understood or being expressed by means of a little suffix thus $D_x y$. Newton just used to write y' and his notation although it affords no means of indicating the 'argument' is often very convenient when we know exactly what we are refering to.

So you must be prepared to see

$$y' = D_x y = dy/dx = nx^{n-1}$$

or even $d(x^n)/dx = n.x^{n-1}$ for we can write the whole function in place of y if we wish.

So again $D_x(x^n) = n.x^{n-1}$.

We refer to the differential coefficient in several ways also.

1) As the differential coefficient.

2) As the "derivative".

3) As the "first derived function".

When we are using functional notation and say $y = f(x)$, then we indicate the 1st derived function by $f'(x)$ (read f dash eks). Thus if $y = f(x)$, then $dy/dx = f'(x)$. This is very convenient if you have a long function and want to talk about it, but can't be bothered to write it out every time.

Molly: 26th March 1923. Hampstead
The chapters aren't the least little bit dis-jointed; they follow each other beautifully. You are quite right – I did start my δs at the wrong end; they don't look so much like 8s if you begin at the other end. I'm awfully sorry to bother you, but I'm afraid I don't know why $x^0 = 1$; I should have expected it to equal x. I understand the other part.

The Binomial Theorem certainly is most awfully useful for multiplying out a binomial expression.

I should have loved to have seen that place all beautiful and sunny on Sunday, with you lying there eating your two oranges, and trying to write a letter with a bee sitting on the end of your pen.

I must stop this, or you'll never get it before you leave Switzerland. All the time I can imagine you thinking that this is your last Tuesday and then it will be your last Wednesday, and then the very last day of all, and I can imagine how sad you must be feeling at leaving it all, and how you are going for your last walks to the very most beautiful places. But I expect you will go back there some time, and you will see that lovely little grassy, flowery place again, and you will remember how you ate two oranges there, and how lazy that bee's buzzing was, and you'll see all the other places, and they'll be much nicer when you remember how you walked to them with the Baron.

We are having an awfully jolly time at Coll., but isn't it hateful – I am not to take Inter Science in June. You see all your professors have to sign a paper saying that you have attended regularly at their lectures, and as I have missed such a lot, they can't do that. Of course, added to the fact that I was away for five weeks, I was very un-brainy to begin with, so there wasn't much chance of my getting through. Still I should have liked to try and I <u>might</u> have passed by some lucky chance. However, there is no use grousing about it.

Shall we see you before you start at Vickers? I don't exactly know what "starting at Vickers" means. Will you be in London?

Chapter 4

BACK IN ENGLAND – ONE YEAR ON

Comforted by getting Molly's letter of the 26th March just in time to speed him on his way home, and after sleeping off the effects of his journey, Barnes lost no time in seeing Molly. Travelling overnight, he arrived home on Good Friday. On Easter Sunday, 1 April, he called in Hampstead without serious ill-feeling or opposition from Mr Bloxam. With both of them in London, Barnes and Molly hoped that Mr Bloxam would relent and allow their friendship to develop more freely and personally. But he stayed firm in enforcing the restrictions intended to protect his daughter. And so Barnes took up the maths again, posting as before.

Barnes: 4th April 1923. The Cottage, Rudgwick.
Your letter couldn't have arrived at a better time – the last evening at the Coll. when I was feeling very sad. Thank you very much for thinking of sending it – I didn't expect to hear from you till I reached home.

I <u>am</u> so sorry about the Inter and think it is awfully hard lines. I never quite understood whether you really wanted to take pure science, and then took medicine because your maths was weak, or whether you started Science by mistake and really wanted medicine all the time? Now that you have this extra year I should think you would find that you could tackle the pure science. Either you could take the Coll. maths course, or else keep to the subjects you have already chosen, (which will serve equally well for a science or medical degree, will they not?) and have a few lessons in maths in the vacs from a really good coach. I am quite sure you would pick up the necessary amount of maths quite easily. Your disability I think is by no means due to un-braininess on your part – it is simply the way you have been taught, and of course being below the necessary standard in maths has made all the rest seem very hard to you.

About $x^0 = 1$. You would not expect it to equal x if you reflected for a moment that $x^1 = x$, only we never bother to write in the index when it is unity. So x^0 could not equal x^1 (read x to the one); as it would do if x^0 and

x^1 were both equal to x. So they have found another interpretation for x^0 and it is very simply done like this:-

First of all you must know the "index laws"

1) when multiplying like quantities add their indices.

2)   "   dividing   "   " subtract   "   "

Examples:-

$$a^3 \times a^2 = a^{3+2} = a^5$$
$$a^3/a^2 = a^{3-2} = a$$

In this way we can readily find a meaning for negative indices, thus:-

$$a^2/a^3 = a^{2-3} = 1/a$$

Explanation:- $a^2/a^3 = a \times a/a \times a \times a = 1/a$

                therefore $1/a$ may equally well be written a^{-1}

Again $x^5/x^7 = x^{5-7} = x^{-2} = 1/x^2$

          (because $x^5/x^7 = x \times x \times x \times x \times x/x \times x \times x \times x$
          $\times x \times x \times x$

          $= 1/x \times x = 1/x^2$

Equally well we can show that $a^2 = 1/a^{-2}$, so that we find, if we want to change a term from a numerator into a denominator or vice-versa, all we have to do is to move it as required and change the sign of the index. Supposing I have then an expression such as a/a. We may write this as a^1/a^1 if you prefer. We know that it is equal to a^{1-1} by index laws. But by simple division $a/a = 1$.

Therefore a^0 must equal 1. And that is the general interpretation placed upon any term to the power 0 i.e. $x^0 = y^0 = a^0 = p^0 = 1$

So you see that when we differentiate

    $y = 3x$ we get $dy/dx = 3 \times 1 \times x^{1-1} = 3x^0 = 3 \times 1 = 3.$

I say Molly, I do hope you don't really get bored to tears by all this maths?

It was very disappointing about the fives final. When I played the semi-final it was a blazing hot day, and I got tremendously hot, and then like an idiot went and had tea and sat about for a long time without changing, and with practically nothing on. Once one is out of the sun its really very cold, and I suppose I got a rotten chill and sore throat, and spent my last Sunday in bed! I also practically lost my voice, but nobody but me really seemed to regret that!

And about fixing another day next week – you see Molly I myself am perfectly vague as to just what Vickers will want me to do. Probably I shall work in London at first, and if so I don't suppose I shall get away from the office till about 6 p.m and perhaps just at first will have to work even later, until I know my way about; so I could not fix a day, only perhaps to have to break my engagement when you had all made your arrangements. You see I am not my own master, once I have started work. But I will write or ring up next week, and I most awfully appreciated being asked. I wish I had decent manners.

I do hope you didn't think I did not want to stay on Sunday? I felt awfully rude and awkward, and probably behaved worse than usual in consequence, but you did understand that I couldn't abandon Fanny. I love the way you speak of all my beautiful places Molly. You imagine things most awfully well, for that is just the way I think and feel about it all. Perhaps I will go back there some day. It would be lovely to see it all again.

Molly: 8th April 1923. Hampstead
Thank you very much for your letter, and for the explanation about x^0. Of course, if I had had any sense I should have thought that x^0 couldn't equal x when $x^1 = x$; but all the same I shouldn't have known why.

Honestly Barnes I'm not the teeniest bit bored by the maths; quite the contrary – I enjoy them, though if you had told me a year ago that I should be saying that, I shouldn't have believed you. At school I looked upon maths as a necessary evil that had to be got through somehow; but when you started writing about them, it made me feel as if you were really interested in them, and that was infectious. But then there are, of course, maths and maths, and I don't believe I could ever get interested in our Algebra book or Miss Shaw's lessons. (Miss Shaw taught us maths at school, and she was most awfully jolly, with a very pink face; but I couldn't understand a word of her lessons).

I really wanted, and still do want, to do medicine, I never had any intention of taking pure Science. You see for the first year Science people (or at least biological science, not maths or physics of course) and medical people do the same work and take the same subjects for Inter.

I am sorry about the fives final, it would have been so lovely if you had won. And fancy having to spend your last Sunday in bed; how awfully rotten for you; but of course if you <u>will</u> sit about in the cold after playing in the blazing sun, what can you expect? You need somebody to look after you and make you change your things and put on dry socks when you have been out in the wet, etc. Nan would be just the one for that; if you have been out in the rain, she gives you no peace until you've changed your stockings and everything that's the tiniest bit damp. Your only hope is to sneak in the back way and keep out of her sight as long as possible till your things have dried.

Nan has been away, and all last week George seemed to do nothing else but tear himself and his clothes. He knelt on the wood of the frames in the garden, and of course his knee slipped, and he went right through the glass. Luckily it wasn't a very bad cut. Then every day he had some fresh thing which he wanted mended. He started off quite gaily with his overcoat, and we gradually worked through all his garments, till on Friday he said to me very sweetly "Molly, would you deign to be humble enough to lower yourself to mend my pyjamas?" and having got down to the very depths in lowering myself to mend his pyjamas, there we stopped.

Barnes: 20th April 1923. New Cross

I am so glad you like the maths. Of course I love them, there's something so very practical about maths. They always help you to do something. The trouble in schools is that they dont give you a wide enough view of what it is all about. They never make it clear that school maths are merely a means to an end, – the necessary mental equipment as it were, which enables you to tackle much bigger problems later on in life.

Calculus: Chapter VII

It is not often that we are lucky enough to get such a simple function as $y = x^n$, in the course of our work.

We are much more likely to get a function such as $s = ut + \frac{1}{2}at^2$ for instance. Now if from this function we could get the rate of growth of distance with regard to time we should be getting the <u>velocity</u>, for that is what velocity really is.

There are two things that worry us here.

1) there are two terms, ut and $\frac{1}{2}at^2$

2) both of these have already got a coefficient, namely u and $\frac{1}{2}a$.

Take one thing at a time, and we will start on the coefficient, difficulty 2).

Take the general case, and say $y = ax^n$, where 'a' is the coefficient that is worrying us. x grows from x to $x + \delta x$.

Then $y + \delta y = a(x + \delta x)^n$

$$= a\{x^n + n.x^{n-1}\delta x + n.\overline{n-1}/1.2.x^{n-2}\delta x^2 + ...\}$$

$$= ax^n + an.x^{n-1}\delta x + a.n.\overline{n-1}/1.2.x^{n-2}\delta x^2 + ...$$

Subtract $y = ax^n$

and $\quad \delta y = a.n.x^{n-1}\delta x + ...$

Or $\quad dy/dx = a.n.x^{n-1}$ (In the limit).

Note that the 'a' here is nothing to do with the 'a' in $(x + a)^n$. This remark you may snort over as quite unnecessary, but I put it in in case you were puzzled by the happy way in which one uses the same letter in different ways, at the same time. 'a' is very very frequently used as a constant – any constant. It is only very rarely that it has special meanings such as acceleration in $\frac{1}{2}at^2$.

Note then that the constant 'a' appears unchanged in the derived function, $dy/dx = n.ax^{n-1}$

This is always true for coefficients.

Try a different form of constant – the constant of addition, thus, $y = ax^n + b$.

$\quad y + \delta y = a(x + \delta x)^n + b$

$$= a\left\{x^n + n.x^{n-1}\delta x + \frac{n.\overline{n-1}}{1.2}x^{n-2}\delta x^2 + ...\right\} + b$$

Subtract $\quad y = ax^n + b$

and $\delta y = n.ax^{n-1}\delta x$ approx.

or $\quad \dfrac{dy}{dx} = n.ax^{n-1}$ (In the Limit).

Note that the b has dropped out. It disappeared in the subtraction, and we get just the same result as by differentiating $y = ax^n$.

Now for difficulty no 1.

Again to be perfectly general let us try

$\quad y = ax^n + bx^m$

where a and b are constants and n and m are the indices of x.

$\quad$ Then $y + \delta y = a(x + \delta x)^n + b(x + \delta x)^m$

You see x has grown to $(x + \delta x)$ and this growth <u>must</u> occur in both terms.

$\quad$ Expanding:-

$$y + \delta y = a \left\{ x^n + nx^{n-1}\delta x + \frac{n.\overline{n-1}}{1.2} x^{n-2}\delta x^2 + \ldots \right\}$$

$$+ b \left\{ x^m + m.x^{m-1}\delta x + \frac{m.\overline{m-1}}{1.2} x^{m-2}\delta x^2 + \ldots \right\}$$

Whence

$$y + \delta y = a.x^n + n.ax^{n-1}\delta x + \ldots + bx^m + m.bx^{m-1}\delta x + \ldots$$

Sub $\quad y \qquad = ax^n \qquad\qquad\qquad + bx^m$

$$\overline{\delta y = n.a.x^{n-1}\delta x + \ldots \qquad\qquad + m.b.x^{m-1}\delta x + \ldots}$$

or $\quad \dfrac{\delta y}{\delta x} = n.ax^{n-1} + mbx^{m-1} \quad$ approx.

and $\quad \underline{\dfrac{dy}{dx} = n.a.x^{n-1} + m.b.x^{m-1}} \quad$ (really)

Just the result we might have anticipated. You simply differentiate each term separately and add the results together.

We are now in a position to differentiate

$$s = ut + \tfrac{1}{2}at^2$$

$$ds/dt = u + 2/2\,at$$

$$= \underline{u + at} \qquad \text{or } v = u + at!!! \text{ Hurrah!}$$

We never bother to employ the little Duke. We know the rule – Diminish index of t by 1 and x by old index.

therefore ut becomes $1 \times u \times t^{1-1} \qquad = 1 \times u \times t^0$

$$= 1 \times u \times 1 = \underline{u}$$

and $\frac{1}{2}at^2$ becomes $2 \times \frac{1}{2} \times a \times t^{2-1} \qquad = 2 \times \frac{1}{2} \times a \times t^1$

$$= \underline{at}$$

But one does it in ones head.

So we say velocity = rate of growth of distance with regard to time

or $ds/dt = u + at$

We have got the following rules for dealing with constants, and sums or differences

1) An added constant disappears.

2) A multiplied constant remains unchanged (except that it may be modified when we multiply by the old index of x)

3) The differential coefficient of a number of terms is simply the diff. coeff. of each term taken separately and added together afterwards,

The following example illustrates all three rules:-

$$y = ax^n + bx^m - cx^p + d - e$$

then $dy/dx = n.\,ax^{n-1} + mbx^{m-1} - p.\,cx^{p-1}$

a, b and c reappear unchanged, while d and e go out.

The reason for the disappearance of the added constants should be plain if you consider the following example.

Let us plot the following graphs – $y = 2x$

$$y = x$$

$$y = 2x + 5$$

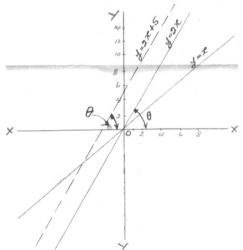

I have written against each of them the function they represent. Notice the following points:-

1) When there is no constant term i.e. in the case of $y = x$ and $y = 2x$, both lines pass thro' the origin (i.e. the point 0)

2) The line $y = 2x$ slopes much more steeply than the line $y = x$.

3) The line $y = 2x + 5$ is parallel to, but separated by a vertical distance of 5, from the line $y = 2x$.

In other words the added constant '5' does not affect the <u>slope</u> of the line, – but only its <u>position</u>. When you remember that dy/dx measures the <u>slope</u>, you will see why the added constant does not appear in the derived function.

Differentiating both $y = 2x$ and $y = 2x + 5$ gives us in each case $dy/dx = 2$. That is to say the trigonometrical tangent of the angle θ is 2, wherever the line may be placed.

Notice again that in these linear functions the <u>coefficient</u> of x really represents the slope of the line. (a 'linear function' is one which when plotted gives you a straight line. You can always tell a linear function, because it contains only variables of the <u>first degree</u> i.e. x or y. x^2 or y^2 or $(x \times y)$ are expressions of the 2nd degree. x^3 or y^3 or xy^2 or xyz are of the 3rd degree. x^4 or y^4 or x^2y^2 or xy^3 are of 4th degree.

All expressions of the 2nd and higher degrees give you a curved line when you plot them.

e.g. $y = x^2$ gives a "parabola"

     $y = x^3$ "   a "cubic parabola"

     $y = 1/x$ "   a "hyperbola"

     $x^2 + y^2 = 1$ gives a circle

     $x^2/a^2 + y^2/b^2 = 1$ "   an ellipse

     $x^2/a^2 - y^2/b^2 = 1$ "   another hyperbola)

Differentiating $y = x$ gives us $dy/dx = 1$, or for every unit that x moves along, y goes 1 unit up.

4) One other point in linear functions:- dy/dx is <u>constant</u>, as x does not in this case appear in the derived function. In some of the functions we have considered, the bigger x was, the faster y grew.

e.g. take $y = x^3$ then $dy/dx = 3x^2$

Obviously the bigger value x has, the bigger $3x^2$ must be, and hence the greater dy/dx or the comparative rate of growth of y.

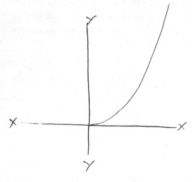

We know $y = x^3$ is a curve like this (its the curve I first plotted for the beanstalk) and the increasing rate of growth is indicated by the increasing steepness of the curve. We might plot this curve, or say $y = x^2$, with an added constant. Let us plot $y = x^2$ and $y = x^2 + 4$ and $y = x^2 - 4$

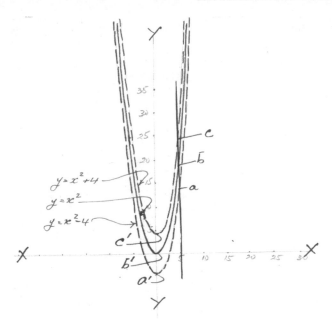

I have drawn the 'father' of the family $y = x^2$ in full line, and the lines $y = x^2 + 4$ and $y = x^2 - 4$ in dotted lines. See how the curves are exactly the same shape, we could telescope one into another if we wished, they are only separated by a vertical distance of 4 units. If we <u>did</u> telescope them we should find that the curves would absolutely coincide and therefore for any given value of x the 'slope' or dy/dx is the same for all 3 curves.

Look at the line I have drawn vertically thro' the value $x = 5$. It cuts the 3 curves at the points a, b, c. Now if I slide the curves vertically until the points a', b' and c' all coincide they you will find the points a, b, and c will all coincide also. Hence the curves are the same shape and therefore also of the same slope, at 'a' and 'b' and 'c'. Looked at another way – if I kept the curves in their present position and drew tangents at a, b, and c respectively, then these tangents would all be parallel. So that the dy/dx would be the same for each curve, and again we see the reason for the disappearance of the added constant in the derived function.

Calculus: Chapter VIII
In this chapter, and indeed in chapter VII, there is not really anything new to grasp; both almost entirely concerned with teaching you how to apply the process of differentiating to certain awkward cases; for it is not always

that we are so fortunate as to get even so simple a function as $s = ut + \frac{1}{2}at^2$ in our physical sciences.

Suppose for instance we had a function like this:- $y = (5x + 3)^2(2x + 6)^2$; and we wanted to find dy/dx? It would be a lengthy business to multiply the whole thing out and then differentiate, and we <u>do</u> get functions where we couldn't do even that. Again, suppose $y = 5x + 3/2x+6$; and we want dy/dx? One doesn't even divide the other exactly – we get a remainder.

How can we tackle these two problems? Notice carefully, that in both cases, <u>both</u> factors are functions of <u>x</u> (i.e. $5x + 3$ and $2x + 6$). That is, although we have a complex function, yet both its component parts are functions of the same independant variable or argument:- i.e. 'x'.

The case where we have two or more independant variables does arise quite frequently, but is dealt with very differently by a process called 'partial' differentiation. At present we are concerned with <u>one</u> independent variable only.

As usual we will take a general case. Let 'u' be one function of x, and v be another function of x – any functions you like, those I have written will do.

Then taking the multiplication first we may write

$$y = uv \qquad (u \times v)$$

Let x grow from x to $(x + \delta x)$ and let the resulting increase in u be δu; and the increase in v be δv; and that in y be δy.

Then

$$y + \delta y = (u + \delta u)(v + \delta v)$$
$$= uv + u\delta v + v\delta u + \delta u\delta v$$

Subtracting $y = uv$ we get

$$\delta y = u\delta v + v\delta u + \delta u \times \delta v$$

Here we see that the 3rd term is the <u>product</u> of two small quantities, i.e. it is a small quantity of the 2nd order of smallness, and therefore as δx, and consequently δu and $\delta v \rightarrow 0$, we may neglect it altogether, Hence

$$\delta y = u\delta v + v\delta u$$

Then $\dfrac{\delta y}{\delta x} = u\,\dfrac{\delta v}{\delta x} + v\,\dfrac{\delta u}{\delta x}$ approx.

and in the limit

$$\frac{dy}{dx} = u\,\frac{dv}{dx} + v\,\frac{du}{dx} \quad \text{exactly.}$$

Now $d\theta/dx$ is simply the differential coefficient of v with regard to x, and du/dx is the d.c. (short for differential coefficient – a frequently used abbreviation) of u with regard to x, for remember that v and w are both functions of x.

If $u = (5x + 3)^2$

$$= 25x^2 + 30x + 9$$

then $\dfrac{du}{dx} = 50x + 30$

And if $v = (2x + 6)^2$

$$= 4x^2 + 24x + 36$$

then $\dfrac{dv}{dx} = 8x + 24$

So that in our 1st case of $y = (5x + 6)^2 (2x + 3)^2$ we get

$$\frac{dy}{dx} = (5x + 3)^2 (8x + 24) + (2x + 6)^2 (50x + 30)$$

which we can leave as it stands or multiply out if we wish.

Try another say $y = (4x^3 + 3)(5x^4 - 2)$

Here $u = 4x^3 + 3$ and $v = 5x^4 - 2$

$$\therefore \frac{du}{dx} = 12x^2 \text{ and } \frac{dv}{dx} = 20x^3$$

$$\therefore \frac{dy}{dx} = (4x^3 + 3)20x^3 + (5x^4 - 2)12x^2$$

$$= \underline{140x^6 + 60x^3 - 24x^2}$$

Let's multiply out first and then differentiate to see if they really do come the same; starting with

$$y = (4x^3 + 3)(5x^4 - 2)$$

we get $y = 20x^7 + 15x^4 - 8x^3 - 6$

$\therefore dy/dx = 140x^6 + 60x^3 - 24x^2$ exactly the same as before!

So now we can differentiate products!

Putting the rule $dy/dx = u.dv/dx + v.du/dx$ into words we say, "the differential coefficient of a product is = the first factor (u) multiplied by the differential coefficient of the 2nd factor plus the 2nd factor (*v*) multiplied by the differential coefficient of the 1st factor".

I always remember it in the other notation like this:- $Dy = uDv + vDu$

And I say it myself when I wish to recall the rule, like this "Dee y equals *u* dee vee plus vee dee *u*. Note that I dont bother to mention to myself the "argument" '*x*', nor do I worry to say "big dee" or "capital dee". The visual memory supplies that detail directly the brain repeats the words, and of course it doesnt matter what letter we chose for the argument in quoting the formula, provided it is the same for Dy, Dv and Du.

By visual memory I mean the mental <u>picture</u> of the shape and form of the letters and the formula generally, which is almost certain to arise on hearing the words – if your memory is at all like mine – I mean, it is not always so – perhaps your memory is more "aural", that is you like to <u>hear</u> the sounds.

Written in the 'capital' notation it is a nice compact little formula to remember, and as we always allocate the letters *u* and *v* to denoting different functions of the same variable, you are in no danger of getting muddled.

The rule holds, however many multiplied functions you may have. Suppose $y = uvw$, where *u*, *v*, and *w* are all functions of *x*. Then, keep two terms as they are, and differentiate the third, doing it in turn to all thus:

$$Dy = uv.Dw + vw.Du + wu.Dv.$$

Note how this is readily written down by inspection from "cyclic order" – you probably know all this – When we have a "symmetrical" expression we can write down all the terms without doing any work, by simply imagining them running round in a circle and then take them as required.

Most simple case –

$$uvw + vwu + wuv$$

are written down in <u>cyclic</u> order.

By simply taking the above expression and differentiating the last factor in each of the 3 terms I get the complete d.c. thus:-

$$wv\underline{Dw} + vw\underline{Du} + wu\underline{Dv}$$

If you care to do the 3 function one for yourself from 1st principles, start like this:-

$$y = uvw$$
$$\text{then } y + \delta y = (u + \delta u)(v + \delta v)(w + \delta w)$$

Then multiply out – reject terms of 2nd and greater orders of smallness and proceed as usual, when you will get the result as above.

Now for the quotients: not quite so simple.

Suppose $y = u.v$ where again u and v are functions of x.

Let x grow to $x + \delta x$

u "   " $u + \delta u$

v "   " $v + \delta v$ all in consequence of the growth of x

y "   " $y + \delta y$

Then $y + \delta y = u + \delta u \ / \ v + \delta v$

An unpleasant looking monster, but let us divide him out, and see what happens. This is just ordinary algebraic division treating δu and δv just like ordinary terms – tho' I admit it looks rather horrid!

$$
v + \delta v \overline{\smash{\big)}\ (u + \delta u)} \qquad\left(\ \frac{u}{v} + \frac{\delta u}{v} - \frac{u}{v^2}\delta v - \frac{\delta u \delta v}{v^2}\ \text{etc. etc.}\right.
$$

$$
\underline{u + \frac{u}{v}\delta v}
$$

$$
\delta u - \frac{u}{v}\delta v
$$

$$
\underline{\delta u + \frac{\delta u \delta v}{v}}
$$

$$
-\frac{u}{v}\delta v - \frac{\delta u \delta v}{v}
$$

$$
\underline{-\frac{u}{v}\delta v \qquad\qquad +\frac{u(\delta v)^2}{v^2}}
$$

$$
-\frac{\delta u \delta v}{v} - \frac{u(\delta v)^2}{v^2}
$$

Notice that the 4th term in the quotient – $\delta u \delta v/v^2$ is a small quantity of the 2nd order of smallness (because it contains $\delta u \times \delta v$ – not because it is divided by v^2), and all terms beyond the fourth will be of increasingly high orders of smallness. Therefore as $\delta x \to 0$ (and consequently δu, δv and δy all tend to 0) we may neglect the 4th and subsequent terms in the quotient and write

$$
y + \delta y = \frac{u}{v} + \frac{\delta u}{v} - \frac{u}{v^2}\delta v \quad \text{approx.}
$$

Subtracting $y = \dfrac{u}{v}$

we get $\delta y = \dfrac{\delta u}{v} - \dfrac{u}{v^2}\,\delta v$ approx.

Adding together the two terms on the left, that is bringing them over the common denominator v^2, we get

$$\delta y = \frac{v.\,\delta u - u.\,\delta v}{v^2} \quad \text{approx.}$$

And dividing thro' by δx both sides of the equation

$$\frac{\delta y}{\delta x} = \frac{1}{\delta x}\left\{\frac{v\delta u - u\delta v}{v^2}\right\} = \frac{v\dfrac{\delta u}{\delta x} - u\dfrac{\delta v}{\delta x}}{v^2} \quad \text{approx.}$$

(Note that I put the δx under the δu and δv because <u>I want it there</u> to form, in the limit, the d.c. of u and v.)

And In the Limit

$$\frac{dy}{dx} = \frac{v\dfrac{du}{dx} - u\dfrac{dv}{dx}}{v^2} \quad \text{exactly.}$$

or in Capital notation

$$Dy = \frac{vDu - uDv}{v^2}$$

In words – "the differential coefficient of a quotient is equal to the denominator × the d.c. of the denominator all divided by the denominator squared".

But I always remember it by the capital formula "Dee y equals vee dee u minus u dee vee all over vee squared".

Notice the points of difference and similarity between the formulae for products and quotients

$$Dy = uDv + vDu$$

$$Dy = vDu - uDv/v^2$$

1) The form is similar
2) The two top terms are made the same way vDu and uDv
3) Products starts with uDv. Quotients starts with vDu
4) Products has 'plus'. Quotients has 'minus'.

5) Quotients has its original denominator squared as the new denominator.

So now we can differentiate Quotients! Our armoury is rapidly increasing.

Let us try a quotient – the one I first wrote.

$$y = 5x + 3/2x + 6 \qquad \text{here} \quad u = 5x + 3 \qquad v = 2x + 6$$
$$du/dx = 5 \qquad dv/dx = 2$$
$$\therefore \ dy/dx = (2x+6)5 - (5x+3)2 / (2x+6)^2 = 10x + 30 - 10x - 6/(2x+6)^2$$
$$= 24/4x^2 + 24x + 36 = 6/x^2 + 6x + 9$$

How jolly simple, but whoever would have guessed that it would turn out like that.

Let's invent another where we know the thing will divide out evenly, so that we can test it by direct differentiation, and see if they come the same. We can easily do it by purposely incorporating the same factor in top and bottom thus

$$y = (2x + 3)(4x - 5)/(2x + 3)$$

Multiply out, and you have the puzzling looking fellow

$$y = 8x^2 + 2x - 15/2x + 3$$

Suppose for a moment we dont know the denominator will divide the numerator exactly, and we decide to work by rule

$$Dy = vDu - uDv/v^2$$
$$\text{Here} \quad u = 8x^2 + 2x - 15 \qquad \text{and} \ v = 2x + 3$$
$$\therefore du/dx = 16x + 2 \qquad dv/dx = 2$$
$$\therefore dy/dx = (2x+3)(16x+2) - (8x^2 + 2x - 15)2 / (2x+3)^2$$
$$= 32x^2 \, 52x + 6 - 16x^2 - 4x + 30 / 4x^2 + 12x + 9$$
$$= 16x^2 + 48x + 36 / 4x^2 + 12x + 9$$
$$= 4(4x^2 + 12x + 9) / 4x^2 + 12x + 9$$
$$= 4 \, !!!$$

(Sorry – I simply wrote down the 1st things that came into my head – they were too simple).

Now from $y = (2x + 3)(4x - 5)/2x + 3$ we get of course

$$y = 4x - 5 \text{ and } dy/dx = \underline{4 \text{ as before}}!$$

Notice, Molly in all my juggling, I perform all the algebraical processes of multiplication, division &c, &c, <u>whilst the Little Duke is still a finite body</u>. That is, I can treat δx's, δu's, δv's and so on as ordinary algebraic terms which I can multiply, divide, or subtract and add at my pleasure. I have to be very careful to do this because it would obviously be absurd to tell you in one breath (or one dip rather) that dx was <u>nothing</u> and in the next dip, to ask you to watch whilst I divided both sides of an equation by dx – that is by <u>nothing</u> and then sweetly smiling assert that the result made sense! No, we must do all the working in finite commensurable quantities and then take the limiting value when the Little Duke fades away. If this distinction is clear to you, you have grasped the only difficult point in Calculus. Do not worry however if you do not quite see it. It will come in time. Stick to the thought that when the little Duke fades, the <u>slope of the tangent</u> remains as a definite property of the curve, and dy's and dx's will always retain their proper meaning for you.

Molly: 8th April 1923. Hampstead
That three function one – $y = uvw$ does come out to $Dy = uvDw + vwDu + wuDv$, because I worked it out to see if it was so. (I'm sure you must be awfully relieved to hear that you are right!)

Barnes: 20th April 1923. New Cross
You amused me very much, or rather I amused myself, by your remark that I must be awfully relieved to hear that the three function differential coefficient was right. The amusing part was, that it wasn't till the fifth or sixth time of reading your letter, that I realised you were making a little joke! Up till then I treated it as a perfectly solemn remark, and said to myself, "By Jove yes, it was a jolly good thing I <u>didn't</u> make a mistake, or Molly would have been awfully puzzled". So you see what a solemn sort of idiot I am. As a matter of fact, although I did at first write the d.c. down 'by inspection', I worked it through as a check, as I cannot risk making any mistakes and so not only wasting your time, but also causing you a lot of worry and confusion (at least, I try not to). You must be getting on jolly well to be able to work that through yourself.

Do you realise Molly its just a year next Monday since we first met? April the 23rd was the 1st Sunday after Easter last year. I wish you were coming again. Do you remember how worried you were over how you were going to earn your living and how you didn't want to grow up.

...Last night we went to see the Beggars Opera. It is perfectly delightful thing. There is the most charming orchestra, with all the instruments they had in Gays time, including a Harpsichord dated 1789, and its jolly sweet to listen to too.... I never go to a theatre, unless I've got somebody to go with, and that doesn't often happen. I once went alone, when I was in the Air Service in 1915, and had to come up to London in a hurry about something, and stayed in a hotel in town, and in the evening went to a show by myself, and felt perfectly <u>wretched</u>!

I seem to be fairly settled in London for the present. They wanted me on Monday, to discuss plans for a new type of flying boat, in metal construction. At present I am going thro' what's called the Estimates and Contracts department where the prices of things are fixed up, and so on. Its most awfully interesting, tho' 10 till 6 is rather a long day. I like the other people there very much. Its absolutely ripping being back at engineering again.

I am most awfully sorry, that I am not ready with my maths chapter this week. You will understand about the maths – I would have done them if I could, only really every moment has been occupied. They are not the least bit of a <u>trouble</u>. I thoroughly enjoy writing them – only it is not the sort of thing I can do in a hurry. When shall I see you again?

Molly: 25th April 1923. Hampstead
I was sorry not to get the maths chapter, but I'm glad you didn't do it if you were awfully busy. I love the maths, but I shall hate them if you are bothering to do them when you haven't time, or if you are tired, or anything.

Yes, you are lovely and solemn; you didn't really think that I'd think it might not be right, did you? Of course you <u>might</u> make a mistake by mistake, (not that you would make a mistake on purpose, but you might be tired or orangey) but that never occurred to me, and I only worked it out for my own satisfaction.

Yes, Barnes, I remember it was a year ago last Monday that I first saw you. I remember after you said "how do you do" to Baba and me, the first thing you said was "Did you have a good night? Oh, that's rather a futile question to ask" (though why it should be, I don't know), and then you smiled, and the first thing I noticed was a whole lot of crinkles round your eyes when you smiled; I've never seen such a lot before. I <u>do</u> hope you don't mind me saying that; it's just to tell you what my very first impression was.

Baba and I went to a dance last Tuesday. A cousin of ours went with us, and it would have been very enjoyable if he could have danced, but he is even worse than I am. .... So Douglas and I trod on each other's toes and got dreadfully mixed up...... I think your partner ought to hold you fairly tightly, and to do the steering himself, not leave half of it to you, and he ought at least to know if he is trying to one-step or fox-trot. Douglas was blissfully ignorant of all such things, and he sailed gaily along, very breathless and very apologetic and always sure that we should get into step in another minute, which we never did.

Barnes: 2nd May 1923. New Cross
I love writing the maths, Molly, only I do like to be able to give you the best I can, that is the only reason why when very rushed I have not written them. Yes, I did really think that you might think the 3-product d.c. might not be right. Why shouldn't you? I often make mistakes, and it never struck me that you would necessarily think it right.

I dont think the new airship scheme will come to anything after all. We had a good look thro' their figures and they are all far too optimistic. The thing simply couldn't be done on the big scale on which they wish to start. And they have grossly underestimated the costs in connection with running a service.

I'm in a way rather depressed about Airships. Until the Government feel rich enough to start experimenting in a relatively small way, I fear nothing will happen. To attempt to start a big commercial service carrying passengers straight away is far too great a risk.

I've just been listening to a lecture by Prof. J.A. Fleming of London University on the wireless. I do hope there's something good on tomorrow. Its the simplest thing in the world to set up a wireless set. You will see when you come.

Of <u>course</u> I don't mind about the crinkles – I like it. They have always been there – must have had a very fat face as a kid I suppose. But you have them too. It will be <u>ripping</u> seeing you tomorrow.

Molly: 9th May 1923. Hampstead
I am so sorry about the Airships; it must be so disheartening when you have worked so hard and are so keen.

Barnes, thank you ever and ever so much for the lovely time you gave us on Saturday and Sunday; the worst of it was that it was all over much too soon. It was topping of you to bother to come and fetch us.

The wireless is wonderful. Quite what I had imagined, I don't know, but I certainly didn't expect it to be like that; it is so beautifully clear and un-scratchy; I think I expected it to be more like a gramaphone.

Thank you very much, too, for the roses, it was lovely of you to give them to us. They are still alive and very beautiful.

Wasn't it fun, Harriet's being away? I love cutting thin (or supposed to be thin) bread and butter. The only thing is, I hope the food lasts out till next Monday. I didn't know you were such an adept in the art of frying sausages and doing friedy bready (how does one spell it?) and making custards; and your salads are topping. I love the way you dance about and get so excited when you can't find the right thing at the right time.

I say, Barnes, do you really think it is horrid of a person not to mind dissecting a rabbit or a dogfish? Of course it isn't nice to think that the poor things have to be killed, but I suppose it can't be helped; and nothing on earth would induce me to do vivisection. It seems to me that you lose sight of the fact that the animals inside isn't very nice to look at, because it is all so wonderful and perfect.

Barnes: 14th May 1923. New Cross
Of course I dont think it is horrid of anyone not to mind dissecting things. Nothing is horrid or repulsive when looked at from the scientific point of view. All things as you say are so wonderful and perfect, that to the scientist they become beautiful. I dont see why any of Creation should ever be considered repulsive. It is only the way in which we look at it that is wrong.... I am awfully sorry, Molly, if I made you feel that I thought otherwise. I was only making fun. I felt quite miserable when I saw the pain I had given you.... Molly, when you are so sweet and open and generous as you are you <u>cant</u> hurt people; there isnt any need to stop to think, and it only spoils the spontaneity and charm of your talk if you do.... So please Molly dont be hurt, and please dont stop to think, or I shall have done you a lifelong disservice.

I'm sorry Molly, I couldn't tell you more about my Mother. You may think it silly for a man, but although its nearly twelve years since she died, I simply cannot talk about it without making an idiot of myself. You see, she was absolutely everything in the world to me.... Molly, I simply worshipped her. Please dont think I dont want to talk about her, – I should love to, only you will understand if one has to stop?

♦ Barnes was fast becoming all things to Molly: friend, host, entertainer, admirer, and a teacher of maths who now produced chapter nine of the Calculus involving Molly's three younger sisters, Betty, Nancy and Pamela.

Calculus: Chapter IX
This chapter introduces you to another and very useful dodge for helping you to differentiate awkward functions. I think one uses it almost more than any other. The dodge is called "Change of Variable", and works like this:-

Suppose we want to differentiate $y = \sqrt{a^2 - x^2}$. Things like this are frequently cropping up, and would completely stump you, as you see the quantity under the root sign is not a perfect square – hence we cannot say "oh well, let's square both sides and say $y^2 = a^2 - x^2$", because we cannot talk about dy^2 ('dee y squared' – be very careful not to mix this with d^2y – dee 2 y, which we have already come across). You see we never want to know the rate of growth of y^2 with regard to x – we never plot y^2 against x; what we are seeking is dy/dx.

So we say "Let $a^2 - x^2 = u$, where u is a function of x, or rather, u is <u>the</u> function of x, $(a^2 - x^2)$. u's, v's and w's are nearly always used for this dodge.

Then substituting, we get $y = \sqrt{u} = u^{1/2}$ and here we have an ordinary function with which we can deal by our first rule.

if $y = u^{1/2}$, then $dy/du = \frac{1}{2} u^{-1/2} = \frac{1}{2} . 1/u^{1/2}$

(you know that root signs can equally well be expressed as fractional indices: e.g. $a^{1/2} = \sqrt{a}$, $a^{1/3} = \sqrt[3]{a}$, $a^{1/5} = \sqrt[5]{a}$ – if not tell me, and I will explain).

Molly: 9th May 1923. Hampstead
No, I didn't know that $a^{1/2} = \sqrt{a}$, but I think I see why it does. Do you remember you said a little while ago that when you multiply like quantities you add their indices? I suppose $a^{1/2} \times a^{1/2} = a^{1/2+1/2}$ and that $= a$. Also $\sqrt{a} \times \sqrt{a} = a$; and therefore $a^{1/2} = \sqrt{a}$. But I don't know if that is the proper way to explain it.

Barnes: 14th May 1923. New Cross
Molly, you are really brainy at maths. Yes you are quite right about $\sqrt{a}$ and $a^{1/2}$. Of course $\sqrt{a}$ is simply a conventional sign, used to indicate the square root of a.

Chapter IX continued:
We have already seen how a negative index may be made positive by moving the term concerned from numerator to denominator, or vice versa. And of course $u^{1/2-1} = u^{-1/2}$ and the original index was $\frac{1}{2}$; so that is

how we get $dy/dx = 1/2u^{1/2}$. Note that $\underline{u}$ is now my variable, in this differentiation.

Now look at our other function, $u = a^2 - x^2$.

Differentiating $\underline{u}$ with regard to $\underline{x}$ we get:-

$du/dx = -2x$ (the constant a^2 disappears)

We thus have two d.c.'s

$dy/dx = 1/2u^{1/2}$ and $du/dx = -2x$

One represents the rate of growth of y with regard to u, and the other represents the rate of growth of u with regard to x.

If I said to you "Betty can walk twice as fast as Nancy; Nancy can walk three times as fast as Pam; what is Betty's rate of walking with regard to Pam?". You would have no difficulty in doing the very simple sum?

Let us put it into miles per hour. If Pam can walk say 2 miles per hour, then Nancy can walk $3 \times 2 = 6$ miles per hour, and Betty can walk $2 \times 6 = 12$ miles per hour. So what I say is obviously absurd, but it will serve as an illustration.

Betty's rate to Pam's rate is then as 12 to 2 or Betty can walk 6 times as fast as Pam. But this was clear from the first for since Betty can walk twice as fast as Nancy, and Nancy 3 times as fast as Pam, then Betty can walk $2 \times 3 = 6$ times as fast as Pam. Or in words, to get Bettys rate with regard to Pam all we have to do is to multiply the two rates together.

We could readily express this in perfect calculus language. I say "perfect" because $d(Betty)/d(Nancy)$ (read "dee Betty by dee Nancy") is as proper a calculus expression as dy/dx and means exactly the same thing – the rate of growth of the distance covered by Betty with regard to the distance covered by Nancy.

So we will say $d(Betty)/d(Nancy) = 2$

and $d(Nancy)/d(Pam) = 3$

Then from our argument above,

$d(Betty)/d(Pam) = d(Betty)/d(Nancy) \times d(Nancy)/d(Pam) = 2 \times 3 = 6$

Notice particularly that I do not 'cancel' the two $d(Nancy)$'s. Remember that $d(Betty)$, $d(Nancy)$, $d(Pam)$, dx, dy, du are not finite quantities, and are not therefore susceptible to ordinary algebraical operations. If you do not understand this point, and it may seem to you a foolishly pedantic one, do not puzzle over it, you will realise the necessity for the distinction later on.

Put shortly d(Betty)/d(Nancy) is a <u>rate</u>, so is d(Nancy)/d(Pam) and all we are doing is to multiply the <u>two rates</u> together, because by so doing we get a <u>third rate</u> compounded of the other two, namely d(Betty)/d(Pam).

So if we multiply the 'rate' dy/du by the 'rate' du/dx, we shall get a third rate – dy/dx. The argument is just the same as that used about Betty, Nancy, and Pam.

But $\qquad dy/dx = 1/2u^{1/2}$ and $du/dx = -2x$

Hence $\qquad dy/dx = dy/du \times du/dx = 1/2u^{1/2} \times -2x = -2x/2u^{1/2}$

or $\qquad dy/dx = -x/u^{1/2}$

But $u = a^2 - x^2$, so replacing this value for u we get

$$dy/dx = -x/\sqrt{a^2 - x^2}$$

How easy, but we could never have done it any other way.
Try another, say $y = (x^2 - a^2)^{4/3}$

put $u = x^2 - a^2$. Then $y = u^{4/3}$

$$\therefore \frac{dy}{du} = \frac{4}{3}u^{1/3} \text{ and } \frac{du}{dx} = 2x$$

$$\therefore \frac{dy}{dx} = \frac{dy}{du} \times \frac{du}{dx} = \frac{4}{3}u^{1/3} \times 2x = \frac{8}{3}u^{1/3}x$$

and substituting the original value of u we get

$$\frac{dy}{dx} = \frac{8}{3}x(x^2 - a^2)^{1/3}$$

Here is one which gives you good exercise in signs:-

$$y = \frac{1}{\sqrt{(a^2 + x^2)^3}}$$

Put $a^2 + x^2 = u$ then $y = \frac{1}{\sqrt{u^3}} = \frac{1}{u^{3/2}} = u^{-3/2}$

$$\therefore \frac{dy}{du} = -\frac{3}{2}u^{-5/2} \text{ and } \frac{du}{dx} = 2x$$

so that $\dfrac{dy}{dx} = -\dfrac{3}{2}u^{-5/2} \times 2x = -3u^{-5/2}x$

$$= -3(a^2 + x^2)^{-5/2} \cdot x$$

$$= -\frac{3x}{(a^2 + x^2)^{5/2}} = -\frac{3x}{\sqrt{(a^2 + x^2)^5}}$$

Chapters 7, 8, and 9 are so important, that before we go on to doing Maxima and minima and some interesting practical problems to which they lead, you can, if you would care to, and have half an hour to spare, try your hand at differentiating a few examples for yourself. But do not bother over it as of course you have heaps of other things to do, and this is only fun. For the first few questions I will give you a hint as to which rule to apply. The last few I leave entirely to you.

For handy reference I will number the rules, and just quote the no. required.

Rule I	$y = x^n$	$dy/dx = n \cdot x^{n-1}$
Rule II	$y = ax^n + b$	$dy/dx = n \cdot ax^{n-1}$
Rule III	$y = uv$	$dy/dx = u\,dv + v\,du$ [denominator dx to dv and du missing in text]
Rule IV	$y = u\,/\,v$	$dy/dx = \dfrac{v\,du/dx - u\,dv/dx}{v^2}$
Rule V	The change of variable.	

I am not always going to stick to x's and y's.

♦ Barnes had clearly envisaged the course continuing with more Calculus, since he had mentioned the next topic to be covered, that of "maxima and minima" and the interesting problems that could arise therefrom. But this was the last chapter on Calculus. At the end of these three chapters VII, VIII and IX Barnes set Molly a test paper of twelve questions, adding the comment "Do you begin to see how delightfully simple this wonderful subject renders the often puzzling parts of physics? The examples I have dug out of a book – I can never make up convincing examples, unless they are actual engineering ones." It was the first time he had given her such a test; indeed it was the only time in the whole course, leaving aside the few odd questions which he threw in along the way, to which he generally gave, or hinted at, the answer.

Molly did not disappoint Barnes in the end, although she did not attempt the questions for some time. Towards the end of September, when the new term was about to begin, she sent the answers.

Molly: 20th Sept. 1923. Hampstead
I have done (or at least tried to do) those examples you sent me last May. There is one that I can't do at all, and I don't know if the others are right.

It certainly does make the examples most awfully easy to do, if you can do them by the Calculus, and it was fun working them out. But I like the way you calmly say that perhaps I'd like to do them if I have half an hour to spare! It took me quite three hours. You can't think how stupidly slow I am at working out sums, and when I go through them again I always find heaps of silly mistakes.

Barnes: 28th Sept. 1923. New Cross
The maths were very good indeed. I will send them back with notes in a day or two. You only had 2 mistakes, which you will easily understand. You mustn't bother to do things if they take too long, but I think the time you took was quite good, seeing that there were 12 sums, and you hadn't done any Calculus for 6 months. Half an hour would only be the time if you were an expert and had had a lot of practice at differentiating.

♦ In April, by the end of this first year of their acquaintance Barnes and Molly knew much more of each other through their letters: likes and dislikes, interests, feelings and opinions, activities both personal and more public, families, friends. In addition to all this, Molly knew more calculus, and Barnes knew more of her mathematical needs. The letters, to say nothing of the maths, had occupied many hours, were eagerly awaited and greatly valued. Mr Bloxam had gained time, but no victory; and time served to harden the intentions of the two correspondents. As the letters continued to go back and forth, the topics that had run through the first year reappeared, all but the calculus. That chapter of the course was closed.

Chapter 5
AIRSHIPS AGAIN

Commander Dennistoun Burney, a dynamic figure, an inventor and an engineer, was causing a commotion in Barnes's life. In 1922 and early 1923 he was pressing for financial support for an ambitious project to construct a system of airship 'Liners' which would link the far-flung parts of the Empire. The proposed 'Trans-Oceanic Airship Liners' would fulfil the function of ocean-going liners at a much greater speed, with benefit to the Empire. Government had blown hot and cold on support for the airship industry, as Barnes knew to his cost. Socialists would not favour assisting private industry; Liberals would oppose any action which could be construed as leading towards a policy on armaments. Action required someone who could reach the ear of those in power. Commander Burney had political influence, self-assurance and money, all of them advantages which Barnes lacked. While Burney's indulgent life-style did not accord with Barnes's ascetic attitudes, the two shared an intense belief in the importance of the Empire to Britain itself, to the Dominions and Colonies, and to the world in general. The opinion that the excellence of Western culture, ethics, political and legal systems benefited all who participated therein was widely held at the time; and the two men shared a determination to develop speedy and efficient communications between every part of the Empire upon which the sun still never set. Airships had a greater carrying capacity and could remain airborne for longer than heavier-than-air craft. Once certain design problems were overcome, airships would bring benefit not only to trade and commerce, but also to the emotional unity of the great family of nations. The reduction of time taken for the mail, and for travelling back and forth, would overcome reluctance to leave Britain and loved ones often felt by those who were needed to work abroad.

Vickers had been investigating the possibility of developing such a service in 1920 but backing from the Government had been firmly refused and the airship programme closed down. Burney, however, was not a man to give up. Still formulating this plan, he began to consider his team.

Wallis, who had been in the forefront of airship design during the First World War, was one of the obvious choices. Until the General Election in the Autumn of 1923, the signs were hopeful but, as Barnes wrote to Molly on November 19th, the election threatened these hopes. A Labour Government, the first to take office, made Ramsay MacDonald Prime Minister, promising policies of Socialism and pacifism with which companies like Vickers Engineering did not comply. Furtherance of airship development was confused and finally confounded by political antagonisms over the control of important industrial concerns, and by in-fighting between Admiralty and Air Ministry claims to expertise and experience. The outcome was the construction of two ships, the R100 under the name of Vickers Engineering, and the R101 under Government sponsorship. The British airship industry ended in tragedy some years later, when the competition for prestige over-hastily launched the R101 on her fatal maiden voyage to India, carrying many important men to their deaths at Beauvais on 5 October, 1930. This disaster sent the successful R100 to the scrapheap.

Molly: 9th May 1923. Hampstead.
I wonder if you would mind telling me a little about Trigonometry next time? After much thought and mentally drawing a triangle, I can remember what the sine, cosine and tangent are. But our Physics man says something about equating the sines to their angles if the angles are small; or have I got hold of the wrong end of the stick? And can you have a sine of an angle when it (the angle) isn't in a triangle?

Barnes: 15th May 1923. New Cross
Oh Molly, such exciting times. Yesterday, at a moments notice I had to attend a conference on Commander Burneys financial scheme for starting a great Imperial Airship Transport Company. He didn't know who I was, but during the night he must have found out, for first thing this morning he rang me up, and asked me to go and talk to him. And then and there he asked me to join in with him in the new company.

I wasn't too keen for a number of reasons, too long to write. Mostly you see my present post in the Vickers company – the biggest engineering firm in the world I think – is an exceptionally good one, with very good prospects of advancement.

On the other hand, Airships have twice stopped altogether, once at the beginning of the war, when I was jolly glad, because I was able to shoot off and enlist, and then in 1921, when also I was not sorry, because I was able to come to London and work for my degree, for which I had matriculated and laid my plans 10 years before!

Now however I want to settle down – suppose I join Burney, and in another few years Airships once more break down? I shall never get another opening in Vickers such as they offered me last Christmas, and wh. I am now enjoying. I should stand the risk of being thrown out of a post for the third time; and it becomes increasingly difficult as one gets more senior, to get posts wh. will suit, as one is not as when a boy, prepared to start all over again at the bottom. Its a big decision. Which shall it be, Molly, adventure again, or safety? I've got to make up my mind in a day or two. Why have all the exciting times in life all come at once.

Some funny fate pursues me, I never seem to be able to settle down, and think I'm safe for life, but something else crops up, and off I go again. I've had 14 different sets of rooms in 16 years. When shall I have a home of my own? What a book I could write on landladies!

Next day. I think it's going to be safety after all, not because I have made a decision, but because I think my particular director of the Company [Vickers] will not let me go. He says perhaps he will be willing to lend me to Burney for 3 months. Ive been having interviews all day, and there are more tomorrow. I simply hate some of these interviews, when I feel I am being "inspected". I always have my oldest suit on, and feel sure theres a smudge over one eye, or my hair isn't tidy. But you must be tired of all this, tho' its such fun for me, having someone to tell it all to. There's lots and lots more, but it would take hours to write. I must save it up to tell you if you care to hear about such things. Perhaps you dont.

Next day. I'm feeling rather heartbroken Molly. It's been one long conference all day. And now I fear I am out of airships – for none of the reasons that I thought – but because under the very complex circumstances, the only decent thing I could do was to stand aside and let Pratt, who has been my colleague for many years, step into the place that Burney offered me.

What an illogical idiot I am. When the thing was undecided, I could see many arguments against joining Burney – tho' even as I wrote to you I said to myself – "You are pretending to be very judicial and cautious, but really in your heart you know there is only one thing to do, and that is to join him". And now that it is all over, I feel perfectly miserable!

Next day. Auntie Fanny and Father left for Wreningham today and I am in lonely state. Do you think I could come to Hampstead on Monday? But perhaps you have something on?

I shall love to do the trig. and am very sorry I have had time to do so very little. I will do more next week and send it on separately if you think I may? No time for more. Please, Molly, forgive this scribbled letter, it <u>has</u> been such a busy week. Shall I ring up tomorrow, Saturday morning, or are you going to a lecture?

I had meant to get this letter off before, but one thing after another has prevented me. Last night Fanny and I spent hours, translating the trade terms on a wretched letter from Danzig asking for engines for a ship!

Molly: 20th May 1923. Hampstead
This isn't really a letter; it is only to ask you to do some Trig. this week, if you have time. Daddy doesn't mind a bit, and I should love it; it is awfully nice of you to suggest doing it; thankyou very much indeed. But Barnes, be sure you don't do it if you haven't time, which will be quite likely if there are a lot of conferences as there were last week; I shall quite understand if you find you can't manage it.

I am already in the middle of writing you a proper letter, and it seems rather silly to write a letter in the middle of a letter, but the proper one won't be finished by this evening.

I do wish you were coming to-morrow, Barnes, but Daddy is absolutely unpersuadable. Aunt Friedli isn't coming after all, and we wanted to ask you here for the week-end since Uncle Charlie and Auntie Fanny are away, but we couldn't. We are all most awfully disappointed.

Please excuse the shortness and scrappiness of this; I wouldn't have written it at all, if I could have spoken to you on the telephone yesterday, but nobody told me that you had rung up until it was all over.

Oh, I do wish you could have come; I was getting so excited. I hope the being in "lonely state" isn't <u>too</u> lonely, Barnes.

I am so sorry about the airships, Barnes. When I first read about it, I thought how lovely it was for you to have a chance of going back to them; and it must have been so disappointing when you decided that Pratt and not you must join Commander Burney. I can just imagine how, all the time you were thinking it would be wiser and safer not to join him, you knew that because it was Airships you must join him. I should think you did hate those interviews, with people inspecting and criticizing you. I <u>love</u> hearing about all the interviews and conferences and everything; it is every bit as much fun for me to hear about it as it is for you to tell it. I like hearing about every single solitary thing you do. Perhaps there will be another chance for you with Airships again soon; I do hope there is.

Barnes: Undated, postmark 23rd May 1923. New Cross.
So very many thanks for your note. I am awfully sorry I have not been able to finish off the enclosed, but am leaving at a moments notice for Germany, where I expect to remain about a week.

I am going with Commander Burney to Friedrichshafen to advise him on the purchase of the Zeppelin Airship Works. Fearfully secret!!

Chapter I. A few notes on Trigonometry.
For the purposes of Inter Physics you do not want to know very much trigonometry. The thing to do is to learn only such parts as are necessary for you to understand fully your other work.

To start right from the beginning, I must run thro' the convention adopted for representing positive and negative quantities by plotting (or drawing) on paper. If you know this, so much the better. It is the basis of the modern method of teaching trigonometry – at least the way I taught it – it seems the simplest.

Just a word first to draw you attention to what I mean by a "convention". When I was learning – generally on my own, there was never anyone to tell me what things were merely conventions, and what things were unalterable (for want of a better word), facts. Consequently I used to spend much time and trouble worrying over the "reason why" of things that had no "reason why" beyond their great convenience. For instance, its no good trying to work out why $\sqrt{a}$ means the "square root of a"; it only means that, because we choose to give that funny sign that particular signification; so you just have to accept the fact without mental argument. In other words, it is a convention. And conventions, in mathematics are almost invariably adopted or created, because they enable us to express in convenient and concentrated form, what is often a very complex idea. Who would bother to write "Ratio of length of the circumference of a circle to the length of the diameter of a circle", when he can express the whole idea for everyone, including himself by writing "π"? Why "π"? I don't know, nor do I very much care! It is merely a "convention". So, much of what follows is simply "convention", and may therefore be accepted by you without further argument or question as being the acknowledged practice of mathematics throughout the world.

The author of the particular convention that I am going to explain was a Frenchman called Descartes. He did it I think in the 18th [sic] century. Hence it is called:- "The Cartesian System of Rectangular Co-ordinates. We are only going to bother about this system for lines or points that all lie in the same plane, e.g. the plane represented by this sheet of paper.

Now, supposing I draw any two straight lines (<u>Any</u> two straight lines, provided they <u>meet</u> – <u>Parallel</u> lines will not do.) on this, (or the next) sheet of paper. You will readily see, that having done so, I am then in a position to write down in figures the exact location of <u>any</u> point on that sheet of paper with reference to those two lines. Neither more nor less that 2 lines are required – See

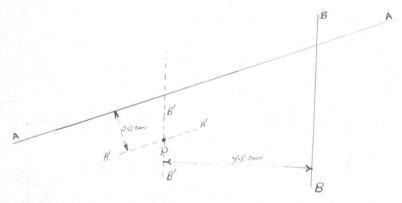

At random, take any point P. My lines are supposed to be capable of extension to infinity in either direction. And suppose <u>you</u> have another sheet of paper, with two lines AA and BB, which lie at exactly the same angle to one another and to the paper. And I wrote and told you that P was situated a perpendicular distance of 2.2 cm from AA, and 7.8 cm from BB. By drawing the two little bits of lines A′A′ and B′B′, you would have no difficulty in obtaining the exactly corresponding spot on your own diagram.

The two lines AA, BB, are termed the "axes of reference", and the distances 2.2 and 7.8 cm are the "co-ordinates of the point P, with reference to the axes AA, BB."

But we are not quite safe yet, for supposing <u>you</u> had marked off your 2.2 cm on the <u>other side</u> of AA? You would still get a point, but it wouldn't be the same point as my "P". Look:-

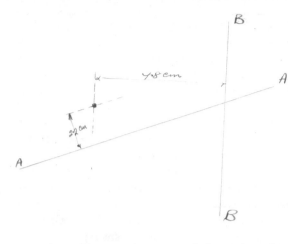

Again, you might have put the 7.8 cm on the other side of BB. and then taking 2.2 cm first on the lower, and then on the upper, side of AA, you get a further two possible positions for P.

How are we to get over this difficulty? Well, I might amplify my directions thus:- The point lies 2.2 cm <u>below</u> AA, and 7.8 cm to the <u>left</u> of BB. That would fix it entirely.

Actually, in maths, we <u>choose</u> to regard distances measured on the upper side of AA, as positive, and those measured on the lower side of AA as negative with regard to AA. Similarly distances measured to the <u>right</u> of BB are regarded as positive, and those on the left of BB as negative with regard to BB.

For the sake of convenience we always take our 'axes of reference' at <u>right angles</u> to one another (hence <u>Rectangular</u> coordinates) and we generally call them XX and YY thus:-

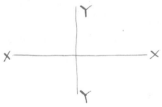

Each is of course supposed to be continued to infinity in both directions. Now, taking one line at a time, and considering the XX line first:-

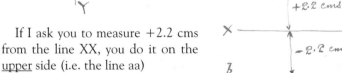

If I ask you to measure +2.2 cms from the line XX, you do it on the <u>upper</u> side (i.e. the line aa)

If I ask you to measure −2.2 cms from XX you do it on the <u>lower</u> side (i.e. the line bb)

Notice, that until we introduce the 2nd axis of Reference YY, we cannot determine the point required. It might be <u>anywhere</u> on line aa if + ve and <u>anywhere</u> on line bb if -ve.

Molly: 23rd May 1923. Hampstead.
I have just got the maths. Thank you so much for them. Are you really going to Germany; how exciting for you. It looks as if you had joined Commander Burney after all. I am longing to hear all about it.

Barnes, I like you even better than ever because of the way you think about your Mother. I know she was absolutely wonderful, and of course I understand that it is difficult to talk about her. Certainly it isn't silly for a man – quite the contrary – it makes him nicer. It <u>was</u> hard for you all, that she died when she did. I would love to have known her.

You know those roses you gave us – the red ones were surprising. Nancy [her sister] threw them away because they were all droopy and she thought they were dead. But I managed to rescue them, and I cut their stems and put them in fresh water in a vase in my room, and they lived till last Saturday, with much stem-cutting and water-changing.

I don't know if "at a moment's notice" means to-day or to-morrow, or when. But I am hurrying this so that it will catch the post to-night and get to you to-morrow morning, in case it means to-morrow. If you have already gone, it will be rather silly of me, because there is heaps more to say.

Goodbye, Barnes. Good luck to you in the "fearfully secret" business. It sounds most awfully thrilling. I do hope everything turns out satisfactorily and that you find time to enjoy yourself in the midst of the secret transactions. I am sorry this is so short, but I do want you to get it before you start if possible.

♦ It is not clear whether the next section, which does unusually carry a date, was intended to be a separate chapter, or whether it was actually the rest of Chapter I as indicated by Barnes. The only guide is that the chapter following is numbered Chapter III.

Trigonometry. Chapter II. [Query] 21/5/23
Turning to the YY axis, we see that distances measured to the right of it are considered positive, and distances measured to the left are considered negative. The line cc is dotted in at +2.5 cms (approx) from YY axis and the line dd is dotted in at −3.3 cms (approx) from the YY axis.

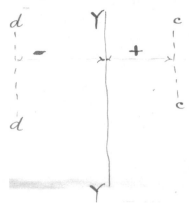

Notice again, that until we introduce the 2nd axis of reference XX (not shown) a point + 2.5 cms from YY may lie <u>anywhere</u> on the line cc, (wh. may be continued parallel to YY to infinity in either direction). Similarly, a point −3.3 cms from YY may lie <u>anywhere</u> on line dd (which may be continued parallel to YY to infinity in either direction)

I emphasise this point because these lines cc and dd are known as "loci" or "locuses" of a point which moves in a path so that it is at a constant distance (namely + 2.5 cms) from YY. (Also of course the lines aa and bb on previous pages.)

<u>Definition</u>. A Locus is the name given to the path traced out by a point which moves <u>in accordance with a fixed law</u>, or laws.

In the case just referred to, the law if "that the point shall always be a perpendicular distance of + 2.5 cms from YY axis".

A locus is not always, or even usually, a straight line. Other common examples of loci are the circle – wh. is the locus of a point moving at a fixed distance from a given fixed point (i.e. the centre); the ellipse; the parabola; the hyberbola; any and every algebraical function. But we need not bother further with them at present.

Now let us again put our two axes of reference together and see what we get.

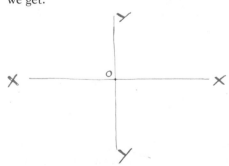

To call the vertical one YY and the horizontal one XX (ecks-ecks) is again merely convention. The point O where they cross is called the "origin". (O for "<u>o</u>rigin" also for "nought")

We have seen that completely to determine (see how beautifully I don't split my infinitives) the position of a point in the plane (in this case the plane of the paper. We assume it <u>is</u>

plane. It never was truly plane, and certainly wont be after going thro' the post!) in which the axes lie, two and only two dimensions are necessary:-
1) Its perpendicular distance from line XX
2) its perpendicular distance from line YY

How do these two dimensions completely determine the position of a point? Each of the two dimensions represents <u>a locus</u> and the intersection of the two loci gives us the <u>point</u>.

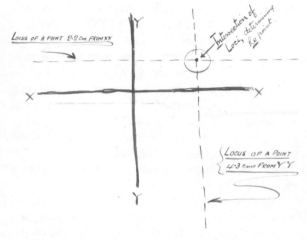

Locus of a point 2·2 cm. from XX

Intersection of Loci, determining the point.

Locus of a Point 4·3 cms from Y Y

Each dimension of course having its proper sign + or − prefixed, to enable us to measure off on the correct side (i.e. upper or lower for XX − right or left for YY) of each axis. Note here that, as is usual in algebra, when a quantity is +ve we dont bother to indicate the sign. We only trouble to write the -ve. Any no. without a sign is +ve automatically.

I call the axes XX and YY because usually in maths, when one wishes to discuss some purely mathematical point, that is to say, when one is dealing with symbols which do not necessarily relate to some particular property of the physical world such as s (space); p (pressure); v (volume) etc; it has become the custom to use x, y, or z. They are purely mathematical you pay them double, and give them any (including no) meaning that you choose.

And again, it has become the custom always to measure values of x <u>horizontally</u> to right or left of the vertical axis; and to measure values of y vertically up or down starting from the horizontal axis.

Hence we call the vertical axis the y axis and the horizontal axis the x axis.

Because you see, if you are plotting the graph given by a function $y = f(x)$ – too puzzling? All right, suppose you are plotting $y = x^2 + x + 5$, before you begin to plot, you choose some length to represent unity (that is the number – <u>one</u>), and by marking off a series of such lengths you get your "scale" by means of which you represent on paper actual numbers. Not very clear I am afraid – I'm not in my usual form to-day somehow. Anyhow lets draw it, that is the best way to make it clear. Do not stop here to puzzle over the graph, read st. on.

Let us choose at random .5 cm. to represent one unit. Marking off on our axes a series of points .5 cm apart we get our scale. (see figure below). Always start from the origin.

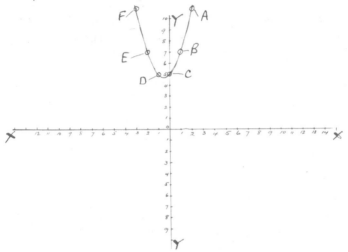

Now as in Calculus, x is always the independant variable or argument.

Let us assign a few <u>definite</u> values to this at present <u>indefinite</u> variable. 'y' the <u>dependant</u> variable is then compelled to follow suit. Why? Because we have <u>chosen</u> for the moment to say that

y shall equal $x^2 + x + 5$ <u>whatever</u> value x may have.

There is no other reason. One might say that "independant" variable is a bad name for x, because the value of x depends on our whim. Just as you please – I prefer "argument" myself, but the name "independant" refers really to the relation between x and y, and not to the relation between x and ourselves.

We will then exercise our prerogative, and make that independant fellow x mean what we choose. And because it is a Bank holiday he shan't have any pay at all! You will see from the size of our paper, that as far as x is concerned, it is no use taking any values of x much over 12 or 14, but since $y = x^2 +$ something and the y scale stops at 9 or 10, it is, as far as y is concerned no use making x greater than about 3, if even as much as that. If we want a greater range of values, we must either increase the size of our paper, or decrease the unit length of our scale. Remember that we may have x either +ve or -ve. Thus when

$$x = -3 \quad -2 \quad -1 \quad 0 \quad 1 \quad 2 \quad 3$$
$$\text{then } y = 11 \quad 7 \quad 5 \quad 5 \quad 7 \quad 11 \quad 17$$

You will see that values of y must be hunted out on the YY axis, while values of x must be found on the XX axis. As a matter of interest the curve here shown is symmetrical about a vertical axis passing tho' $x = -\frac{1}{2}$ and it is a parabola, shooting off to infinity. Note again however great (numerically) we may take the -ve values of x, y is always +ve.

Now we can name each of these points very neatly and easily.

The point marked A is known as the point (2,11)

    "   "   "   B " " " " " (1,7)

    "   "   "   C " " " " " (0,5)

    "   "   "   D " " " " " (-1,5)

    "   "   "   E " " " " " (-2,7)

    "   "   "   F " " " " " (-3,11)

We always write it in brackets, with the x (or horizontal) value quoted first, the two figures being separated by a comma.

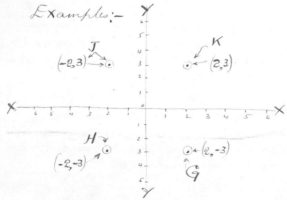

Examples:-

See how, with the same <u>numerical</u> value for x (i.e. 2) and for y (i.e. 3) in each case, we can get 4 different points by varying the signs.

I want to draw your attention to the fact that we can have a +ve value for x, in conjunction with a -ve value for y (see point G above), or 2 -ve values (see point H), or a -ve value of x with a +ve value of y (see J) or 2 +ve values as at K. <u>This is important.</u>

The 4 right angles made by the two axes are termed Quadrants, and are numbered for purposes of reference thus:-

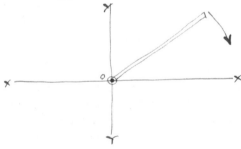

Note that the quadrants are numbered the opposite way to that in which the numbers on a clock go.

NOW, this convention has been extended to enable us to represent <u>angles</u> on the same diagram. For supposing we imagine a rod to be pinned at the point O so that it can rotate about the pin, then if we always make the rod start from the same place we can swing it round in <u>either direction</u>, thro' any angle we choose, just like the big hand of a clock, and the place where it stops indicates the angle thro' which it has turned from the starting point.

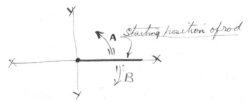

The convention is, always to make the rod start from the positive XX axis thus:-

If we swing it upwards, in direction of arrow A, we say it describes a pos-itive angle. If we swing it round downwards, as arrow B, we say it describes a negative angle. The first is called rotating anti-clockwise, and the second, clockwise, the names being self-explanatory from the way the hands of a clock move.

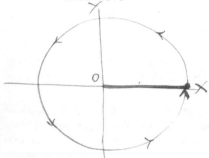

If the rod starting from OX, goes anti-clockwise thro' a com-plete circle, we say it has described or passed thro' an angle of + 360 degrees (written 360°). If it went round clockwise, and again came to rest along OX, it wd. have passed thro' −360°.

♦ Barnes wrote from Friedrichshafen as soon as he arrived, describing in amazement the financial crisis in Germany. Tension in Europe was high. Early in the year French and Belgian troops had occupied the German industrial area of the Ruhr valley in an attempt to force payment of repa-ration for war damage laid down in the Treaty of Versailles in 1919. Burney had made a previous visit to Friedrichshafen for discussions with Dr Eckener, who had succeeded Count von Zeppelin at the head of the Zeppelin company. The Friedrichshafen airship shed had been reprieved from the destruction of the German airship industry ordered in the 1919 Peace Treaty since a rigid airship for the U.S.A. was under construction there; but the shed was to be destroyed on completion. The Zeppelin company, however, would not be inactive as a consequence, since their expertise was sought after in other countries in Europe, and more impor-tantly by the U.S.A. Negotiations between Zeppelin and Goodyear were already advanced; German skills and technical knowledge were to be incorporated into Goodyear, for which Zeppelin would receive a third share in the company. Burney realised Britain was being left out of the race.

Barnes: 26th May 1923. Friedrichshafen
I cannot resist writing a line to send you <u>one penny</u>, in the form of a thou-sand mark note. One wanders about with ones pockets simply bulging with sheaves of these things running up to one hundred thousand (8/4d)

notes. Our dinner last night at the hotel, if the mark had its proper value, cost us £13,000. As Commander Burney said, even Rockfeller [sic] couldn't keep it up long at that rate.

Sunday I wrote the above, under the belief that I should be starting for London today. Now it is all changed and I am to stay here for I fear at least another week. They tell me that letters to London take about 5 days to get thro', so that if I post this tomorrow, it should reach you approximately on the right day. So that I shall not be breaking rules after all; nevertheless I leave the above, for it was indeed my intention to do so. But it was too interesting to miss, as I also wanted you to see 300 marks in stamps on the envelope. Nominally of course this is worth £15. Actually it (the stamps) are today worth but little over a farthing!! Why ever is it that every letter which I sent from Switzerland cost 4d at least – many of them more of course, as one cd. only send 20 grammes for that. No wonder many English firms who have to send out lots of circulars, are having them posted in Germany.

We got here at 5 o'clock on Thursday afternoon, after travelling continuously since Wednesday morning. Very tiring, as being in such a rush we could get no sleeping accommodation on the trains, and so got no sleep for 40 hours. I used to be able to do this easily years ago, and have even been to play tennis after 36 hours continuous work; but now I find I get very tired and sleepy.

Nevertheless we went straight to the Zeppelin Works and started negociations and got no tea! Friday we conferenced from 10 in the morning till 11 at night, and yesterday, from 10 a.m. till after midnight, and today we were at it again till about half an hour before Burney was due to leave at 1.40. He has to be in London for a Cabinet Committee meeting on Tuesday, and so has had to leave me in charge to draw up the draft agreement between the English and German companies. As the whole affair involves capital to the extent of over eight million pounds (English) it is no small affair, and the inter-related companies are very complex. I hope to goodness I dont make a mess of it, but do not feel much worried.

There is now, in my mind not the least doubt that in 2 or 3 years from now, we shall have the most wonderful series of air lines running from England to India, twice every week and back, in $2\frac{1}{2}$ days! Carrying 200 passengers each, and mails and parcels. I don't care if I die after that. It will only be the start of course, and the line will subsequently be extended to Australia and New Zealand, and then there will be other lines.

This is a very beautiful place on the N. side of L.Constance, or the Bodensee as they call it in German. It is bigger than L.Geneva, I think, wh. is about 50 miles long and the mountains stand further back, giving one a much more open impression. Three nations have their frontiers on the lake, Austria, Germany, and Switzerland. Curiously, although it is in Germany, (i.e. Friedrichshafen) the quickest way from London to here, is thro' Paris, then down to Basle on the Swiss frontier, and then thro' Zurich to Romanshorn, a little Swiss village on the opposite (South) side of the lake. From there a service of steamers runs twice a day to here. It takes about an hour to cross. I feel quite an old hand at Continental travelling now. We were going to fly to Paris, but found it did not save any time, much to my disappointment.

My German is very small – I knew much more when I was at school, but have never kept it up since. The first morning I had no end of a job to make the Zimmermadchen understand that I wanted a <u>cold</u> bath. I'm sure she thought I was mad. When I got it, she had filled it to the brim, with <u>icy</u> water. I had to let some out, as it was the sort of water one either goes into with a flop, or stays out of altogether. I'm not one of those strong minded iron willed heroes who can gently lower themselves into cold water without a murmur. Are you?

I've just been for a walk along the lake, for about an hour – the first exercise I have been able to get. It feels a bit lonesome without Burney. I hate being alone in a big hotel. Everyone else seems to have a particularly charming and jolly companion, and you sit alone and glum at your little corner. This a most luxurious hotel, for Friedrichshafen is only a little townlet of 12,000 inhabitants – very picturesque and quaint. It was only a small village before Count Zeppelin chose it to build his first Airship. We went thro' their Airship museum the other day – most fascinating and interesting, with relics of his earliest ships, dating back to 1900. He was a very wonderful old man.

Oh, the food, and the men, and the women! I have seen some of the most disgusting men I have ever seen in my life; huge rolls of fat hanging over their collars, and their tummies! One couldn't help wondering whatever shape their coffins would have to be! I shall starve till I am as thin as a rake when I get back. Starve here one simply cannot. KEWS [he had tried Queus Qu Qeues, crossed them out, and given up] of obsequious waiters all but push food into ones mouth. I'm sick of being waited on. When one asks for the mustard they wont let you help yourself! And the food! Molly I'm not a pig, really I'm not, but I have consistently eaten too much ever since I came. We all have. Its the most delicious, most tastefully served, food I have ever eaten. Generally, I dont care a hoot what I eat, you know I dont, as long as it doesn't give one tummy-ache, but here one actually thinks what new 'delicatessen' they are going to spring on one next.

Burney and I actually used to consult solemnly with the head waiter as to what we should have for dinner, and used to feel quite interested in it, – a performance which usually bores me to tears.

<u>Monday</u> I have now definitely offended the head waiter, and order any old thing I want much to his disgust and disapproval. He clearly thinks I'm mad. I simply couldnt, after Burney left – stand their long drawn out and weary meals. I fear I must finish this, as I do not care to risk leaving it until tomorrow. If I get time to write any trig. I will send it on separately. It is too aggravating to think of a letter from you sitting waiting at home for me. Thank you ever so much for your note, which I got safely on Monday. It was very kind of you to write. I daren't ask Harriet to send on your letter, as I should probably miss it.

Molly: 7th June 1923. Hampstead
Just fancy a thousand marks only being a penny, and I suppose a hundred marks is worth absolutely nothing! And the stamps on the envelope too – I am glad you sent me a letter from Germany. Poor Germany, I do feel sorry for her – at least I suppose it is bad for a country to have such a low exchange, or do I mean high? I haven't the remotest idea which it ought to be, and I don't even know if exchange is the right word; but I am trying to be very learned and intelligent.

You must have been tired on Thursday evening – fancy, 40 hours without any sleep! And all those long conferences must have been very tiring too. What a responsibility for you having to draw up that agreement; I hope it was satisfactorily arranged? Do you really think there will be air lines from England to India in a few years? Won't it be exciting; I am so glad it will be so soon, because I'm always afraid things like that won't happen for about fifty years, till after I'm dead, and I do so want to fly in an Airship. And won't it be lovely for people in India to be able to come home to England so quickly.

Goodness no, I most certainly don't gently lower myself into my cold bath – quite the contrary – I jump in with the greatest of haste, and out again even more quickly. Our water in the summer is horribly warm, but in the winter of course it is plenty cold enough. Why should your zimmermadchen be so surprised because you wanted a <u>cold</u> bath; don't

Germans have cold baths? I've never learnt German; is it easier or more difficult than French? A client of Daddy's came to dinner the other evening, and he and Mother talked German to each other; it sounded like nothing on earth – all sort of right back in your throat, and the way he rolled his r's. He was enormously fat like your German men; he had only been in England for a week, and so I suppose hadn't had time to get any thinner, or rather, less fat.... It is queer how much Germans like eating; and it must have been a bore for you to spend such a long time over their weary long meals, specially when you were by yourself. I have never stayed in a hotel in my life, but I should think it would be rather lonesome if you are alone and everyone else has a friend. I would like to see an Airship on the ground right close up to me. I can only remember seeing one once when I was young, a long time ago at Elvaston and it was in the air then. Are German Zeppelins called after the name of the man who made them, then? I always supposed they were a different kind of Airship.... What made you first start on Airships, Barnes? No – don't tell me now, I'd rather hear you tell it some other time.

Do you know, I don't know if you have come back from Germany yet; Mother had a letter from Auntie Fanny this morning, but she has been so dreadfully busy all day that the poor dear hasn't had time to read it yet.

Barnes: 14th June 1923. New Cross
It was simply delightful to get home after more than 2 weeks absence, and find your ripping letter waiting for me. Thank you most awfully. And then 2 days afterwards to get another, I hardly knew myself. Dont you sometimes find "thankyou" an awfully inadequate expression? I do, and this is one of the occasions. Still I must be content. There's such heaps to write about and so little time to write – I've spent nearly all my time on the maths, as I thought you would rather have that, at least I dont quite mean that, for I dont think you would. I wouldnt like it very much if you wrote me an article on say dog-fish, instead of a letter. But its really much more use to you, especially as I have left such a long interval, but I couldnt get time in Germany, and the trains are so bad that I cannot write decently while travelling. I wish I had had your Whitsuntide letter before I left, still it was all the better to come back to because I knew it must be there. It was very nice of you to try to get it done in time. Like you, I think I shall have to take to writing letters in bed. Molly, I dont really think you ought to, you know. It sounds as if you were not sleeping very well, and that must be perfectly wretched for you, because I know how awful it is. I do

hope it isnt that. Just after the war I had a dreadful time – the result of overwork and strain, and couldnt sleep for weeks, but I'm all right again now, except for a night or two when one is extra worried perhaps. So I can jolly well sympathise. I think the only way is to train oneself to lie quite still, and to try to feel happy. But its a <u>miserable</u> business and I am most awfully sorry, Molly. Often it helps if you get up and have a drink of cold water, but I certainly wouldnt write or read. The Docs say that one rests much the same whether asleep or awake, provided you resign yourself to it and dont worry.

I love telling you things Molly, and it is such a joy to feel that you are interested. When I come to think of it you never write much to me about yourself and your work, you always seem to write about other people and things. I hope it isn't because you dont think I am interested, because I am, most awfully. Oh Molly it is nice of you to understand so well about Mother. I've never talked to a soul about her and she filled so large a part of my life. During the latter years I used to have a letter from her <u>every day</u> of the week. Ive got them still treasured, hundreds and hundreds. You see I went away to school when I was 12, then was at home just for 2 years from 16 to 18, and have never been at home since, except for an odd time or two like the present, and never during her lifetime.

Yes, exchange is quite the right word. One says high or low according to which country one belongs to. An Englishman speaks of a low German rate of exchange, whereas a poor German speaks of a high English rate, because he has to pay highly if he want to buy English money.

From the chambermaids horror, I could only argue that the cold bath is practically unknown in Germany. German is I think harder than French, because all the nouns have cases, and "the" and "a" and all adjectives, are declined also, and there are 3 genders, masc, fem, and neuter, all with different declensions. Awful. But the construction seems to me to come more easily.

I got quite heart-achy when it was time to leave. I wish one didn't get so fond of places. Some places do seem to appeal to one so. On the whole-holiday Saint's day the directors of the Zeppelin Company took me out sailing in a yacht which used to belong to the late King of Wurtemburg. We had the most topping time, and I got as brown as anything – and still am – I burn most awfully easily – do you, I wonder? There was Dr Eckener and Frau Eckener, and Herr Lehmann and Frau Lehmann (another of the directors) and Herr Friker, a banker, who used to live in London and has rowed in Henley Regatta. He went back to Germany on the outbreak of war, and was gassed.

I didn't know whether I should take any food with me, as I didn't want to offend my hosts, so I consulted the Head Waiter at the Hotel; he is a most important person. He said yes, he thought I had better, and some wine too, and so he sent down to the "Jachthafen" a basket like a small washing-basket containing tons of sandwiches – a dozen hard boiled eggs and two bottles of wine and some mineral waters!! When I arrived at the yacht there was no one there – this was about 11 in the morning. Presently in the distance I saw the Lehmann's (I always want to say the Lehmen!) plus servant, carrying 2 large baskets. We stood and looked at each other. Presently again came the Eckeners, accompanied by an even larger basket. At this we all burst out laughing! We had enough food for a regiment. Lehmann had also brought an <u>enormous</u> accordion, on which at intervals during the cruise he would play slow and very harmonious German airs, while the old Doctor stumped up and down the deck, humming the words. They are funny people, just like great children. Somehow I felt absolutely at home among them.

The two men retired to the cabin soon after we had started, and began to prepare a most elaborate meal. This was the menu:-

Speck mit Ei – (Bacon and Eggs)
Wurstchen   – little sausages that you hold in your fingers
White wine
Cigars

Then about 3 oclock we had coffee and cakes their substitute for afternoon tea. And then about 6 we had my eggs and ham sandwiches! And then about 9 I got back to my hotel and had dinner!!

Getting thro' the customs at the German frontier coming back the man searched my pocket book. He was most fearfully curious about your letters – couldn't make out why I had several I suppose and thought they must conceal gold or notes or something. I thought the brute was going to try to read them!

I must stop now Molly, to catch the post. I will send on a further chapter on trig. in a few days – real trig this time.

Chapter 6
POOR OLD Y

Some time in the third week of June, Barnes sent Chapter III of trigonometry.

Trigonometry. Chapter III. On Zero and Infinity.
At first sight this wont seem to have very much to do with trigonometry, but have patience with me.

We have dealt to a certain extent with the ideas of very small quantities in the introduction to the Calculus. I must now go a little more fully into this question, because it leads to some very important conclusions which we make constant use of in physics.

What is zero? To the non-mathematical, I suppose it means nothing – void, blank. (I dont mean by 'nothing' here, an absence of meaning). To the mathematician it means as you know the <u>limiting value</u> of a quantity which is being made smaller and smaller, smaller than anything that can be imagined or conceived. We do not in general treat zero as if it were the blank nothing or void – this is sometimes called 'Absolute zero' to distinguish it from the mathematical conception; when we are manipulating any functions or mathematical expressions in which zero occurs, or may occur, we nearly always regard it from the small quantity point of view, only introducing (if necessary) its limiting value of absolute zero at the end of our operations. The reason for this is obvious, if you think for a moment. It is just the same reason as makes us use the little Duke in the Calculus. We must have finite, tangible quantities if we are going to do any work with them; it is only at the end that we can take the 'limiting value' i.e. 0, for otherwise it is absurd to ask us to multiply <u>nothing</u>, or divide <u>nothing</u>. It is important that you should get this (after all) very common-sense way of looking at zero firmly into your mind, so that the idea becomes your own.

At the opposite end of the scale we have infinity. The symbol generally taken to represent infinity, to save us the bother of writing the word is ∞. Note how this differs from the symbol for "varies as" ∝ which you may also have met. I always think the infinity sign is a good one, for it gives one the idea of being able to let one's pen run on and on over the same line for ever and ever. Infinity is often defined as being a quantity which is greater than any quantity which can be imagined or conceived. Anything that partakes of the "infinite", whether infinitely large, or infinitely small, is from its very definition beyond the grasp of the human mind. But you will find this essential difference between the large and the small; – that whereas one, the small, has a limit (i.e. 0); the other, the large, has none.

And although it is true that you are unable to conceive anything infinitely small, your mind is yet able to pass easily to the conception of the ultimate limit – absolute zero. But you are quite unable to get any idea of the infinitely large, because directly you have formulated a thought in your mind it becomes <u>finite</u> and therefore is not what you are seeking, namely infinity.

The best that we can do to accustom ourselves to these ideas is to replace infinite values by finite values of such magnitude (the word magnitude includes smallness as well as bigness) that the finite values so chosen represent (in relation to the quantities with which they are associated in our thoughts or operations) the infinite values which they replace. That is a beast of a sentence and really wants working out again. But perhaps an example will make my meaning clear.

If we are dealing in millions, and want to get a finite representative of something exceedingly small, 1 will probably be quite small enough; but if we are working in units or tens then 1/1,000,000 may have to be taken as our small quantity. Again, if we are working in units or tens, one hundred million is quite a fair representative of infinity; but if we were working in millions, then millions of millions of millions might not even be great enough to bring us into the realm of the comparatively infinite.

Let us apply the ideas we have gained by joking for a while with friends zero and infinity, and see what jokes we can contrive.

You have perhaps if you have ever had occasion to think of it at all, always thought of the value of say 5/0 (five divided by nought) as 0. Nothing goes into 5 no times. It sounds reasonable. We are quite <u>positive that</u> $5 \times 0 = 0$. Then why not the division too. Well, let us replace the 0 by a finite (and therefore manipulateable. (<u>I</u> have invented this word on the spur of the moment, therefore the spelling is right, $\therefore$ I say so. How very convenient!) quantity, which in relation to 5 is approaching towards the very very very small.

Such a quantity would be say 1/10,000,000,000

Then our sum looks like this 5/1/10,000,000,000

Which equals $5 \times 10,000,000,000/1$ {Because to divide by a fraction turn it upside down and multiply}

$= 50,000,000,000$

Now 50,000,000,000 is a quantity which in relation to 5 is approaching the very very very big.

Let us chose our small quantity even smaller, that is even nearer the limit 0.

Then our sum might look like this:-

5/1/1,000,000,000,000,000,000,000,000,000,000

Which equals 5,000,000,000,000,000,000,000,000,000,000

The conclusion you have probably jumped to for yourself long ago. Put mathematically we say

Lt $5/h$ as $h \to 0$ is infinity

or $\underset{h\to0}{\text{Lt}}\ 5/h = \infty$ where h simply represents our small quantity

More shortly still – tho' not quite so good but so general as to be quite allowable, except in a reasoned mathematical argument, we say $5/0 = \infty$. ← A Which startling conclusion shows you the difference between mathematics and common sense.

I dont think however that your reason will rebel against this conclusion, especially if you will train yourself to regard 0 or zero as the limiting value of an infinitely small quantity.

I need not spend time over $5/\infty$. You will see at once that $5/\infty = 0$. But please note that as a matter of mathematical "style" it does not follow from equation "A" above. Why? Because "A" is an equation involving infinite values:- 0 and ∞ which are not susceptible to algebraical manipulation. Properly to draw our conclusion, we must start the argument all over again, by writing 5 ÷ by something very large = something very small ∴ in the limit $5/\infty = 0$

Having given you the key, you can reason out any other combinations for yourself. I will write down as many as I can think of under 3 headings:- (here 'a' represents any finite quantity).

Important.	*Interesting.*	*Indeterminate.*
(1). $\dfrac{a}{0} = \infty$	$\dfrac{\infty}{0} = \infty$	~~~~
(2). $\dfrac{a}{\infty} = 0$	$a^{\infty} = \infty \,(\text{if } a>1)$ $a^{-\infty} = 0$	$\dfrac{0}{\infty} = \,?$
(3). $\dfrac{0}{a} = 0$	$a^{\infty} = 0 \,(\text{if } a<1)$	$0 \times \infty = \,?$
(4). $\dfrac{\infty}{a} = \infty$	(10) $a \times \infty = \infty$	$\dfrac{0}{0} = \,?$
(5). $\infty + a = \infty$	(11) $a \times 0 = 0$	$0^{0} = \,?$
(6). $\infty - a = \infty$		(Compare $a^{0} = 1$)
(7). $a + \infty = \infty$		By indeterminate I mean
(8). $a - \infty = -\infty$		that one cannot assign
(9). $a \pm 0 = a$		any value to such an expression.

Do not bother to puzzle over them if you do not see them clearly.

Of the ones under the "important" heading 1 and 3 are the only ones that you need to grasp. 5 and 6 give a good illustration of what I mean when I say that these infinite quantities are not susceptible to ordinary algebraical processes. For subtract 6 from 5:- Then if we try to do it by algebra we get

$$2a = 0$$

An absurd result due to handling these quantities in the wrong way. If we took the precaution to substitute a finite representative of ∞ we should see the impossibility at once.

One further very important point:-

You will note that I have not hesitated to write $-\infty$ when necessary. Infinity can be either +ve or -ve.

Now consider the fraction $y = 1/x - a$

when $x = a$, then $x - a = 0$ and $y = 1/0 = \infty$

But if x is less than a, then y is finite and -ve.

And if x is greater than a, then y is " and +ve.

So that you see, from a negative value, y passes <u>thro'</u> infinity to a positive value or in other words, a function may <u>change its sign</u> while passing thro' infinity.

Perhaps if I roughly plot the function for you this will be clearer.

In order to plot, we must assign a definite value to the constant 'a'. For simplicity make $a = 1$. Then $y = 1/x - 1$.

If $x = -3, -2, -1, 0, \quad 1, \quad 2, \quad 3, \quad 4, \quad 5$

then $y = -\frac{1}{4}, -\frac{1}{3}, -\frac{1}{2}, -1, \quad \infty, \quad 1, \quad \frac{1}{2}, \quad \frac{1}{3}, \quad \frac{1}{4}$

Take a few additional values in the region of the infinity value:-

$x = \frac{1}{2}, \frac{3}{4}, \frac{7}{8}, \frac{99}{100}, 1$ {positive change to +infinity see next row}

$y = -2, -4, -8, -100, -\infty$

$x = 1\frac{1}{100}, 1\frac{1}{8}, 1\frac{1}{4}, 1\frac{1}{2}$

$y = +100, +8, +4, +2$ etc. etc.

Look at the curve. You see that y shoots away very rapidly as x approaches the value 1 to $-\infty$ and then as x <u>passes thro'</u> the value 1 y reappears from $+\infty$

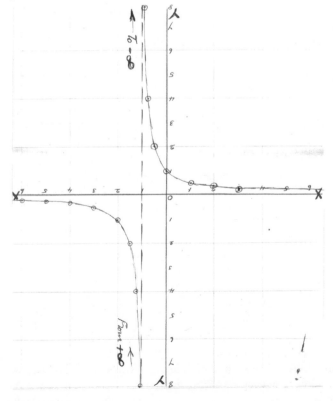

I have dotted in the line passing thro' $x = 1$ so that you can see clearly how y disappears in the direction of $-\infty$ for $x = 1$ and reappears from $+\infty$ for the same value of x.

We say that "y changes sign from $-\infty$ to $+\infty$ (or vice versa if you like) as x <u>passes thro' the value 1</u>".

This is most important in trigonometry. We could arrange any number of functions to behave in this way, by artfully contriving that for some value of x, a 0 appears in the denominator. Then poor old y has got to shoot out to infinity and back the other side whether he likes it or not. I always imagine him getting more and more annoyed as x approaches the "critical value", muttering to himself, "I <u>dont</u> want to go, I was just settling down so comfortably, – why I haven't even packed my bag, and the things aren't back from last week's wash". And then you give him a kick, and off he goes, via the sun, Jupiter, Urianus, to Infinity, reappearing punctual to the second, a little breathless perhaps, to crawl over the top edge of your paper, and flinging his bag on the floor exclaims angrily "I <u>wont</u> be a 'discontinuous function' any longer".

It looks "discontinuous" to us and so we call it so; but poor old y knows better, for he has to tear along like anything to get round again the back way.

Molly: 21st June 1923. Hampstead
Isn't infinity queer? It seems as if poor old y gets turned inside out or back to front when he goes off the negative side and comes back on the positive.

Yes, thank you is often a very inadequate expression; but I guess I find it much more so than you do, and I never can thank you enough for all the Maths. and everything. I wouldn't rather have maths. than a letter.

Barnes, you are ripping; but I do sleep well, really. What I meant was that there is time to think before going to sleep for about an hour. You see when Betty and I slept together we always used to talk when we were in bed, but somehow Baba likes to go to sleep pretty soon, so I have got into the habit of just lying awake for a little while.

I never work or read in bed when I am at home; if I am busy and want to write letters, I get up early and do it – at least it isn't really early, only about half past five or a quarter to six, but it gives one an hour and a half before breakfast. As a matter of fact that is what I am doing now. Last summer before Matric. I used not to be able to go to sleep. You know how it is before an exam., you keep on saying things over and over to yourself, and when you are in bed you simply can't stop. Then I used to get out of bed and sit by the open window on the window sill, and it was so lovely and dark and quiet and scented that it soon made one feel sleepy, listening to the faint little rustlings of the trees; and occasionally a bird would give a feeble little chirp, only to subside, feeling very much squashed because it wasn't yet day and nobody joined in. Oh dear, I'm so sorry Barnes, I didn't mean to dash on like that; only I forgot, and night-time in the summer is so lovely.

How I should have loved to have seen you all the day you went out sailing in the yacht, when each one arrived with his provisions. I did laugh when I imagined you all standing staring at the baskets!

Barnes, I know you are interested in what I do, only there never seems anything very exciting or unusual to tell you. You always do such lovely things – like dashing off to Germany in a hurry.

I <u>do</u> love your Mother.

Barnes: 21st June 1923. [Note in Chapter IV]
Someone saw the opening sentences and simply shrieked with laughter at the pompous and elaborate style in which it is written. I didn't tell what it was for, except that I was writing a book! I'm sorry if it reads like that – it isn't meant a bit. If I stopped to consider style and grammar I fear you would never get it at all! If anything requires further explanation let me know.

B.N.W.

If there are discrepancies or repetitions, you must forgive them, as it is hard to bear in mind just what one has said when it goes back 5 or 6 months, and you always have the great remedy – SKIP!'

Trigonometry. Chapter IV: Angles and their Measurement.
I have spoken before I think about the difficulties of fixing standards for the measurement of time, space and mass. For each of these three fundamental physical properties mankind has had to fix artificial standards, – actual bits of stuff in the case of space and mass, and an elaborate system of observation and record in the case of time (Greenwich Observatory etc).

Angles are of course included in the general class of spacial [sic] measurements, but they are in the happy and exceptional position of having two natural and easily derived units for their measurements, the right angle and the radian.

The right angle you are already familiar with, the radian may require some explanation. You must be familiar with both systems of measurement, for whereas the R. angle and its subdivisions is <u>exclusively</u> used for measuring angles in real life (e.g. in astronomy, engineering, chemistry and physics when we wish to perform an actual measurement), yet the Radian is equally exclusively used for all mathematical and physical <u>calculations</u> (excepting of course such calculations as may be necessary after making measurements in the 'right angle' units) of a general nature.

Taking them separately:-

a) <u>The Right Angle</u>, is too big a unit for ordinary use, and actually we always measure in a sub-division of the R angle called the degree. All nations (excepting I think the French, and they only for limited purposes) work on a 'degree' which is obtained by dividing the R.angle into 90 equal parts. These again are divided into 60 equal parts called minutes, and these minutes again into 60 equal parts called seconds.

A degree is a 'small quantity' compared to a R.angle; so the minute is a 'small quantity' of the 2nd order, and a second is a small quantity of the 3rd order, i.e. a second is 1/90th of 1/60th of 1/60 of a R.angle, = 1/324,000 of a R.angle. Nobody much beyond astronomers and navigators bothers about seconds in real life. You will find most tables of angles will not go below blocks of even 6 minutes. The signs are ° for degrees, ' for minutes and " for seconds.

The French did try a decimal subdivision of the R angle, but I dont think anyone worries about it now.

Four R.angles are formed by the intersection of any two (or more) lines (i.e. all meeting at the same point); if the lines are perpendicular to each other, the angles are of course actual R.angles. If the lines are not perp^r then the sum of the angles = 4R.

Thus $\hat{1} + \hat{2} + \hat{3} + \hat{4} = 4R$

I write angle thus $\wedge$, it is not so liable to misinterpretation as the sign generally used "$\angle$"

All this you have known since childhood. You are now perhaps [the word "probably" here is crossed out and replaced] for the first time going to realise its use and importance.

Do you – but I expect you must – use the word "subtended"? The angle subtended by a given line, or portion of a line at a given point, is simply the angle formed by joining the extremities of the given line to the given point. Thus the angle subtended by AB at C is the angle ACB, or we can write it the other way and say AB subtends an angle ACB at point C.

It doesn't matter whether AB is straight or curved. Thus arc, or Curve AB, subtends an angle ACB at point C.

With this brief explanation we can go on to consider the second natural unit of angular measurement.

b) <u>The radian</u> The radian is the angle, subtended at the centre of a circle, by an <u>arc</u> equal in length to the <u>radius</u> of the circle.

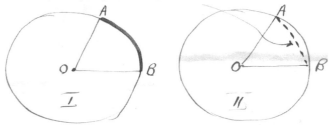

In the figure I the length of the curved line AB, would if stretched out straight, exactly equal the radius OB or OA.

Be very clear in you mind that is not the <u>chord</u> shown dotted in figure II, to which we refer.

It is obvious that the distance along the curve AB, (in either figure) is greater than the straight line joining AB.

Hence if in I <u>Arc</u> AB = Radius

and in II <u>Chord</u> AB = Radius

then ∠AOB in I must be smaller than ∠AOB in II.

Now in making this comparison I have purposely omitted to postulate that the two circles must be of equal radius. They need not be. Why? Because all circles are "similar figures". That is the ratio (circumference/diameter) is constant (i.e. the same, or "does not change") whatever size the circle may be drawn. Of course you know this. It is capable of mathematical demonstration, which is too advanced for us to consider here. But if you want to convince yourself that it is so, you can readily do so in a rough way, by cutting out 2 or 3 different size circular discs of cardboard, of some convenient diameters and then, making a fine ink mark on one edge, roll them over a sheet of paper, until the ink mark again comes to earth.

roll carefully

A

B

This distance is equal to the circumference, if you have rolled the disc without slipping

Divide the distance from A to B by the diameter, and you should, if you have worked accurately, get a very fair approximation to the value of π. You have not done well if any of your results lie above 3.2 or below 3.1, the true value for π being of course 3.14159.

This merely means that in <u>any</u> circle the circumference is 3.14 times as long as the diameter; and since the diameter is twice the radius, it follows that the circumference is 2×3.14 times as long as the radius. Of course we generally say,

circumference = $\pi \times$ diameter, = $2\pi \times$ radius

Now since Circum = $2\pi r$ where 'r' is the radius it follows that, when we come to mark off the radius round the circumference of a circle, we shall always get 2π divisions thus:-

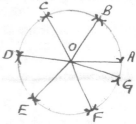

Each of the arcs AB, BC, CD, etc. is equal in length to the radius of the circle, and therefore subtends an angle of one radian (by definition). You will see that there is a little bit left over, – the sector AOG, there being 6 whole parts.

From the above equation ($C = 2\pi r$) we get $C/r = 2\pi$, or in other words the no. of radians subtended by the whole circumference is $2\pi = $ 6.28... radians. The little bit left over is the odd .28... of a radian.

The radian is a difficult thing to measure off, as you cannot step the distance out with dividers; one would have to use a flexible tape measure if one really wanted to do it. Fortunately one never does.

Now all the angles at the point O [the centre] make up 4 right angles, and since each R.angle = 90 degrees, ∴ all the angles at O together = 4 × 90 = 360°

∴ 6.28 radians = 360°

or 1 radian = 360/6.28 = 57° 19' 12" approximately.

By means of this relationship we are able to convert radians into degrees and vice versa, when we want to.

As a matter of fact, we never speak of radians as one, 2, 3, 4 ... radians; we always speak of them as multiples of π, for the simple reason that nobody ever wants to work them out, (that is in the course of a calculation. We always leave the actual conversion till the end), and it is much more convenient to write "π" than to write a dreadful string of incommensurable never-ending figures.

Hence we say

$360° = 2\pi$ radians

$180° = \pi$ "

$90° = \pi/2$ " (read "pi by two"radians)

$60° = \pi/3$ " (pi by three)

$45° = \pi/4$ "

$30° = \pi/6$ "

and so on.

You will do well to get the equivalents of the principal angles fairly well in your mind. I dont mean learn them by heart, but remember the "key value" $360° = 2\pi$ radians, and all the others follow by simple division.

There is a sign for radians you may sometimes meet – 2π radians is written $2\pi^c$, the little c standing for "circular"; because radian measure is sometimes referred to as "Circular Measure".

<u>Convention</u>. When we are talking about an angle and we wish it to be understood that it is to be measured in radians, we always give it a greek letter thus the angle shown is the angle "theta" and you would know that I am dealing in radians.

If we are talking about an angle in

degrees we always give it a capital letter from the ordinary alphabet.

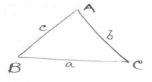

While on this subject I may as well mention the standard method of lettering a triangle.

Each angle has a capital letter, while the sides are named by small letters – little a is the side opposite angle A, and so on. This is much quicker than speaking of angle BAC and side BC.

Molly: 21st June 1923. Continued
<u>Friday</u> I have just got the maths. Thank you so much. It never appeared to me to be pompous or elaborate, and I don't think it is now. That is one of the marvels to me – how you can make each one follow the last so beautifully without having anything to refer back to. I have just been rolling cardboard circles along and they act beautifully. With one $\pi = 3.15$ and with the other 3.143. We didn't do anything about similar triangles at school, but they are one of the things I have been trying to learn about lately. Our Physics man is always talking about similar triangles and symmetry.

Barnes: 26th June 1923. New Cross
Yes, I know exactly the pre-exam feeling. Poor Molly, I can so feel for you. How can I say what I think of you and the little bird who tried to chirp too early. Why dont you always write me things like that? Molly, just the same, I love hearing about every single solitary thing <u>you</u> do. I dont <u>want</u> to hear exciting things. I want to hear quiet, beautiful things like that. You cant think how much I love your letters. I've just spent the afternoon reading thro' every one, from start to finish from the one on May 1st last year. I dont think exciting things are the most interesting. You never told me for instance that you had a new dress, (in which you look ripping); and lots of things about work, and swimming and cricket and tennis are all much more interesting to me than journeys to the continent which anyone can make who buys a Cooke's ticket. But I never knew anyone else who could understand the rustling of the trees and the feelings of the little bird when no one would join in; and who could think of fairies resting under the quivering noon shadows. I dont know why you should keep that side of you all to yourself, and then be sorry because it happens to have bubbled out. I <u>know</u> the night time in summer is so lovely. Many and many a time I have slept out of doors, so I know. I've got the most topping little tent, just holds two, and only weighs 6 lbs, and goes in a rucsac. I

simply love camping. There is no holiday in all the world – away with Switzerland, France, Germany – to compare with packing your rucsac and your little tent and going off on your own to tramp and camp when and where you please, in ENGLAND.

♦ A letter from Molly intervened, breaking the rules to return a booklet on Count Zeppelin's life. Her parents were in Swanage on the Dorset coast, searching out lodgings for a holiday for their large family and Nan. Barnes had unhappy memories of Dorset, memories of an incident from the early part of the war which he did not at that moment describe to Molly. He had taken his father to Weymouth for a much needed holiday and, being prevented by his work on airship design from joining up as he desperately wanted to do, he was not in uniform. He suffered the despising looks and cruel comments that were frequently meted out to those not in the forces and, being unable to retaliate or justify his position, he ran away.

Molly: 27th June 1923. Hampstead
Here is the book. Thank you so very much for it; it is very interesting. Count Zeppelin was a wonderful old man, and I do like his face in the picture at the beginning. I'm afraid I have a confession to make to you – I was sitting on the floor in the schoolroom reading the book when Nan brought down some gooseberries for us to top and tail. I shut the book up and put it on the floor beside me, and George promptly trod on it – quite by mistake, of course. It really wasn't his fault at all, it was entirely mine, I ought not to have been so silly as to put it on the floor. You can see the mark on the cover. I am so awfully sorry, Barnes, please forgive me. I'm not really careless about books, specially other people's. Please don't never lend me another book because you'll be afraid I may spoil it.

Barnes: 26th June 1923. Continued
The only time I've seen Swanage, I was tramping by myself at the end of a week's holiday with Father in Weymouth (before he married Fanny) from Weymouth to Southampton. He went back of course, by train. I had no tent then, and after stopping to bathe in Lulworth Cove, I spent the night at a little inn above Corfe Castle somewhere – it was dark when I got there and dont know where it was now. Next day I stopped to bathe at Swanage, and then on to Bournemouth where I had to pawn my gold shirt links for 8/6, because I had no money. The next night I slept somewhere near New Milton, and the next day or evening rather, got lost in the New Forest, only getting to Hythe about 10 at night. (I hate sticking to roads,

and without a really good map tried to cut my own way across the Forest). I found a friend next day in Southampton who lent me enough to pay my fare back to London, as I had to start work next day. But I'm tired of being lonely now, so I spend my holidays with friends.

The book with your note arrived safely this morning. My dear Girl, if it wouldn't have looked odd and made you feel uncomfortable, you should have had every treasure I possess, from a gold watch to a few leather bound volumes, by special messenger this morning, with instructions to drop them in your cold bath or spread butter on them, if you wished to do so! I was quite prepared when I got as far as the gooseberries to find one squashed all over the dear old Counts face; and as for the footprint I declare I had to look several times before I could find it! Molly, dont <u>ever</u> be worried over a thing like that with me. I'm not even one of those people who assume an angelic air of resignation and thereby make you feel far worse than if they raved at you. Your little "Please <u>dont</u>" at the end makes me feel more miserable than I suppose you would feel had you accidentally dropped the book in the fire so Please <u>dont</u> ever worry again, and at the first decent excuse I will lend you anything and everything I have. Of course I know you're not careless. And anyhow scars of use and battle only render things to me even more beloved.

I have joined the Territorial Army again. Oh, for lots of reasons Molly, – I dont know what you'll think. This time as a gunner, in a mobile anti-aircraft battery. I had to enlist for 4 years, and am not taking a commission, just going in the ranks – its much more fun. It doesn't really take much time, one has to do 40 drills a year and go into camp for 15 days a year, and of course you are liable to serve anywhere if there's a war or anything abroad. I'll tell you all about it when next we meet – it helps to keep one fit and hard, and what weighed with me quite a lot – the best way to learn how to make airships invulnerable is to find out all about the methods available for shooting them down! Also one never knows, with Europe in its present state, what may happen, and the next war will be sure to start instantly with a huge aerial attack, so that the A-A batteries will be the very first people in action. All the Pater did when I told him was to grunt. Fanny was quite enthusiastic. We go to camp the middle two weeks of August.

Molly, I haven't said about the book very nicely – I'm awfully clumsy at that sort of thing, but you <u>couldn't</u> do anything that would hurt me, except by being unhappy yourself. Good night Molly, and have the best of holidays.

Molly: 4th July 1923. Hampstead

The first time I saw your writing was that Monday evening at New Cross when we were playing what our family calls "getting words out of", and I remember thinking how nice and strong and firm it was, and I specially admired the way you printed "tergiversation" at the top. Are you brainy at drawing? your writing and specially your printing makes one think you are. Of course you must be – how stupid of me – you took drawing and design in Inter. didn't you? and you must draw for engineering, and also there are your diagrams. The thing that strikes one most about your writing is the fact that you join so many of your words together. I like your writing very much; it never seems to me to be untidy. I can't think how you manage to write so nicely when you are in bed; unless I am sitting at a table, my writing is always dreadful – witness now, when I am writing with the pad on my knee in the garden. As a matter of fact it is getting too dark to see any longer. Hasn't it been a gorgeous day, and now it is so delightfully cool.

Barnes, I don't know what to say. I didn't know you'd <u>like</u> me to write things like that. [Letter 21st June] I was thinking after I had written it, that that was an exceptionally silly letter, and I was in two minds as to whether I should send it or not; but I shouldn't have had time to write another one, so I did. You can't think how glad I am that you don't think it is stupid, and that you love all those beautiful things. You know, it is most awfully exciting not knowing everything about you, because in every letter I discover something fresh that I didn't know of before, and that makes you nicer.

Barnes, I am most awfully glad you have joined the Territorial Army; I think it is simply splendid. Though it makes me hope that there won't be another war. That sounds rather inconsistent, but I can't help it.

It is distracting writing out-of-doors. To begin with the wind keeps on blowing everything about. I don't remember ever having such a strong wind with such a perfectly cloudless sky. And when you are sitting on the grass there are such lots of funny little creatures rushing busily about, and you have to stop and watch them. Also there are the great, fat, lazy bees and the thin, energetic butterflies, and all the shadows on the leaves and on the grass. I do think shadows are interesting things, don't you?

Do you really call what you said about the book clumsy, Barnes? I consider that nobody else could have said anything a half, or a quarter as nicely as you did. It is so lovely of you not to mind; and I won't worry about a thing like that again. Do you know, Barnes, sometimes it makes me almost uncomfortable to think of everything you have done for me and very little I have ever done for you. I can't understand it. I do think I'm lucky; and the worst of it is, I haven't anything interesting to lend you or give you. I do hope you'll have a perfectly lovely holiday.

♦ The Territorial Army involved more than just playing at soldiers. The experience of the Boer War had shown that the military strength of Britain was inadequate. Supplementary troops were drawn from the militia and the yeomanry, raised locally as for centuries and with no obligation to serve abroad; a voluntary, often poorly trained and heteregeneous force. Lord Haldane, Secretary of State for war in the decade before the First World War, carried out extensive army reforms. The standing army remained relatively small, and the disparate elements of the militia and backup forces were coordinated into an organised body, voluntary but well-trained, required to go through specified periods of training, and to go abroad for service as necessary. The men who volunteered had the advantage of keeping their civilian jobs while they experienced the life of a soldier and served their country. Barnes, who longed to serve, valued this opportunity. The life of discipline, physical hardship, gruelling activity and male companionship suited a need in his character and filled a gap in his emotional life. He felt at one with the masculine traditions of history, in the knowledge that this was what his forebears had done from time immemorial. Throughout his life he was glad to harden his physique and subdue his will; there was an element of monastic austerity about him.

Chapter 7
REAL TRIG. AT LAST!

Barnes completed the next instalment before leaving on 14th July for his summer holiday, which he spent in Wales with a group of friends from wartime days.

Barnes: 5th July 1923. [Enclosed with Chapter V]
This instalment is perfectly <u>rotten</u>. I am awfully sorry, but I have been very pressed for time, and it has been written in numberless little bits wh. made it hard to think it out properly.

Next time we <u>really do</u> start on Trig. I daresay you are far ahead of this really, but it may be of some help.

Your results for π are <u>very good</u> indeed. [Molly 21st July].

Did I make it quite clear last time that since all $\odot$s are similar figures $\therefore$ the radian is a constant angle? Why. Because the radius always divides the circumference into 6.28... parts $\therefore$ the 4 rt angles at centre are always divided into 6.28... parts, whatever may be dia. of $\odot$.

Chapter V. Real Trig. at last!
The object of this chapter is to couple up for you the work we have done in all the previous chapters. So far we have seen how we can:-
   a) Fix the position of a point by means of its coordinates.
   b) Represent an angle by means of a rotating line.
   c) Measure an angle by 'degrees' or 'radians'.
   By joining up these ideas we come to a consideration of what are called by the somewhat puzzling names of "Trigonometrical Ratios" and "Trigonometrical Functions".
   First a few definitions:- (Poor Molly!)

1) <u>Scalars and Vectors</u>
I dont know whether you have ever done any "graphic statics", or graphics of any kind. In case you have not this explanation may be useful – We can very conveniently represent such quantities as distances, forces,

weights, etc, etc, on paper by means of lines. For instance any map worth the name is drawn to a certain scale, say 1 inch = 4 miles. On such a map we can represent a distance of 4 miles by drawing a line 1 inch long. Such a line is termed a "Scalar" because it represents to some scale a physical quantity or property.

Now suppose you were asked to measure off 4 miles in a northwesterly direction from a certain point, your resulting line would be 1 inch long and would slant upwards and to the left at an angle of 45° thus:-

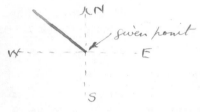

Such a line is termed a "Vector" because it represents not only the magnitude, but the <u>direction</u> also, in which the measurement was made. Direction here is a little loosely used. If we were really going to do some graphics, I should have to point out fully that "direction" includes two ideas – 1) the first is sometimes called the "sense" – you will be more familiar with the term "slope" and 2) the second idea is in which direction was the measurement made, i.e. from the top downwards or from the bottom upwards? – In the example given above there is no doubt, but suppose I said to you "this line represents the distance travelled by a motor car in 10 minutes, to a scale of 1" = 4 miles" you would at once say, "<u>oh yes but which way was it travelling</u>?" We can show this idea too, by putting a little arrow head on the line thus, which at once shows the complete thing:-

1) The distance travelled. 2) The sense or slope. 3) The direction or way in wh. the car was going

In this case car moved from top left to bottom right

In this case car moved from bottom R to top L

All clear?

So:- A <u>scalar</u> is a line representing magnitude only.

A <u>vector</u> " " " magnitude + direction.

2) <u>Ordinates and Abscissae</u> – This is easy – you already know it.

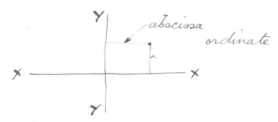

The abscissa is the horizontal co-ordinate
   " ordinate "   "   vertical   "
The two together are called the co-ordinates of the point.
So that x distances are all abscissae and
      y "   "   "   ordinates.

3) <u>Radius Vector</u> – This is a special application of the vector defined in 1) to the purpose referred to in b).

When a line OP (see figure below) of fixed length rotates about a fixed point O, to any position or positions we chose, it is called a "Radius Vector". I suppose because

1) it is virtually the radius of the circle which the end P sweeps out and
2) it will represent any slope we like, hence it is a vector.

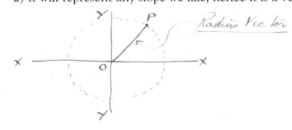

There now, I do think I have covered all the preliminary ground which is essential for a clear grasp of what is to come. I have found it extraordinarily difficult, why I dont know, unless because it is to hard to tell what to put in, and what to leave out. Many things you probably know already. I have however assumed that you did not know them, as it was better for me to write them than for you to be puzzled in case I omitted something that was new to you.

Trigonometrical Ratios.
Let us look again at our radius vector.

For a convenience we always consider OP to be of unit length, – it doesn't matter what the units are – one inch, 1 cm one anything.

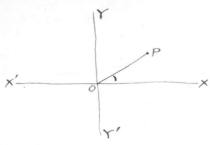

Now suppose I asked you to make the angle XOP equal to 30° – unless you had a protractor you would be rather stumped. And even with a protractor it is very difficult to measure accurately within 1/2 or 1/4 of a degree.

There are however several other ways of laying out this angle.

For let us drop a perpendicular from P onto XOX′. (Note that I have introduced a ′ on the negative X and Y, so that when I talk about the angle XOP you cannot mistake it).

1) You will see that by dropping this perp^r we obtain the x and y (i.e. the ordinates) of the point P where OM is x and MP is y; so that if we could get the value of OM and MP from say a book of tables, for the particular angle –30° which we require, and in the particular unit we want, we could by reversing the process of <u>dropping</u> a perp^r (i.e. by erecting a perp.) obtain the position of P and so, by joining OP get the desired angle.

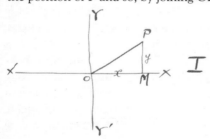

2) Again, instead of getting both x and y from the tables, we could get say y and then striking a circle with OP for radius, draw the <u>locus</u> of y, when the intersection again gives us P, and joining OP gives the angle [*below*].

3) Or again, striking the circle, and drawing the locus of x will give us P and hence the angle [*below*].

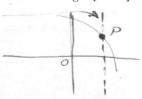

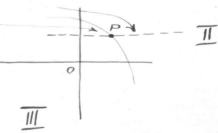

I have drawn, for convenience, all my figures in the 1st quadrant. If we had the necessary tables we could however draw any angle, in whatever quadrant it fell. It would only mean that some of the values of x or y would be negative, according to the quadrant.

This is a good place to say that the radius vector OP is <u>always</u> considered <u>positive</u> in whatever quadrant it lies.

Now whichever method we employed, would necessitate looking for 2 values in our book of tables, for every size of angle:-

method   I) – we want x and y

"   II) – "   "   OP and y

"   III) – "   "   OP and x

This would mean big and clumsy table books, but a convenient fact comes to our rescue here.

Let us examine method 1). (see fig. on [p. 120])

Perhaps in geometry you have got as far as what are called "Similar Figures"? In case you have not, I will explain them, or at least the simplest form – the similar triangle. The idea of similarity is a familiar one, I think, tho' textbooks often seem to make it terribly hard.

Let us have a short digression, to make quite sure that we understand "similar figures". I am not going to try to prove the fundamental proposition on which the similarity of triangles is based. To tell you the honest truth I dont remember how it is done; all that remains in my mind is a recollection that it seemed appallingly difficult, and that I never understood it properly, and consequently for years used to regard the whole subject as one quite beyond my mental capacity. So here is a dull fellows method of grasping at least the idea, although the methods employed may not be quite valid.

Let us cultivate a due sense of proportion. A correct idea of ratio and similarity is of the greatest use to you in physics, and everyday life – being able to prove academic problems is a worthless accomplishment. Someone tried to teach me the proof before I understood even the idea.

If I have 4 slips of paper A,B,C and D (they are not drawn strictly to scale) and I want to know 1) how many times (<u>not</u> how <u>much</u>) longer is B than A and 2) how many times longer is D than C:-

You would find no difficulty in doing this. One simply says –

length of B divided by length of A = B/A = $2\frac{1}{4}/1\frac{1}{2} = \frac{9}{4} \times \frac{2}{3} = \frac{3}{2} = 1\frac{1}{2}$

And similarly D/C = $2\frac{5}{8}/1\frac{3}{4} = \frac{21}{8} \times \frac{4}{7} = \frac{3}{2} = 1\frac{1}{2}$

In words	– B is half as long again as A
or	– B is one and a half times the length of A
or	– B is fifty per cent longer than A
Similarly	D is half as long again as C
or	D is one and a half times the length of C
or	D is fifty per cent longer than C.

The expansion B/A simply expresses the number of times the length A is contained in the length B. This is given the short and convenient name of a "ratio". When written B/A it means the "ratio of the length B to the length A". Notice that a ratio is <u>always</u> "non-dimensional" that is, it is a mere number; (i.e. simply the <u>number</u> of <u>times</u> A goes into B. Hence nothing to do with the units of measurement) – we should get the same figure $(1\frac{1}{2})$ if we measured B and A in any units of measurement whatever, (i.e. cms, millimetres, decimals of a foot, decimals of a mile).

We must of course measure both B and A in the same units (e.g. you must not say B millimetres/A inches!).

Similarly D/C is merely the ratio of the length of D to the length of C.

Ratio is thus simply a comparison between two quantities <u>of the same kind</u>.

Do not be puzzled if you are asked to compare say the <u>weight</u> of 1cc of alcohol absolute with the <u>weight</u> of 1cc of water. Here the substances are different but the <u>property</u> or <u>quality</u> of each that you are comparing is the same, i.e. the weight. This particular comparison or ratio, is as you know called the "specific gravity" (or "specific weight" is a term I prefer but it is nearly always referred to as Sp. Gr. Gravity here simply means weight) of the alcohol, the weight of 1cc of water being taken as the basis of reference or unit of weight. This gives us a good example of the non-dimensional quality of a ratio, because if we work in English units, and compare the weight of a cubic foot of alcohol with that of a cubic foot of water we shall get exactly the same figure for the Sp. Gr. of alcohol. Ratios are thus (and we may be duly thankful) absolutely international. The value of π is the same to a Frenchman, a Russian, a Turk or a Chinaman as it it to an Englishman, although the 1st may be measuring his diam. and circ. in cms, the 2nd in versts (or whatever Russians measure in) and goodness only knows what Turks and Chinks measure in; because π is a mere number, representing the comparison between the length of the circ. and the length of the diameter of a circle.

All clear as to the perfectly simple meaning of "ratio"?

Now here is a useful and interesting fact. First one or two definitions:-

1) Two or more lines radiating from a point are termed collectively a "pencil".

2) Each individual line composing the pencil is termed a "ray". Very simple and very apposite, – they look just like a beautifully sharpened pencil point.

3) Parallel straight lines, – there is no need to define these for you.

NOW, let us state our fact first in mathematical language, so that you can't understand it, and then show how clever we are by explaining it – I do <u>try</u> not to obtain a fictitious reputation by means of this very simple dodge.

When a pencil is crossed by two or more parallel straight lines, all the rays are similarly divided. Ha! ha!!

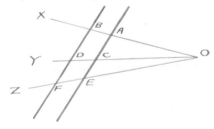

The ray OX is divided at A and B

" " OY " " " C and D

" " OZ " " " E and F

If I am considering the line OB, and A is a point within it (he! he!!) then I say OB is divided internally at A, and the ratio of the lengths into which it is divided is the ratio AB/OA where AB and OA are measured in any units you please (or OA/AB of course)

If I am considering the line OA, and B is a point on OA produced, then I say OA is divided externally at B, and the ratio of the lengths into which it is divided is the ratio OB/OA, or OA/OB if you like to put it the other way up.

Now whichever kind of section (internal or external) we take doesn't matter, – we can make up our ratios to suit our purpose – our fact talls us simply this:-

Ratio OB/OA = ratio OD/OC = ratio OF/OE

or taking the other series:-

Ratio AB/OA = ratio CD/OC = ratio EF/OE

We can of course write both series inverted:-

OA/OB = OC/OD = OE/OF

and OA/OB = OC/CD = OE/EF

The Proof – take a ruler and measure them! You were content to accept the value of π on an actual demonstration, so why not this? It is only the useful fact itself that we want to use.

Look again at a pencil of only two rays, crossed by two parallel st. lines.

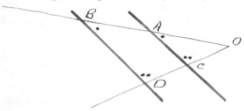

As before we can write down the ratios:-

OA/OB = OC/OD

Now draw separate figures so:-

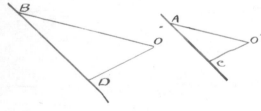

I have lifted off as it were, the small triangle O'AC. But still, since it once formed part of the original pencil I know that the ratios above are unchanged.

So that I can say; In the $\triangle$ O'AC and OBD

O'A/OB = O'C/OD

I omitted to add [on p. 123] that the pieces of the parallel lines cut off by the rays were also in proportion to the lengths of the rays.

That is to say, taking the figure [on p. 123], we also have the ratios:-

OB/BD = OA/AC   and   OD/BD = OC/AC

And of course, (writing O' for O in triangle O'AC) these rations also remain true when we lift off the $\triangle$ O'AC.

Looking again at the figure [p. 123] we see, from the properties of parallel st. lines that

1) $\angle$B = $\angle$A

2) $\angle$D = $\angle$C

3) $\angle$O is common

and hence the $\triangle$s OBD and OAC, are equi-angular, or as it is more generally termed are similar.

The sides OA and OB are called <u>corresponding</u> sides so are OC and OD and so are AC and BD.

When the $\triangle$s are nicely placed as I have drawn them, it is very easy to pick out the corresponding sides. It is not so easy if one of the $\triangle$s is twisted round, or even turned over. For instance these triangles are similar although it is not very apparent unless DEF is lifted up and turned over so that F lies on the left of DE.

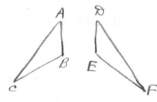

In this case the corresponding sides are:-

AC and DF

AB and DE

BC and EF

<u>Similar</u> triangles have no <u>equalities</u> except:-

1) Their corresponding angles

2) The <u>ratios</u> of the corresponding sides.

If one side of one triangle is <u>equal</u> to the corresponding side of the other in a pair of already-known-to-be similar $\triangle$s then the $\triangle$s are equal in all respects, and would come under one of the three famous propositions dealing with the identical equality of $\triangle$s.

All we have to do in order to be able to make use of the "ratio" properties of similar triangles is to prove that they are equi-angular.

If we can do this from any given figure, then we are entitled to use the ratios I have referred to in any way we please.

Conversely, if we happen to be able to prove that the ratios are equal, then we may say the triangles must be equi-angular.

Consider 2 $\triangle$s

 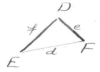

Given $\hat{A} = \hat{D}$ then   c and f are corresponding sides
$\hat{B} = \hat{E}$ b and e   "   "   "
$\hat{C} = \hat{F}$ a and d   "   "   "

Then we know

$$\frac{c}{f} = \frac{b}{e} \quad \text{etc, etc,}$$

Now this is simply an equation, and we may juggle with it algebraically in any way we choose:-

For instance c/b = f/e would be one variation obtained by multiplying both sides by the factor f/b.

You will note that this gives us the ratio between two sides of the Δ ABC equal to the ratio between the <u>two corresponding</u> sides of Δ DEF.

And similarly c/a = f/d and a/b = d/e

or in words:- Given two similar Δs the ratio between <u>any</u> two sides of one Δ is equal to the ratio between the <u>corresponding</u> sides of the other Δ.

To illustrate:- here is a series of similar triangles

  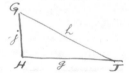

and we can write at once

$$\frac{c}{a} = \frac{f}{d} = \frac{j}{g}$$

and $$\frac{c}{b} = \frac{f}{e} = \frac{j}{h}$$

and $$\frac{a}{b} = \frac{d}{e} = \frac{g}{h}$$

How many angles of one Δ must we show to be equal to those of another Δ to prove similarity?

Two only. Why?

because the sum of all the $\angle$s in a triangle is constant and equal to

2 right angles

or 180° degrees

or π radians.

∴ If 2 angles of one Δ = 2 angles of the other Δ the 3rd $\angle$ of the one must = 3rd $\angle$ of the other.

Hence if we are dealing with <u>right angled</u> Δ's we only need to show equality of 1 other angle to prove similarity.

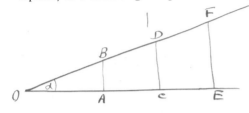

∴ All the Δs OAB, OCD, OEF are similar since angle α is constant and $\angle A = \angle C = \angle E$ = R.angle. (α = greek letter for a – called alpha)

Therefore:-

AB/OA = CD/OC = EF/OE and upside down.

and AB/OB = CD/OD = EF/OF   "   "

and OA/OB = OC/OD = OE/OF   "   "

♦ The chapter broke off here, and Barnes sent this first section before he had finished the whole. A few days later, the second instalment of Chapter V laid the foundation for an explanation of the trigonometrical ratios sine, cosine and tangent; although these were not explicitly named for another three chapters. He re-drew the final figure of the previous section, using P, P′, P″ and M, M′, M″ as perpendiculars in place of B, D, F and A, C, E, reiterating that coordinates can be used to determine any angle; and that in right-angled triangles with one other angle equal, the ratios between any two corresponding sides is constant.

Chapter V continued

We can call this ratio say "k", where "k" has some numerical value, <u>independant of where we measure OM and MP</u> but <u>dependant</u> on the value of "α" (or size of the angle).

So that $\dfrac{MP}{OM} = \dfrac{M'P'}{OM'} = \dfrac{M''P''}{OM''} = k$

What is the use of this? Why, instead of my having to give you two separate coordinates to determine an angle, it is only necessary for me to give you the <u>ratio</u> of MP/OM or "k".

Supposing I say, "draw me the angle whose MP/OM = 1.5" (this is an inverted comma not an inch " mark). We can rewrite this equation MP = 1.5 × OM. Then choosing <u>any</u> value for OM that is convenient – let us say in this case 2 inches we get MP = (1.5 × 2) inches = 3" and this is the resulting angle,

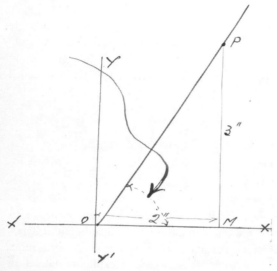

There is another possible solution to this, for suppose feeling orangey I said "Ha ha I will choose OM equal to –2 inches". What then? Simply follow the rule for signs and write MP = {1.5 × (–2)} inches = –3"

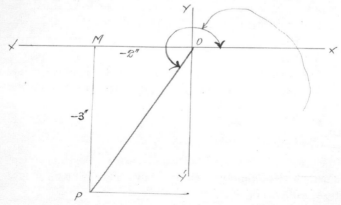

Then <u>this</u> is the resulting angle, and you know that $-3/-2 = +1.5$ as before.

There are therefore at least 4 separate and distinct angles that we can draw whose MP/OM = 1.5. Here they are.

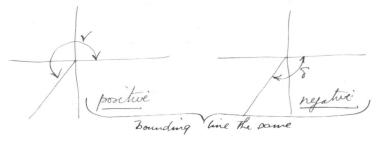

Putting them all on one diagram:-

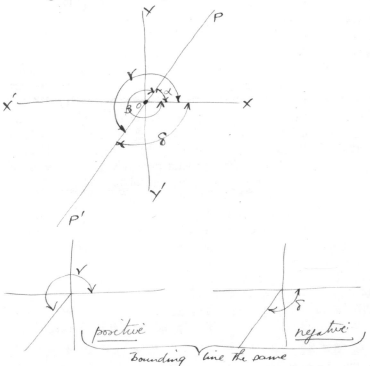

When MP/OM = 1.5 then the angles are:-

$\alpha \approx 56°\ 20'$ $\quad \beta \approx -303°\ 40'$

$\gamma \approx 236°\ 20'$ $\quad \delta \approx -123°\ 40'$

$\approx$ means "is <u>approximately</u> equal to", I haven't a book of tables here, so cannot say exactly. And here in case it may be useful is a Greek alphabet, tho' very likely you may know it already – if one doesn't – (and I dont) it is a great handicap when trying to read maths.

Note that β and δ are negative angles, the radius vector having started from OX and rotated clockwise thro' 123° 40' and 303° 40' respectively, the finishing position being coincident with the radius vectors for the positive angles α and γ.

These four angles are referred to as "principal angles". For this reason:-

The radius vector OP might have spun round any number of times before coming to rest at the angle α. Suppose, starting from OX it turned anti-clockwise thro' a complete circle to OX again. It would have turned thro' 2π radians. If it then goes thro' the angle α and comes to rest on OP as I have drawn it, it will have turned thro' $(2\pi + \alpha)$ radians, whatever α may be. In this actual case α is 56° 20', or nearly one radian.

If the radius vector had turned thro' <u>two</u> complete circles, and then on to α it would have turned thro' a total of $2 \times 2\pi + \alpha$ or $(4\pi + \alpha)$ radians; – three circles + $\alpha = (6\pi + \alpha)$; n circles + $\alpha = (2n\pi + \alpha)$ radians.

So we see that there are an infinite number of angles whose boundary line is coincident with OP the boundary of α. So that α is referred to as the principal value, or principal angle. Similarly with OP′; and in both cases the radius vector might go round positively or negatively. The expression $(2n\pi + \alpha)$ is (read – two enn pi plus alpha) called the "general solution" for an angle whose MP/OM is 1.5.

Can we write down the general negative solution, supposing the R.V. (radius vector) had rotated negatively? It is simply $(-2n\pi - \beta)$ or more neatly $-(2n\pi + \beta)$. But in giving general solutions the positive angle only is usually considered.

Sorry, I have not given you the complete general solution even now which must include angles bounded by OP′ as well.

OP′ bounds all angles expressed by $(2n\pi + \gamma)$.

But $\gamma = (\alpha + \pi)$ (i.e. $\alpha + 180°$ if one could add radians and degrees)

$\therefore 2n\pi + \gamma = 2n\pi + (\pi + \alpha) = \{(2n + 1)\pi + \alpha\}$.

So that the complete general solution is

$2n\pi + \alpha$ and/or $\{(2n + 1)\pi + \alpha\}$.

Now we can combine these two very neatly. Note that whatever value we may give to "n", the expression $2n$ must always be an even number; it being understood of course that n is always an integer i.e. a whole number. For example – take $n = 53$, then $2n = 2 \times 53 = 106 =$ an even no.

Since $2n$ is always even, it follows that $2n + 1$ must always be odd.

Let us see what effect this has on our radius vector. Starting from OX and rotating $2n$ times (because $2n$ pi's must be an even no. of pi's) the R.V. must always come to rest on OX again, whatever value we assign to n, (<u>integral</u> value of course), whether +ve or -ve.

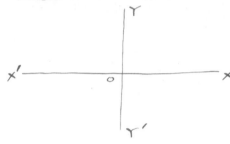

On the other hand, starting again from OX and turning thro' $(2n + 1)\pi$ radians, the R.V. will always come to rest on OX', because the $2n\pi$ part brings it to OX and the π turns it thro' an extra half circle (since $2\pi = 1$ complete circle) to the position OX', whatever the value of n. Looking again at the diagram [p. 129] we see that the angle X'OP' = XOP = α (vertically opposite angles)

So that whatever the value of n, the expression $(2n\pi + \alpha)$ always gives us the halting place OP while $\{(2n + 1)\pi + \alpha\}$ always gives us the halting place OP'.

In other words – an even no. of π's $+ \alpha$ gives OP, while an odd no. of π's $+ \alpha$ gives OP'.

Now <u>all</u> angles having OP and OP' as their boundary, or put another way all angles whose "principal values" are α, β, γ, and δ have their MP/OM = 1.5. (α being the particular angle ($56°\ 20'$) referred to on p. 130).

Hence we dont really care whether n is even or odd, in either case we get an angle whose MP/OM is 1.5.

So we can write the complete general solution for all angles whether bounded by OP or OP' as $(n\pi + \alpha)$.

When n is even, boundary is OP

" " " odd, " " OP'.

♦ Barnes's explanations confused rotations through a full circle, and rotations through 2π radians. 'Rotating though $2n\pi$ radians' and 'Rotating n

times' are equal: the '2' and the 'π' both go with radians, neither with counting rotations. He put 'starting from OX and rotating $2n\pi$ times', intending to continue 'the R.V. [Radius Vector] must always come to rest on OX again'. He realised this was wrong, but made the inconsistent correction: he replaced '$2n\pi$' by '$2n$' and left the 'times'. He made a number of attempts, and crossed them out with such exceptional vigour that it is hard to see what he had at first put. At one point he actually put the basic relation that made it all clear – '2π = one complete revolution' – and then struck it out. First making clear that whatever the integer value of n, $2n$ is always even, and $2n + 1$ odd, he then combined the two. Whether the value of n is positive or negative, if the radius vector rotates $2n$ times [sic], it must reach the +ve abscissa OX; alternatively, if it rotates through $(2n + 1)\pi$ radians, it must reach the -ve OX'. Molly made no comment on all this until September when the vacation was over.

Molly: 20th Sept. 1923. Hampstead

I am very glad you wrote out the Greek alphabet; I only know one or two letters, and it was awfully puzzling when in physics the man put a sort of squiggle on the board, and you didn't know if he meant it for an = or a number or a letter or what. δ puzzled me for a long time, he did it just like an 8, and very foolishly it never occurred to me that he was writing a Greek letter.

I'm afraid there's one thing I don't understand – why are the angles β and δ negative? You say they are negative since the radius vector has rotated clockwise, so I suppose that if the radius vector rotates anti-clockwise, then the angles are positive. But I don't see why it should be so; perhaps it is only a convention? Also I'm curious to know why you chose that MP/OM should equal 1.5. Perhaps there was some very deep mathematical reason which I shall learn later; or did you merely take the first number that came into your head?

Barnes: 28th Sept. 1923. New Cross

β and δ being negative is only a convention – see early chapters on Cartesian Coordinates as used to measure angles – I ought to have explained that, but may have left it out. If R.V. rotates or is considered to have rotated anti-clockwise, angle swept thro' is +ve. If R.V. is looked on as having rotated clockwise, angle swept thro' is -ve. 1.5 was the first convenient no. that came into my head, just for an example. I'm so glad you write about these little points – it is often only a little thing like that, that holds one up.

Chapter 8
HOLIDAYS

Barnes left for Wales on 14 July, joining his war-time friend Dr Boyd. 'Doc's' wife and little son Leslie, and his brother Norman made up the party. Norman Boyd was, like the headmaster at Chillon College, the kind of man that Barnes greatly admired: a Christian and a sportsman. The 'muscular Christianity' suited him. He also enjoyed the company of little Leslie and the fun of family life. Since he left home as an apprentice, 20 years before, his life had been without such cheer. While he was away, the Wallis household was disrupted by the faithful Harriet's absence, and Auntie Fanny invited Molly to keep her company. Before he left, Barnes arranged his room with loving care, leaving a request with Fanny.

Fanny to Barnes: 23rd July 1923. New Cross
There is only one topic you want to hear about in this letter so I won't waste time in preliminaries. The roses turned up all right on Sat. morning – and I arranged them in her (your) room – white and red they were with some asparagus fern – and very nice they looked. Nan-nan brought her – and came up to the room to carry her suitcase – so I couldn't say anything about them then – but I expect she guessed at once – and I told her afterwards they were from you and she declared it was "topping" of you.

Barnes: 14th July 1923. Borth, N.Wales
And a Friday too, what a dreadful day! Thank you ever so much for your very welcome letter. I've just wasted 2 sheets of this precious notepaper, as I started my letter to you 3 days ago, and then didn't like it, so started all over again.

Oh Molly, how on earth did you manage to remember that awful word "TERGIVERSATION"? Now that you have said so, I remember that that was the word, but I hadn't the faintest recollection of it before. You really are fearfully brainy. Do you remember everything? Because I've several times noticed that you seem to.

[Norman] is Vicar of St James Bethnal Green; he's an awfully fine fellow and went as a Naval Chaplain during the war and incidentally won the boxing Championship of the Navy one year. Doc. [Barnes's close friend Dr Boyd] boxes even better I think. I've only once had a go with him, as I used to box a bit when I was an apprentice, – thats what has made my wretched nose all crooked inside, and I think given me such colds, – but he simply knocked me about like anything. After a cup of tea I went out and played a round of golf with Norman. Fortunately he's just about as bad as I am, so we have great fun. Then after supper we all drove into Aberystwith, about 10 miles and went to the theatre, and after that sat on the pier and eat ices before coming home. I <u>was</u> tired, as I had had an awfully busy week, and hadn't been home much before 9 or 10 oclock any night except one, partly with drilling [for the Territorials] and partly with work.

They all got up early today, and went to Early Service, but I stayed in bed and slept, until Leslie [E.J.'s little son] invaded my room, and got into bed too – fully dressed! And I had to start showing him how to tell the time by my wrist watch. Goodness only knows what the time is now, as he insisted on learning how to turn the hands round, by pressing the little thing in the side – only fortunately his little nails are too soft to work the thing without my assistance, or my poor watch would soon come to grief. We insist on sitting together at table, and misbehave most shamefully until Mrs Boyd gets cross – I always get the blame, but really am much more easily kept in order than Leslie – when she's really cross.

We all played on the sands in the morning doing stupendous water works, damning streams and making lakes and channels, and after lunch played golf until it came on to rain about tea time. The correct dress here is a tennis shirt and old grey flannels, no socks, and any sort of shoes that come handy, tho' I've been wearing a sweater ever since I arrived.

<u>Monday</u> We went out after supper last night and played golf with one club each, till after 10. It was quite good fun, only my left hand was so sore from gripping the clubs that I had to stop after a bit. You see I hadn't played since Christmas. If ever you get a chance Molly do play golf – even if you only mess about with a club, you can learn to swing, and the younger you learn the better you will play. Its the most fascinating game in the world, and you can play it all the year round, where ever you may happen to be. Doc and Mrs Boyd are awfully keen players and play a lot together. It makes a tremendous interest in common for a married couple who are both keen.

I dont think you can really learn very much of a persons true nature from letters Molly. You could hardly expect me to start telling you all my faults. I suppose without meaning to be hypocritical one naturally keeps them in the background. Thats the worst of only writing letters. And anyhow its too beastly egotistical to start attempting to analyse oneself even if it were possible, which I dont believe.

Look here Molly you jolly well know that doing maths for you is one of the greatest pleasures I have. I am in the middle of another chapter on trig. only it is difficult to get it done here, as the others always are wanting me to do something, and we practically live out of doors. I never can do anything out of doors. As you say, there is such heaps to look at, even if it is only clouds and shadows. I wonder why a bee, who probably is fearfully busy should always seem lazy while a butterfly who does as far as I know nothing except eat ones cabbages should seem energetic. I always fancy the poor little creatures rush so wildly about, because they are permanently lost in a huge forest of grass stems. You simply couldn't make any land marks out of the grass on a lawn. Not when you lived <u>in</u> it I mean.

You know there's not the least need for you to feel uncomfortable. Surely an older cousin may do lots of little things like that, without making you feel as if you were under any sort of obligation to me. Why Molly, you have repaid me a hundred times by your letters and no such feeling must or need exist. You needn't even think that you must go on writing to me. Whatever you choose to do is absolutely right and just and I shall <u>perfectly</u> understand and be content, as long as you are happy.

It will be lovely to think of you at home. I have left my watch for you, in the extreme right hand corner of the 3rd shelf up of my cupboard. It does the watch good to be kept wound. To set the hands pull the knob out hard till it gives a click and then turn as required, and then push in. You may find many things to browse on in the cupboard on mechanics and so on. Only please dont think its always in such an awful state. Since my return I really simply haven't had a moment to put it straight. There are a collection of rather good sea stories in the bookshelf over the chest of drawers. I do hope the wireless will work all right.

Molly: 22nd July 1923. New Cross
I love this room of yours, and this is a delightfully comfy bed; I slept like a top last night. When you are lying in bed, you can see that picture of Chillon and the Dents [du Midi] reflected in the glass; I have been lying and looking at it for ever so long, trying to imagine what it must be like to live in such a beautiful place. No wonder you felt sad when you had to leave it. I do hope I shall be able to see it some day. Also I love that

beautiful picture of Windermere, and the World's Cow, and the little cottage on the mantlepiece, and the calendar under the bookshelf, and heaps of other things. Though why I am enumerating all the things in your room when you know them off by heart, I don't know.

I found the watch quite easily; thank you so much for leaving it there for me. The watch I wound up and put right according to instructions. Isn't it a beautiful aristocratic-looking one – so thin and elegant.

The very first thing I noticed when I came in here was the roses – two vases of gorgeous red and white ones, one on the mantlepiece and one on the table by my bed. (Bath time now)

Monday. All last night I had them right close up to me, and every time I woke up for a few seconds just to see what sort of a night it was I smelt the sweet scent of those roses. I have just been cutting their stems and changing their water. Auntie Fanny told me that you had ordered them to be sent here for me. Barnes, I don't know how to thank you. They are so sweet and beautiful and it was lovely of you. The white ones are very much out and very big, but the red are buddier and littler, though they have grown a good deal since Saturday; and they all have beautifully long stems. There is also some asparagus fern with them, so you can imagine how glorious the vases look. I have just been smelling them, and I can't decide whether the red or the white ones smell sweetest. I think they are about the same, though perhaps the red ones are a teeny bit stronger, I don't know. Thank you very very much for them.

Yesterday we went to Church in the morning, but the chief thing we did was to have a grand WASH UP after tea. It wasn't just a wash up, but it was a collection of all the things we used, beginning with supper on Saturday night. We had to collect plates and cups from the highways and hedges for tea on Sunday, there were such a few things left.

Yesterday, also, Auntie Fanny showed me some photos of the family in its infant stages. You were a funny, solemn, fat little boy. Poor Barnes, I know how you hate seeing and hearing things about you when you were small, so I'll stop. But one thing Auntie Fanny told me, and that is that you can knit, and you have knitted a sock on two needles. I have yet to discover the thing you are incapable of doing.

Monday – evening. This is fast degenerating into a sort of "dairy" as George calls it; I am so sorry, but I had to write some this evening because I wanted to see what writing by candle light is like. I undressed in the dark so as not to waste too much candle, and now I am writing all nice and cosy by its light. All the same it isn't very easy, it's too flickery.

You see when you are a housemaid you have to write whenever you get the opportunity!; and we came to bed extra early to-night. That sounds as if I were dreadfully hard-worked, but I'm not – quite the contrary – Auntie Fanny is much too energetic; she won't let me do half enough. I am learning quite a lot in the way of cooking. I helped Auntie Fanny to make some savoury eggs to-night. You must understand that when I say "help" I mean that I sat on the edge of the table and watched, fetching the necessary ingredients when they were required.

The wireless worked beautifully at first on Saturday evening, but after a bit it got fainter and fainter, so that the people seemed to be speaking from a great distance. It started getting faint in the middle of a lovely thing – it was in a Somerset inn about 100 years ago, and you heard all the old labourer men talking and laughing and singing songs. Auntie Fanny thinks the accumulator (is that right, or it is something else?) wants re-charging, so she telephoned to Mr Sparkes and he came round for it this morning and will bring it back to-morrow.

How topping the new wireless arrangement is. It is so nice to be able to hear it all at once – no, I don't mean that – I mean for everyone to be able to hear it at the same time, and without having to put those things over your ears.

No, I don't feel uncomfortable now, Barnes, I only feel jolly glad that I've got such a cousin. I daren't start another page, for I should never finish, though I could fill it easily. Once again, thank you for the roses – they are as lovely as ever.

Don't you love Uncle Charlie and Auntie Fanny. They are so kind, and I am enjoying myself so much.

Barnes: 26th July 1923. Borth
It is simply topping of you to write, when you are so busy. I am sure you <u>are</u> hard worked, because I've tried being housemaid myself, and I thought it was jolly awful. I always think Auntie Fanny is simply wonderful; she's always so bright and energetic. Both she and Uncle Charlie worry me, because they <u>will</u> not ask me to do things, but start off on their own, so that I have to be continually on the alert to try and stop them before they have got too far. Of course, I can see their point of view; they say they like to be independant; but I tell Fanny it is so fearfully bad for me, because I so easily get lazy and selfish, and it is good for me to be made to do things. But she never will.

I <u>am</u> glad you have enjoyed your stay at home. I know they have simply <u>loved</u> having you. Auntie Fanny sometimes I think hesitates to ask you all, because she fears so much that you will find it too dull and slow; as there really isn't much to amuse you at home. And it gives her such infinite pleasure, when you all write such topping letters to her, and seem to enjoy yourselves so much.

Yes, I love them both, more than I can say. Of course between myself and Father there is a very great bond. I dont want you to think that because I often chaff him, I am rude. He enjoys the joke as much as anyone, and you see we're more like friends than father and son. When John and I are together, we have the most enormous jokes and we can make him laugh till he cries almost – and all about himself. His mother died when he was born, and my grandfather would never have him at home, but sent him away to be brought up by friends, so he never had any home life at all and was treated as quite an outside person by his halfbrothers and sister when my grandfather married again.

I loved to hear about all the things in my room, and I am so glad you liked it. The worlds cow Victor sent me not long ago. He is always chaffing me because I often say "I am all over jam", or paint or oil, or whatever it may be. He will say in his joking way – "What you really mean Wally is that jam is all over you"; and I received that cow with just a slip inside to say "As you would say, this cow is all over world". For I expect you noticed that the markings did form a map of the world?

The calendar Mrs Lushington [the headmaster's wife at Chillon College] painted for me and sent at Christmas. And all the china is the china which we used to have at my old school. And the bookshelf I made myself, in fact I used to keep the house supplied with bookshelves and picture frames when a child, tho' my best frames John has stolen and has in his home. Also some of my pictures. But all my best pictures are at my rooms at Dalton [in Furness] in the North, where I still keep many of my things, as that room isn't really mine of course, and I dont suppose I shall be at home very much longer. I've got about a dozen pictures, water colour and oil. I hate landladies pictures, they are generally awful.

Father gave me the watch. Poor dear, he had the one Mother gave him stolen, but it was insured, and so out of the money, he got me that, just 9 years ago, almost to a day. I dont suppose you saw what he had had engraved inside. I think it was very wonderful of him. Most men would have bought another one for themselves. I think I value that more than anything. I've nothing of my Mother's and in a way that comes from both her and him. (Yes I have got something but not of that sort).

I will be very cross with Fanny when I get home, for giving me away over the knitting and early childhood. Its probably all untrue. All I can remember is generally being in a towering passion – I was very ill tempered, and once stabbing John in the eye with a pair of scissors; what a charming child! (age then about 3).

I am awfully sorry the wireless went adrift just when you were there. Accumulator is the word; My dear Miss Inter Sci, where is your electricity gone to? You know quite enough to build and operate a wireless set yourself – in fact you probably know much more really than I do. I do wish you wouldnt think me so brainy, it makes me feel as if I were going about under false pretences. I do hope Sparkes sent the battery back in time for you to hear it properly.

You are funny – why ever undress in the dark to save a candle when there is a gas in the room. And how could you see well enough to squeeze the toothpaste out of the tube onto your toothbrush? I do hope you got it all over your fingers! I read your letter again in bed to make the good night come right.

<u>Friday</u> Dont you simply hate washing up? I do, it seems so endless, you no sooner heave a sigh and say "Well <u>thats</u> done", than you have another huge meal to deal with. But the lack of conveniences, and the old fashioned sink at home all tend to make it more than usually arduous. I <u>do</u> think it was topping of you to go and stay.

Before I forget, I want to try to cheer you up over Inter. I think you were awfully brave to be so cheery about it, and I know just how it must have felt. But I am perfectly sure that even if you had got thro', and from what Uncle Arthur told Fanny, it seems very likely that you would have done so; you would not ultimately have done so well, as you will now, by spending a further year on this fundamental work. It forms a foundation on which you can build anything afterwards and I do think you will be very glad later on. And probably the Professors take this view and wouldn't let you sit for it <u>in case you passed</u>.

I am returning home tomorrow (Saturday) and feel very sad. We've had continuous bad weather – just two really fine days out of a fortnight. The first fine day was last Friday the 20th and a party of 8 of us went to Aberdovey on the other side of the river, to play golf. There is rather a celebrated course there. It was quite hot and we took sandwiches and had lunch on the course. Norman Boyd and I bathed twice – we had no towels or bathers, but just slipped away into the sandhills, and ran about to dry ourselves. Oh such gorgeous bathes. We played two rounds of golf and after our final bathe we two missed the train home! So we got ferried across the river and then had about 3 or 4 miles walk into Borth, along the sands. I should think we must have walked about 15 miles altogether as it is a very long course. So with the bathes and the golfing we had really a good day. Doc. is most tremendously energetic and I haven't sat down to read once since I came. We are always in the sea by 8 o'clock in the morning, even when its pouring with rain and blowing a gale. Its great fun bathing in the big seas. Its very queer that whereas one crawls down the beach miserable and shivering and feeling cold in the wind, one comes up sauntering along feeling warm as toast in wet bathing things....

<u>Home again Sunday</u> I made up my mind that you would probably catch the 10.20 from Paddington, and so kept a careful watch for your train. We must have passed within a few feet of each other. We crossed just before you got to Welshpool, about a quarter to three. We had to wait for your train as there was a bit of single track, and we were pulled up just outside a little station called Bunnington. You came by puffing hard with two engines, but you were going too fast to see more than a blur of carriages.

Molly, I do believe I'm going back to Airships again. I daresay you have seen in the papers that the Cabinet have approved Commander Burney's financial scheme for the formation of a subsidised Imperial Airship Company. I went to see Burney's secretary this morning to find out how things are going. He told me that Burney is very anxious for me to join him; so I said that my particular director at Vickers, Mr Sadler, did not want me to go, as they had just offered me the managership of their midland depot. However the Secretary said that Sadler was going to become one of the directors of the new Airship Company, and she had little doubt but that he would now want me to go there instead. I don't know what they would offer me, but I should think not less than £1000 a year. I didn't much want the other post and cannot tell you how pleased I am. Airships are the only thing I really care about. I've given up all my time and energy to them for the last ten years nearly, staked my future on them as it were. I don't think anyone knows quite how hard I've worked, and what a strain it has been. The year before the war I hardly ever had even a Sunday off, and never a Saturday afternoon and never had a holiday. I suppose Molly you may think it wrong to work on a Sunday? But it wasn't for the money I got – I didn't get any more for it; but simply I felt in a way that that was my special work, and it became a way of worshipping one's Creator to work to the utmost limit of one's strength and ability. Do you think you could see it that way?

Tonight I have been to drill and then met Sherard and Hugh [his brother John's sons] at Victoria to take them home for the night as they broke up today. Their train was awfully late, and didn't get in till after 9 p.m. I swelled with importance at mingling with a crowd of Fathers and Mothers, and swaggered about just as if they had been my own (children I mean).

♦ Molly had joined her friend Mary and family, also in North Wales, for the next part of her holiday.

Molly: 6th August 1923. Barmouth
It is jolly staying at a hotel, and this is not a very big one, so it isn't too alarming. I was horribly homesick at first – not so much because I was away from the family but because I couldn't feel that I was at home. I don't know if you understand what I mean – but I like to feel that I am in my own home; it would be the same where-ever or what ever the home was. Though of course I do miss the others and I am longing to see them on Sunday.

It all depends for whom I am washing up, if I like it or not. If I were a servant and had to wash things up, I should probably hate it, but you can't help liking to make things nice and clean for those two to eat and drink from. (I'm afraid that's not a very polite way of saying it, but I mean Uncle Charlie and Auntie Fanny). I do wish I could have stayed till Harriet came home on Saturday.

Of course I know that when you chaff Uncle Charlie you aren't rude. Auntie Fanny was telling me how much he enjoyed it, and how he loves to have you and John joking with him. She also told me about his mother and father; poor little boy – it must have been sad for him. I quite agree – it was wonderful of Uncle to give you that watch, and I'm not surprised that you value it so much. No, I didn't read what was written inside; for one thing I didn't like to open it because I thought I might spoil it, and for another I thought there might be something engraved in it, and I didn't want to pry into something that you might not want anyone to see. Now I know its history, I think it was even nicer of you to lend me such a precious thing. I love to think that it comes from your Mother as well.

My dear Barnes, I most certainly <u>didn't</u> get the toothpaste all over my fingers. If you take a tube and a tooth brush to the window you can easily see to squeeze the paste out, even although there isn't a light in the room. No, the paste went nicely and neatly onto my brush. I didn't use the gas, firstly because Auntie Fanny never suggested it, and I thought it might be unusable; and secondly because it is much more fun to use a candle.

My dear Mr B.Sc. I never did any electricity except such as I gleaned from a rather involved book; at least I suppose the book was alright but I never saw or did any experiments. The only thing I did was a little about conductors and non-conductors and electroscopes. I was away for all the rest.

Barnes, you are cheering about Inter. I couldn't possibly have passed this year, unless it was by a great fluke, but it is nice of you to say that, it certainly never occurred to me that it could have been like that. And I am so glad that you know how it felt.

I do wish I had known that we would pass you on the way down from here, and then I would have waved my handkerchief when we got to Bunnington. We were looking out of the window most of the way, so I probably saw your train.

Barnes, I am so awfully glad about the airships. I think I know a little of how hard you have worked – Auntie Fanny was telling me a little, and I can guess a good deal. I hope more than I can say that you will join Commander Burney. No, I don't think it wrong to work on a Sunday as long as it isn't for money. I quite see that it is a way, and a very beautiful way, of worshipping God. Only I suppose one ought to leave time to worship Him a little some other way – at Church, I mean. I don't know why you should go to Church every Sunday, but Uncle Charlie gave many good reasons for it, and I suppose he's right, though it's awfully difficult, and I never know what to think. Anyway, I certainly don't think your working on Sunday was wrong, – quite the contrary, it was very wonderful of you, and I like to think of you doing it.

On Saturday Mary and I walked to a little place called Llanaber and saw its Church – a dear little old place. We went to a service there yesterday. After we had seen the Church, we turned off the main road and went up a hill to try to find a Neolithic burial ground which it speaks of in the guide book. We saw a lot of big grey stones dotted about over the side of the hill, but as there are big grey stones on every hill in Wales, I don't think they were much to go by. When we got to the very top, we lay full length on the grass with the sun pouring down on us and ate our lunch. It was a gorgeous place covered with bracken and heather and thyme; and there was a wonderful view of the sea – which sparkled and glistened so that it almost hurt your eyes to look at it, – and of distant mountains. We came down the hill by a brook, and we took off our shoes and stockings and came down partly walking in the water and partly on the soft green grass and moss beside it. Rather an undignified proceeding, I am afraid, but the water was so delightfully cool and fresh, and it was so lovely to have it trickling over your feet, and to stand on a slippery, mossy stone under a tiny waterfall. And the grass at the side was softer than any velvet. Of course I slipped and made my dress all wet, but it didn't take long to dry in the sun. I did enjoy that walk. I think I like best going alone with Mary for beautiful, solitary walks like that. We never know where we are going until we get there, and what we generally do is to ask somebody in the village if they know of a very nice walk of about 7 or 8 miles, and they usually tell us something very beautiful. Then we wander about just as we please, allowing enough time either to get back to the station or to walk home so that we shall be back at the time Mrs Turner [Mary's mother] expects us. As neither of us is any good at finding our way about, we usually get lost, but one can always come back the way one came. I do think the hills and woods and streams are so beautiful, don't you?...

We want to climb Cader most awfully, but they won't let us go alone and Mr and Mrs Turner can't climb. We are trying to find some reliable person who will go with us. I don't see why we shouldn't go alone; if one goes on a clear day, can one lose the way?...

You will let me know about the Airships won't you? I am simply longing to hear. I wish you could tell me by wireless; I don't want to wait a week.

Barnes: 8th August 1923. New Cross
This isn't an answer to your letter, but just to tell you that so far there are no further developments in Airships and I do not expect to hear anything for some weeks. Anyhow they haven't sent me North so apparently they are marking time on the other appointment.

I go to camp at Hunstanton on Sunday (12th) with the battery, and I will write from there. We are there for a fortnight.

Molly, I cant thank you enough for your letter. It is too good for me. The marvel to me is that you should care to hear from me, and spend your time in writing me such simply charming letters. Its quite incredible – in fact I never <u>do</u> believe it until I see your letter. But I mustn't go on or I shall have to tear this up.

Of course, regarding toothpaste – I forgot you went to bed by <u>daylight</u>.

Molly: 9th August 1923. Barmouth.
<u>I</u> go to bed in daylight! My dear Barnes what are you thinking about? Never once at Auntie Fanny's did I go to bed before 10.30 p.m., and quite

often I went <u>much</u> later. Also I do my hair and wash my face and hands before I clean my teeth. So it certainly couldn't possibly be daylight. You try it yourself at about 11 o'clock and you'll see I'm right. Daylight indeed!

I hope you will have a good time at Hunstanton. I haven't the least idea what you will do, and I very much want to hear about it. Do you have to wear ~~khaki kahki~~ [both attempts crossed out] bother it, where does the h come, there is one, isn't there?

Barnes: 10th to 16th August 1923. New Cross – Hunstanton
How on earth did you ever manage to write in bed with a candle? I meant to start this letter days ago, and simply haven't had a minute, so now in despair, I am trying your device; but its most difficult, and as you said, very flickery. Its nearly half past eleven, and I didn't get home till nearly ten, and Im too hot and sticky to stay up any longer. Molly I did so enjoy your letter, thank you again ever and ever so much. I always feel as tho' every letter you write me is the most delightful I've ever had. I <u>am</u> so glad you are having good weather.... I can enjoy myself in the rain – I mean a man can, better than a girl, because it doesn't matter getting wet. One day we were wet thro' three times, and I was nearly reduced to appearing in my pyjamas!

I'm jolly glad I was able to cheer you up a bit over Inter. Perhaps I can understand what you have felt, because like you I am the second child of a family. You see the eldest always does things first which seems to emphasize things when you <u>dont</u> do them, and I never could entirely suppress, not a feeling of jealousy, but a sort of feeling that one was left behind as it were. Everyone gives you the impression that you are expected to be more or less level with your elder, and sometimes it makes one feel a little "out of it". Why Molly, I failed for Matric when I left school, when I was nearly 17 in September 1904 and didn't get another chance to try till I was 23! My degree has a queer history. Fancy spreading one's efforts to get a degree over nearly 17 years, counting my 1st failure at Matric. I suppose its one of the hardest things I've ever done. You see I'm 36 next September.

I'm glad you think like that about working on Sunday. I dont do it now, but would if it were right and necessary.

About going to Church, I dont know either Molly. It is very difficult, when one is hard at work indoors all the week, to go and sit in a stuffy Church when one might worship in another way, by seeing the beautiful country or getting exercise to fit one to do ones work better. I'm afraid I don't often go, but I think if one had children it would be a duty to them. All life must seem different when one has the care and responsibility of young lives. But I'm not good enough to have an opinion worth listening to.

I think perhaps if I had been responsible for you and Mary I would scarcely have let you climb Cader alone, although it is only walking all the way. Still its a long climb and you might have lost your way.

Oh yes, I know that home feeling. Do you know that even now I always have to overcome a sort of dread of going away. Even for instance to go to Borth, tho' I know I am going to friends, at the last moment I feel as if I didn't want to leave home. I seem to differ from you there – I get my homesickness before I leave. When I've taken the plunge and get there, I'm happy enough. But prior to leaving I get perfectly wretched and go mechanically about my packing like a condemned prisoner. Were you homesick at New Cross?

On the Monday [Bank Holiday 6th August] I went to my old school, Christs Hospital. Of course it was holidays but I called at the school Sergeants house, and had a long chat over old times with him. He always remembers me, because I used to be very keen on gym. and we are great friends. I begged the key of the swimming bath off him and had a topping swim – tho' it did seem so lonely, absolutely alone in a bath that one had never bathed in before except with a crowd of 50 or so house fellows. About 8 of the boys that were in my house along with me were killed in the war. And many more that were there after my time.

Then he told me that my old science master [C.E. Browne] had not yet gone away for the hols. so I went to his house for lunch. His wife's awfully decent – still calls me Barnes. She used to be ripping asking one in to tea on half hols, and I suppose I was something of a favourite being head of the science school in my year. I love rambling over the physics and chemy labs with him. They've got the most magnificent labs. We discuss methods of education – he was awfully amused to hear I had got a degree and had tried my hand at teaching. Its the most wonderful school in the world I think – they've got over 800 boys! and about oh, I dont know 1200 acres of land perhaps or it may be more. I love it and so does everyone who has been there.

Oh dear, however shall I tell you all about a military camp. It's all so familiar to me, having lived in several during the war, and it seems funny that anyone shouldn't know just what they are like. But I suppose a girl never sees or hears about such things naturally.

Well, I suppose to me, the outstanding feature is always one of acute discomfort coupled with hard physical work. That is just at the start. One soon gets broken in, and begins to enjoy the thing immensely. But just at first everything is too awful for words. The roughest of clothes, ill fitting and coarse, with no lining at all, enormous boots, weighing nearly 3 lbs each; living and lying on the ground; having to pack all ones belongings into a small canvas kit bag, barely big enough for a boy scout; the inability to wear clean things and clean collars – one cannot wear a collar at all – all combine to make one feel miserably uncomfortable just at first. Washing arrangements are a great trial – there is one out-of-doors tap for 180 men, its quite impossible to have a bath, or even to wash ones body piece meal. Today Tuesday I have had a bathe for the 1st time. One's hands get coarse – nails dirty and broken – large lumps of skin off my knuckles where I have struck the breech of the gun in loading and so on. Here's a sample timetable

Reveille	6 a.m.
Parade for Physical training	6.30 7.15
Breakfast	7.15
Main parade	9 am till 1 pm
Lunch	1 pm
Parade	2 – 4.30 or 5
5	Bathing Parade
5.30	high tea

After that games or if one is detailed for guard

Parade for guard	7.20

All the intervals are filled in with inspections, drills, marching or gun drill etc. Then one has lots of equipment to clean, tent to tidy, plates to wash, Kit Bag to pack etc etc. Tonight I am in charge of the guard which protects the guns and ammunition, situated practically on the sea shore, about a mile and a half from the camp. Lying in the guard tent – the sentries off duty are stretched out on the ground, I am writing to you by an inch of guttering candle stuck on my upturned enamel plate. The door of the tent is open its ten to twelve, and pouring with rain, simply beating on to the tent. The sentry on duty comes every few minutes and stands in the door to get partly sheltered. All I can see of him is a pair of enormous boots, the bottom of his puttees and a huge khaki great coat, and his rifle, which in rain he carries with the lock protected under one arm. We've just had our supper, dry bread and cheese – and a mug of beer, thousands of moths are in the candle and the beer; and as I drained my mug – there was a large earwig drowned in the bottom. Oh my left arms gone to sleep from leaning on it.

The flickering shadow of the stacked rifles – tripod fashion is thrown high up on the inside of the tent. We are doing Molly something that men have done from time immemorial I suppose. That does appeal to me so much. One reads of guard room scenes in so many historical novels – the guard is always an interesting and romantic feature.

Ive just changed the sentries (12 oclock midnight). The very terms we use must be hundreds of years old – "Sentries Pass" is the order to get one sentry off his beat and the other one on to it. Now I can rest till two. I have been up since 6 this morning working and drilling hard all day so I must stop – Goodnight or rather good morning Molly. I wish I could describe things like you do – I could make it all so interesting for you. But perhaps your imagination can fill in something of what I have failed to show.

4 pm Wednesday afternoon. After a very wild night and wet morning we have got a simply baking afternoon. All our necks and arms are very sore from sunburn and the rough blankets – no sheets – and nights are almost unbearable. We had a very busy morning – a general has been down inspecting the guns, so Ive been paying compliments highly technical things they are too, all the time. The etiquette connected with a guard is very complex, I suppose built upon their old history. Some day I must tell you all about it.

I am so tired – practically no sleep last night, and we dont get off till 8 p.m. tonight. They made me a Bombardier which is a non-commissioned officer, although I didn't much want to be. Still its awful fun. I am no.7 on our guns crew, the one who loads and fires. It takes 11 men to work one gun.

We had a good bathe yesterday, and I started a swimming class!

I say Molly please forgive me for this letter – its such a rotten one compared with yours. I suppose when ones been up for just on 36 hours on end one isn't quite at ones best tho' I'm not really tired. On reading thro' I seem to have made a great fuss about ones trials – but really I enjoy the whole thing immensely, its a great game and I love it and one gets most awfully fit and keen.

I'm so sorry if this is late. No one seems to know when the camp post goes and I cannot get out.

Yes khaki!

Terrific thunderstorm just started. Tent nearly blown away!!!

♦ Molly by now had joined her family in Dorset. Mr Bloxam enjoyed the company of his children; they joined in energetic activities which his wife could not share. They walked for miles, and hired bicycles. With his three elder daughters he set out along the Dorset coast, but Molly came to grief skidding on a gravelled corner in the road, cutting and bruising both hands and her right arm in the fall. Never one to admit to illness or incapacity, which in her mind amounted to deplorable weakness, she dismissed the episode as trifling. She had seen her mother's hypochondria feed upon her father's endless support, and she would not allow herself even a step along that path. Barnes wholeheartedly admired physical strength and endurance, but did not usually hide ill-health in himself.

Molly: 19th August 1923. Swanage
I am awfully interested in your camp life, and you describe it all beautifully. I can easily imagine you lying there in the guard tent with the sentries lying on the ground, and that little bit of candle throwing great, queer shadows on the sides of the tent. Barnes, I think you are a wonder to be able to write anything after 36 hours with practically no sleep, and to write a lovely letter like that – I can't think how you did it.

I expect you are wondering what on earth I have done with this writing. I am doing it with my left hand. The result [of the accident] was that I had to have everything done for me for two or three days, but my left hand is heaps better and I can use it now. I am becoming quite brainy at doing things with my left hand, except for writing. Your pen keeps on trying to run away and it is impossible to write a line straight; also it takes such a long time to do – I have been $1\frac{3}{4}$ hours over this so far. It is Nan's fault that I am not using my right hand; it's the arm and not the hand that is bad really, and Nan is so awfully careful and says I am not to use my hand until my arm feels better. Anyway, it's pretty sure to be all right to use tomorrow. The only other result of the tumble is that I can't bathe, partly because of my arm, and partly because I am still stiff and bruisy. It's a great nuisance, but still I'm jolly thankful I didn't hurt my leg or foot, because then I should not have been able to walk.

I have just been watching the most wonderful sunrise I have ever seen. As a matter of fact I have never seen the sun rise out of the sea before. We got up just in time to see the top of his rim appearing above the horizon, and soon he turned all the little red clouds near him to gold, and he made an orange pathway across the sea. He has just been hiding his great red face behind a cloud and when he came out he had done something to himself because he had absolutely changed, and he is now so yellow and bright that we can't look at him.

I should think it must be wonderful to feel that you are doing the same thing that has been done for hundreds of years; I can quite understand how it appeals to you, and I think it's ripping that it does. Christs Hospital is an old school, isn't it? What a huge school it must be. Is there a Science and an Arts school, then? Fancy you being head of the Science School, how jolly. You must have enjoyed Monday there.

Yes, 17 years does seem a long time to spend over getting one's degree. I think it was all the more persevering of you to go on till you got it, seeing what a lot of long interruptions you had. Doesn't it seem funny to think that that September when you failed in Matric. I was just born. I do wish our family were 10 or even 7 years older, and then I might have done something useful in the War instead of just going to school and trying to knit mittens. It would be funny – George would be nearly 17, and I should be 26 – how nice.

It is very interesting for me having that sample time-table, because now I can always imagine what you are doing at any time. You have your breakfast just at the time they go down to bathe (I mean the time our family goes down to bathe – I've only bathed with them twice). You certainly have a very strenuous time, and I should think you must be most fearfully tired at night. I wonder what you look like in khaki (thankyou); I'd love to see you. Are the other men in the camp nice? and did you know any before? How is the swimming class going?

Oh dear, Nan wants to do my arm, so I must stop for a little while. I do hate having it done.

<u>Tuesday</u>. Nan won't let me use my right hand again because yesterday she thought my arm looked redder or more swollen or something – I'm sure it doesn't really – and she says it will heal more quickly if I keep it quite still. I'm sorry – it is horrid for you to have to read writing like this. I didn't know whether you'd rather I write now when the writing is all bad, or if you'd rather I'd waited till it was a bit better. Perhaps I ought to have waited a week or so, only then you might have wondered what had happened.

I have just read this letter through, and it is just about as horrid and untidy as it can be. I am most awfully sorry. I say, Barnes, you won't think I haven't bothered, but have just written anything anyhow, will you? Honestly, I've done it as well as I could, but I can't do it fast, and my left hand is still a bit stiff (from the bruise – the graze is all right).

Barnes: 21st August 1923. Hunstanton
I am so awfully sorry to hear from home about your arm and do hope it is better? Did you get my letter last week? I only ask because I had to entrust it to a gunner to post, and cannot now identify him to ask if he has forgotten it or not. We are drowned out – everything soaking wet. It interferes very much with firing as the clouds are so low.

Barnes: 28th August 1923. New Cross
I think your letter is simply wonderful, and you the very brainiest person I know to be able to write like that when you were all shaken up, and had to use your left hand. Molly, you must have spent hours and hours over it. Why <u>did</u> you do it? Of course I would wait and if you had felt that I should have worried, a postcard would have done. I can't bear to think of you so patiently writing hour after hour, when you must have felt simply rotten. It is a very precious letter, and thank you very very much indeed for it. And as for thinking you hadn't bothered, why Molly, you know that the smallest line from you is more than I expect or deserve. How much more then a letter so long and written under such conditions. And it <u>isn't</u> untidy or muddly. I think the way you write with your left hand is absolutely <u>wonderful</u>. Seeing how you did it, I thought I would try and the result is simply awful as you can see. [The last 10 words are untidily scrawled]. That is rather queer – your left hand writing retains all the characteristics of your right hand and is recognisably yours. Mine looks like that of a child of 4! It's a <u>ripping</u> letter and I was most awfully interested in all you had been doing. I <u>am</u> so sorry for you having that wretched accident, its such hard lines when you were on a holiday, and not being able to bathe, or anything. I do love the way you always take things so bravely when they go wrong with you. I always grumble like anything, and go about feeling injured and miserable. (Every other word seems to be underlined, but thats just how I felt).

Such heaps of things happened at camp after my last letter. To begin with we began shooting with live shell at aeroplanes flying above us. The cartridges are loaded with a reduced charge so that the shells burst about 700 or 800 feet below the plane – near enough to judge whether the shot would have hit, but sufficiently far to make the pilot quite safe. The guns make a tremendous row – most unpleasant at first, but one soon becomes so accustomed to it that it seems un-noticeable. One has to wear special plugs in the ears, which while they allow orders to be heard, prevent the gun blast doing any harm. Its very exciting especially when the whole battery is firing at once. There is tremendous competition between the guns crews to see who can load and fire the fastest. It takes 12 men to work one gun, everyone going top speed the whole time. You can imagine that when shooting at an aeroplane moving at perhaps 120 miles an hour it is within range for so short a time that the number of rounds which one can fire per minute becomes of the greatest importance. Our gun attained a rate of nearly 20 rounds a minute, that is one every 3 seconds and the average for the whole battery of 4 guns was 15.3 rounds a minute. A huge cup has been presented to be held by the battery which does best in an annual competitive shoot, and this year – the first that the cup has been open, our battery has won it. Our battery – the 155th – is recruited entirely from men in Vickers and is commanded by one of the firms directors – Colonel Bouverie. So I knew a good many of them before, and we are all very happy and friendly. The firm are most awfully pleased at our success. Everyone had the jumps on the morning of the competition shoot, as we are closely watched by officers and instructors from the Army School of Gunnery who stand behind one with a notebook and pencil and note every fault. Its all right when once one starts firing, but its awful waiting, and thinking of all the mistakes one may make. The loading, when the guns are nearly vertical is very difficult as the shell and cartridge are about 2'6" long and weigh about 25 lbs.

I had one more guard during the second week – we had rotten weather all the time and lots of cloud. We shot for the cup last Thursday; and Friday was spent in packing up the guns – they are raised by means of huge screw-jacks and then gigantic wheels are fitted as you can see, so that they can be towed by a heavy lorry. On the road home, the guns have to be accompanied by a guard of 12 men, and I was put in charge of this party. The lorries are simply hired and driven by civilians. We set out at 8 a.m. on Saturday morning, with 7 lorries and 6 guns and a Ford van carrying provisions. Each gun weighs about $6\frac{1}{2}$ tons. We arranged to spend the night – Saturday night at Cambridge and to go on to London on Sunday. The lorries were awful and kept breaking down every few miles which much delayed our journey. Finally one broke down completely about 8 miles from Cambridge, and at the same time it came on to pour with rain. So I and a sergent (I dont think thats right) (sergeant?) mechanic who travelled with us walked on to some cottages to try to borrow a rope for towing. We always travelled in the rear, in case of any trouble. The cottagers directed us to a farm, about $\frac{1}{2}$ mile away, where they might have one, so we plodded on, but I smiled at an old woman in one cottage, and asked her if she could make us a cup of tea against our return. We got a rope, from a farmer who looked like a man off the Stock Exchange, – grey striped trouser like one wears at a wedding, black coat, fashionable collar, and curley greased moustache.

We found a gorgeous cup of tea waiting for us, the first drinkable tea I had had for a fortnight. And then proceeded to tow a lorry plus a gun with the seventh lorry. By this time the rest of the column of 5 guns were miles ahead and our pace was very slow – 4 or 5 miles an hour at most. So we didn't get into Cambridge until nearly 7.30 – 60 miles in $11\frac{1}{2}$ hours, some rapid movement. I had sent the Ford van with one man on ahead to prepare a meal and we parked the guns in a long line outside Kings College. The men had their meal in the open street, but I appointed some sentries and took an hour off to get a wash and some food at a hotel, as I have to stay up all night again for the 3rd time. The remainder of the men had only 2 hours guard duty each thro' the night, that is 2 would be on from 8 till 10, two more from 10 to 12, 2 more from 12–2 and so on. There were two covered lorries, but the civilian drivers promptly occupied most of the space in these. I had no authority over them, so most of my men had to sleep in the open. This was rather unfortunate as it rained continuously all thro' the night. We had brought a supply of blankets and waterproof sheets, but when one actually comes to try it is exceedingly difficult to keep dry under a narrow waterproof sheet (about 6 feet long by 2 feet wide) when in pouring rain. If you cover your head your boots stick out and the sheets are not large enough to tuck in so that turning over is a manoeuvre of the utmost delicacy. Also with head covered if you doze off, there is nothing to keep the sheet off your face and you wake half suffocated with mackintosh. If you make sure of covering your toes, and bare your head, the rolled up blanket which you use as a pillow gets simply soaked and the water goes down your neck. Two of the younger fellows got wet right at the start, and I could not persuade them to lie down. I told them it did not matter how wet they were, if only they would keep warm, and said they could have plenty of blankets each. That is quite true and is a good hint to remember. Temporary exposure to damp and wet is only harmful when the evaporation of the water causes a big loss of body-heat, so that if you can only wrap up so much that you really feel too hot, the experience will leave you quite unharmed. Particularly with men as we were, in fine physical condition. But they were young and would not listen, and preferred to tramp up and down.

Having had a hard day they were soon exhausted and about 2 in the morning one of the sentries called me to say that one of them had been very sick and was feeling ill. So we saw him safely into a corner and wrapped him up. A few hours later, about 4 the next sentry came, and said "Please Bombardier, Skinner is lying in a heap against the railings I dont think he's very well". I found him very nearly in a state of coma – he couldn't answer me coherently, and at first I felt worried. Still I knew he was fit and sound, and decided that it was probably only nature putting the boy to sleep in her own way, so we picked him up and I put him in my blankets and covered him over – he never spoke or moved.

Of course I could have ordered them both to lie down right at the start, but after all one must under such conditions give a man some latitude to get thro' in the way they think suits them best and they must learn by experience for themselves. I felt like twelve different sorts of father and mother before the night was thro'. I dont think I've ever welcomed the dawn more heartily. But personally I enjoyed the whole experience. To lie in an open lorry, in the main street of Cambridge, with the muzzle of a gun poking over ones head and the wonderful pinnacles of Kings College Chapel, built I think by old Henry VIII, and to listen to the innumerable chimes coming from every quarter of the town was very wonderful and impressive. One distant chime seemed to play an old Gregorian chant every quarter of an hour – at least it sounded like a 4 line chant and one got an extra line as each quarter went on. I tried to learn it, but it always got so mixed up with all the others sounding at the same time that I never quite succeeded in disentangling the entire chant.

I went about 10 oclock to clean my teeth at a public lavatory in the market square where I had previously noticed that there was a wash place. Alas when I got there, although the lavatory was still open, they had turned off all the water and taken away the towels. Fortunately I use Kolynos, which works best on a dry brush, so I cleaned away vigorously and then went to a coffee stall and had a cup of tea to wash my mouth! I had to have a 2nd cup, before it began to taste like tea again. Swallowing ones tooth wash is rather a drastic proceeding. I dont mind missing a wash, but I cannot <u>bear</u> going thro' the night without cleaning my teeth. Just a habit I suppose.

We were away from Cambridge soon after 7 on Sunday morning, and without further serious breakdown reached Chelsea about 5 in the afternoon. I didn't get home till 7 as we had to unload all the lorries. I was really rather tired this time, and rather foolishly after having a of tea and talking to Father and Fanny, I had a very hot bath – (and I jolly well needed it too). The effect was almost magical for I nearly fell asleep in the bath itself. I just managed to crawl back to my room and flopped into bed, – never said goodnight to anybody, and went straight off to sleep. The only thing I can remember is thinking that Fanny must have changed my pillow because it seemed so soft after a fortnight on the floor. Wasn't is heroic of Harriet to keep up a blazing fire on a Sunday, just to give me a hot bath? I love Harriet and always want to kiss her, but can never summon up the courage, in case she wouldn't like it. But we always shake hands most enthusiastically. On Wednesday we had a swagger dress parade at the Chelsea Barracks, and Colonel Gill presented the Cup – a huge silver two handled one, about 15 inches diameter to our commanding officer. Afterwards it was filled up and everyone had a drink out of it.

A letter came while I was at camp from Victor Goddard [friend from war days in Barrow] to tell me he is engaged. He's been perfectly wretched for the last year as he proposed at Easter (or just before) last year and she wouldn't make up her mind. Instead of going on resolutely, he fell into a state of doubt as to whether, if she didn't love him, he could really love her, and got quite upset and ill over it. Anyhow they've both made up their minds now and I hope they will be very happy. Only I fear she doesn't like me very much, which means that I shall never see so much of him again, and we have been very intimate friends, spending our holidays together, flying in the war together and so on. It makes me a tiny bit sad, for he is the last of my bachelor friends. Did you know I have been a best man three times? They say its perfectly fatal to ones own prospects of getting married!

I fear this is a very interrupted letter, but really I never seem to get a minute – all my days are so full. I suppose you find the same. I think its the happiest way to live dont you Molly? Simply fill up all your time and work and play to the limit of ones energy.

♦ With this letter Barnes sent two photographs. One, taken at Kent's Bank while he worked at Barrow, shows him in a smart blazer sitting nonchalantly sideways on a wall with his pipe in his mouth, not smiling but just about to when he had removed his pipe. The second shows him in Sunday suit, debonair and well groomed, in the bows of a boat on Windermere with the waters of the lake behind him. With these he included a snap of Charles and Fanny. In return Molly sent six little holiday snaps. One was of herself on a garden seat gazing out to sea from the top of a cliff above

Swanage bay, in clear profile with a bun smooth and full at the back of her head. A year ago she had complained about the problem of 'putting up her hair', but it was now an accepted sign of her maturity. The profile shows the regular, shapely outline of her features with their strong and determined aspect. Summer it may have been, but in spite of the sun she was properly clothed in black stockings and shoes; and her hands, with the poor bandaged right arm, were neatly folded in her lap.

Molly: 5th Sept. 1923. Hampstead
I was most awfully interested in everything you had to tell me, specially about the night you spent at Cambridge. I can just picture you lying in that lorry, listening to the chimes and trying to learn that special chant. I do think men are lucky being able to do things like that.

I do love to think of you feeling like twelve different sorts of fathers and mothers the night when you had to look after that boy. Poor boy, I expect next time he will take your advice.

Congratulations to the 155th Battery on winning the cup. I am so glad you did. I had no idea it took so many men to work a gun; and you have the loading and firing to do, it sounds as if that were the most responsible part. And fancy your gun doing 20 rounds a minute. Weren't you jolly proud of it? You can't think how I love hearing about all these things. It is all absolutely new to me; I've never heard anything like it in my life before, and it is all so strange and different and interesting, specially when I know that you are doing everything that you describe.

I like the photo of you sitting on the wall smoking your pipe best. I think it is a topping one, and I am most awfully proud of it. I think the one of Uncle Charlie and Auntie Fanny in the garden is lovely. Auntie Fanny looks the picture of comfortableness, as if she were just feeling at peace with everybody and everything after a good cup of tea; and Uncle looks as if he thought he ought to look very serious and stern, but all the time he wants to smile.

I'm sending you 6 (not all of me) if you would like to have them. Mary expressed a great desire to have a photo of my "bun", but the one she took didn't come out. So I sat in the garden feeling (and looking) like a martyr one Sunday, while Betty photographed my bun; and the joke is that the precious bun is nearly out of the picture!

Yes, it was heroic of Harriet to keep up the fire for your bath. I do think she is a dear. I shouldn't think she'd mind you kissing her – quite the contrary. I guess she'd rather like it. I know she is as proud of you as the rest of the family is, in fact as everybody is.

I've only been to a wedding once in my life when I was very small, and the only thing I can remember was that we had to eat a tiny piece of wedding cake afterwards, which everybody said was awfully nice, but which I thought was horrible. I haven't the remotest idea of what a best man is supposed to do. Poor Barnes, it must be rather lonesome for you to feel that all your friends are married and only you aren't.

I too think it is by far the best way to live, always to have heaps to do. Yes, my days are always full – either working or sewing or practising or reading to the kids or playing with them, and crowds of other things. Since we have come home, I seem to have done nothing but darn. It is surprising how socks and stockings and things wear out on a holiday. And Nan has gone away to-day.

Barnes, I don't want you to think that I didn't want to write that last letter, or that I wrote it as a sort of duty, or anything like that. I enjoyed writing it as much as I always do. My only trouble was that you might rather have a respectably written letter late than a badly written one at the right time. I don't know if I can explain myself or if you will understand – but I liked to feel that I was taking a little bit of extra trouble (not that it was a trouble at all – quite the contrary) over something – even although it was only a letter – that was meant for you; and I liked spending a long time over it and doing it as carefully as I could – I just loved it; and I knew that being a friend like you are, you would have understood if I had only written a card, the same as I should if ever you couldn't write, but I didn't want to send a card, I wanted to write a letter. Goodness, what a terrible sentence. You have to go ever so far back to find the beginning.

I found a teeny bit of white heather, at least I thought it was white, but I'm afraid it looks a little mauve now. Anyway, would you like it? But perhaps you don't care about things like white heather? I can quite understand it if you don't.

Chapter 9
STALE BREAD AND MARGARINE

Barnes: 11th Sept. 1923. New Cross.

Many very very happy returns of the day to you. And all my best wishes for your great happiness and success in this new year. Heaps of games and dances and a triumphant Inter. next July.

Do you mind my sending the accompanying book? If you have read it, I can easily get it changed. If you have not, you will love it I am sure. I think it is Stevenson's best and I have read it over and over again.

Alas, we are in sad trouble. Poor dear Harriet is not well again and Father thinks she must stay in bed for some days. So Fanny and I have been washing up and laying supper. I'm going to insist on Harriet coming down to my room, tho' Fanny nobly says she wont hear of it. But its such an awful drag up all those stairs that I shall have my way. I would clear out and go and live in town, only I think I can more than make up for the trouble I am, by dashing about in the evenings. But I must say you are simply noble to <u>enjoy</u> this housework business. As you said, however, it is trying to help Fanny and Father that is the great compensation. I love them both more every day, but never can manage to tell them so. Men are perfect idiots at that sort of thing. Oh dear, just opened door to urgent parish case, Pater's got to go out late. Everything happens at once.

Now this isn't my proper letter, its an extra, surely a permissible one. Nevertheless, thank you very much indeed for your letter, Molly and specially for the end. Yes I do treasure the white heather, very much. It has found a safe resting place.

Barnes: 12th Sept. 1923. New Cross

I admire your "bun" most tremendously Molly, how on earth do you make it so smooth and neat, and keep the insides inside?

Thank you for your congratulations on the Cup. It looks very fine in the big hall of Vickers House in Westminster.

Yes, we all 5 slept in the same tent. At least one is absent – there were 6 really. Six in a tent is quite comfortable, but during the war we had 13. That is really a squash, and you always had 2 or 3 peoples legs lying on top of yours. I am a very quiet sleeper and hardly ever move, but some men are dreadful. Even if you can persuade all the men to start the night lying absolutely radially like this [he inserted an aerial view of 13 matchstick men lying in a circle with their feet round the central pole of a bell tent], there is not room for everyones feet at the tent pole, for the tent is only about 13 feet diameter, so that when your head is touching the brailing (or skirt) your feet are practically touching the pole. And when they go to sleep – the men I mean and turn over and so on one soon gets rather mixed up. And then there are the invariable two or three who snore. Why isn't there a law compelling everyone with adenoids to have them removed. It must be so awful for their wives. Before I accepted a man I would certainly say "Do you snore?"! if I were a girl. Snoring is so frightfully unromantic.

I've had such a busy day. Molly, Commander Burney has asked me to join him in the new Airship Company this afternoon.

<u>Friday</u>. I had no time to write more yesterday as there was so much to do at home.

The new Airship Company is being floated almost immediately, and Burney wants me to go down to Pulham Air station in Norfolk to survey two old airships and report to him as to whether we can re-commission them in order to start a training service. I am most awfully pleased, as I am quite certain it is going to be a huge success. In two years we shall be running a regular weekly service to India – total time 60 hours against 25 days now taken. I expect in 5 to 10 years time we shall be carrying practically all the first class trans-oceanic passenger traffic of the empire.

I like Burney very much and shall love working for him. We didn't fix up anything about salary, but he said he would give me as much as he could but that just for a year or two we should have to be careful, while the new company was finding its feet. In any case he would get me more than I am getting now, which is £650, so that things look very hopeful. It relieves me of a good deal of anxiety, in case anything happens to the Pater.

Besides, Molly, its the most wonderful work in the world. Some day perhaps I shall be able to take you all over a ship, and then you will see for yourself. You see its all new, and one is thinking out new things all the time. I think when we have done enough trig. I shall have to start writing you a book on Airships, so that I can talk to you about them freely.

Speaking of trig, I am sending off some more, [Chapters VI and VII] as perhaps you will like to start maths again before the new term starts? But dont bother, Molly, if you are too busy. I'm afraid its an awfully muddly instalment; its been written at all sorts of times and places and lacks continuity. Somehow one can not make any fun out of trig. its all so matter of fact – its difficult to say just what I mean – Calculus is an art – it endows you with wonderful powers; you can let your imagination go to all sorts of lengths and not pass out of the realm of reality – Calculus is like chocolate meringues – elementary trig is like very thick stale bread and margarine. It improves a lot later on, and like everything else, merges into calculus, but alas you will never want to go as far as that, for a medical degree.

Oh dear, there's so much to tell you and I really haven't time. For instance:-

How I found Commander Burney doing experiments with "Exploded Tea" – very confidential, and how he and I solemnly sipped tea – one lot exploded and the other not, out of two large (and dirty) tumblers, while he talked about his Airship plans – then we changed glasses to see if they tasted any different.

How I won a battle with Kiddy [the New Cross cat] this morning when she wanted to put the kittens in the coal hole.

How wonderfully Auntie Fanny now gets the breakfast – I am just an assistant and do as I am bid – a post I very much prefer.

How I've been twice to the dentist this week who's been dead for a year, and I didn't know it.

All about a best man at weddings, and why you should stop away from them whenever possible.

Did ever idiot write letters like this? I doubt it.

My dear Molly, if you choose to write to me in Hebrew (which I cant read), it wouldn't make any difference – I love your writing, however its done. It was very very sweet of you to take so much trouble for me. I only get afraid sometimes that you may feel as if you couldn't stop writing even if you wished to, without hurting me. I do hope you understand that no such thought must cross your mind. If you felt like that it would worry me most awfully, and much as I love writing to you, it would be better for you to stop. As I have said so often I <u>never</u> expect you to answer – that is really genuinely true, and you must trust me, and not say you wish to go on, or anything like that, because now you have so many growing interests in other directions that you yourself are quite unable to predict that what interests you now will do so in the near future.

Trigonometry. Chapter VI. [12th Sept. approx.]

I am tired of the last chapter, and its so long ago that I cannot now remember its number. In fact I've had a feeling for a long time that I have missed a number out of the sequence. You must forgive me if I have. Pages [127] to [131; second half of Chapter V] seem to me an awful muddle, and I did start to rewrite them, but concluded I hadn't time. It was written while I was on holiday and everyone was interrupting, and consequently is rather confused. If it is not clear you must let me know, as it is rather important.

Page [130] starts rather a new subject, [on "principal angles"], and I should really have made the new chapter begin there. Going back to [p. 127–130] you will readily see that instead of confining myself to talking about OM and MP, I might equally well substitute OP and MP or OP and OM. For we saw previously that <u>any two</u> of the three measurements OP, OM, and MP were sufficient to determine the angle 'α' (see [p. 127]), it being understood that OMP was always a right angle.

Just to make this quite clear, what I mean is that we might have started [p. 127] by saying

$$MP/OP = M'P'/OP' = M''P''/OP'' = k_1 \quad \text{(read "kay one")}.$$

I say k_1, so that you shall not confuse it with the k we used on p. 127. The ratios are all constant but all 3 are different that is MP/OM $\neq$ MP/OP. The sign $\neq$ means "is not equal to".

Or we might have said

$$OM/OP = OM'/OP' = OM''/OP'' = k_2 \quad \text{(read kay two)}$$

Let us just summarise. From all that has gone before we now know that <u>whatever the position of the bounding line</u> (radius vector) we can write for any given angle

1) $MP/OP = M'P'/OP' = M''P''/OP'' = M'''P''' / OP''' = \ldots k_1$

2) $OM/OP = OM'/OP' = OM''/OP'' = OM''' / OP''' = \ldots k_2$

3) $MP/OM = M'P'/OM' = M''P''/OM'' = M'''P''' / OM''' = \ldots k_3$

NOTE:-

a) I have put the 'equals ... equals' (= ... =) above like that to show that we can go on taking as many fresh positions for M and P as we like.

b) I have changed the plain 'k' used for MP/OM on [pp 127–130] to k_3 to bring it into line with k_1 and k_2.

c) All the above ratios can be inverted, e.g. – OP/MP $= 1/k_1$ etc etc.

I particularly want you to realise what I mean by the words underlined [as above]. Do not get into the habit of thinking of your radius vector as lying only in the first quadrant. At the risk of wearying you, I will show you the R.V. lying in all the quadrants. I have already drawn and discussed the angle in the 1st quadrant. Here is the second.

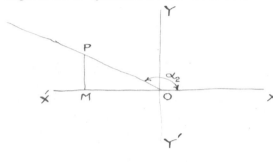

The angle about which we are thinking is the new angle α_2. I call it α_2 not because it has anything to do with the α on [p. 127] but because it is a similar discussion. You must try to get used to this very general way of using symbols, it isn't really confusing if properly employed. (I try to, but you must excuse failures, due to inexperience and difficulty of revising etc.)

Note that OM is now a __negative__ quantity. We do not write it as –OM; but suppose I give it a definite value (say 2") __then__ we write

OM = –2" Clear?

As before, __any two__ of the three measurements OM, MP, and OP will enable you to draw the angle α_2 and we can still say:-

1) MP/OP = k_1 (Of course k_1 here has not the same value as k_1 on [p. 160] (see note about α_2 [above])

2) OM/OP = k_2

(Note that since OM is -ve, OP is __always__ +ve ∴ if we give a definite value to OM, it follows that k_2 is __negative__ but as before, the -ve sign does not appear until the values are assigned. For example, if I say OM = –2"

$$OP = 2\tfrac{1}{2}"$$

then $\quad k_2 \quad = –2/2\tfrac{1}{2} = –0.8$

But you do __not__ write OM/OP = $–k_2$ for k_2 is __in itself__ a negative quantity in this instance. You will readily see the importance of this in the following examples. Supposing I say to you "In the equations OM/OP = k_2 and OM/OP = $–k_2$ substitute the value k_2 = –.8. The first equation would come

OM/OP = –.8

whereas the 2nd would come

$$OM/OP = -(-.8) = +.8$$

which is a very different thing. Wait a minute – I will just deal with the 3rd ratio, and then I will write a separate page on this negative quantity question.

3) $MP/OM = k_3$

Here again MP is a positive quantity and OM is a negative quantity so that k_3 will be a negative quantity.

I have somewhat loosely spoken of OM as negative and OP as positive – what I should really say is "OM is a negative quantity". This may help to make [the discussion] more intelligible. When we thoroughly understand what we mean, there is no harm in using the loose and convenient form above, as long as you are quite sure that you understand that we do not mean that OM must have the minus sign before it.

If you can understand the way I have put all this, you must have the brain of a Huxley – I'm sure I couldn't!

Before I deal with the 3rd and 4th quadrants, just a word to make negative quantities quite clear. I expect you <u>are</u> quite clear already, but I do not care to risk your having any uncertainty on the point – if you are satisfied – skip!

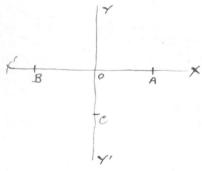

In the figure, I have drawn OA numerically equal to OB. That is to say, the same number – (say 20 millimetres) i.e. <u>20</u> represents the size of each length measured in mm's. One however, according to our convention, is measured in a +ve direction and the other in a negative direction. So that OA represents a positive quantity, and OB represents a negative quantity. But we do not talk about –OB. OB is <u>itself a negative quantity</u>. (Because we made it so!) This negativity (sorry) becomes apparent however whenever we wish to express OB in connection with any other quantity.

For instance OB = –20 mm (because we made it so).
But   OA = +20 mm   "   "   "   "   "
Hence   OB = –OA

So that when we come to equate a negative quantity to some other numerically equal quantity, we have to be very careful to look to see whether this other quantity is a negative or a positive one. If for instance I draw in [downwards, i.e. negative, on the Y axis] OC = –20 mm, then OB = OC, both being negative quantities in addition to being numerically equal.

The same applies of course if I chose to call

$$OA \text{ say } a \qquad \text{then} \qquad a = 20 \text{ mm}$$
$$OB \text{ " } b \qquad \text{ " } \qquad b = -20 \text{ mm}$$
$$OC \text{ " } c \qquad \text{ " } \qquad c = -20 \text{ mm}$$

and
$$b = -a$$
$$c = -a$$
$$b = c$$

If we wish we for convenience we can invert the 1st two and say

$$-b = a$$
$$-c = a$$

Its equally true; for if we substitute for b its value of –20 mm we get

$$-(-20) = a$$

or $\qquad a = + 20$ as before.

I have made a great deal out of a very simple point, but it enters so largely into what follows, that a full explanation now may save much confusion later.

Going back to where we were on [p. 161]:- Here is the 3rd quadrant.

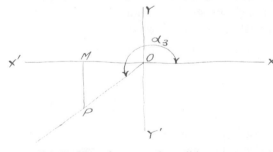

We are now thinking of the obtuse angle α_3 (alpha three). Both its OM and its MP are negative quantities, and as before, wherever we choose P we get the constant ratios:-

1) MP/OP = k_1 (k_1 will be a negative quantity as MP is neg. and OP is pos.)

2) OM/OP = k_2 (k_2 will be a neg. quantity as OM is neg. and OP pos.)

3) MP/OM = k_3 (k_3 will be a positive quantity as MP and
                        OM are both neg.)

Here is the 4th quadrant:-

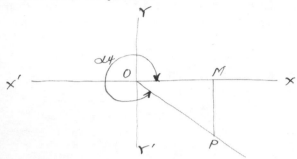

and again we can say:-

1) MP/OP = k_1 (k_1 will be a negative quantity as MP is neg.
                        and OP is pos.)

2) OM/OP = k_2 (k_2 will be a pos. quantity as OM as both
                        OM and OP are pos. quantities.)

3) MP/OM = k_3 (k_3 will be a neg. quantity as MP is neg. and
                        OM is pos.)

I hope all this time you have not forgotten that
1) k merely stands for "some constant";
2) that k is non-dimensional, that is, it is a mere number, being the
number of times the length MP is contained in the length OP, etc;
3) that for any given angle its several k's are all of different value
(except in one or two special cases of which we shall speak later).

But for the same given angle its k_1 is always the same
  "   "   "   "   "   "   " k_2 "   "   "   "
  "   "   "   "   "   "   " k_3 "   "   "   "

NOW, the Great Conclusions which we draw from all this terrible bother
are simply these:-

1) that there are more ways of measuring angles than the two (degrees
and radians) which I mentioned in the beginning chapters.

2) to be precise there are 3 more ways, for I can give you for any angle
   a) its k_1 and having given one of these you, by
   b) its k_2 referring to a book of mathematical tables,
   c) its k_3 can find out the angle.

3) If I chose I can make this into a total of 6 more ways for I can give the reciprocals of k_1, k_2 and k_3 thus

    d) $1/k_1$

    e) $1/k_2$

    f) $1/k_3$

4) Having given you either the k_1, k_2 or k_3, then (provided you have a book of tables which give you these ratios) you can translate them easily into degrees or radians as you may require ONLY you must bear in mind that the tables will give you the <u>principal value</u> (see [pp. 127–130]) of the angle whose k's you are looking up. It is left to you to remember (tho' it is not often necessary) that all the angles expressed by the "General Solution" will have the identical k's that you are looking up.

Trigonometry. Chapter VII.
A little more about OP, OM, and MP.

    Suppose we draw at random any angle θ (theta).

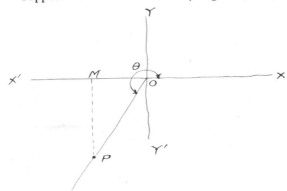

As you see, its Radius Vector has fallen in the 3rd quadrant. Give θ an OM, MP and OP by the simple process of choosing any point P on the R.V. Drop the perp^r on X'X to fix M.

    OM and MP are now the co-ordinates of the point P.

    OMP forms a right angled triangle whose right angle is always at M.

    Pythagoras Theorem (Euclid I 47) tells us that:-

$$\overline{OP}^2 = \overline{OM}^2 + \overline{MP}^2$$

($\overline{OP}$ means the length OP hence $\overline{OP}^2$ means the square on the length OP.)

You probably know this theorem well. The proof has a figure like this:

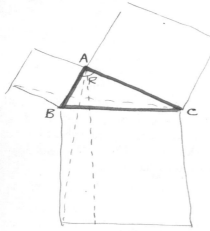

In words:-

The square on the hypotenuse of a R. angled Δ is equal to the sum of the squares on the other two sides.

Since the R. angle in our case is always at M, then OP is always the hypotenuse.

Hence both OM and MP are always shorter than OP.

$\therefore$ the ratios MP/OP (k_1) and OM/OP (k_2) are always <1.

That is to say k_1 and k_2 are <u>always</u> fractions.

$$k_3 = 1 \text{ according as MP } \begin{array}{c} > \\ = \\ < \end{array} \text{ OM}$$

You read the above thus:-

k_3 is greater than, equal to, or less than one, according as
MP is   "   "   "   " " " "   OM.
($<$ means "less than" and $>$ means "greater than")

1. k_1 and k_2 are connected by virtue of Pythagoras' theorem thus:-

$$k_1^2 + k_2^2 = 1$$

See how it is done:-

$$k_1 = \frac{MP}{OP}, \quad k_2 = \frac{OM}{OP}$$

$$k_1^2 = \frac{\overline{MP}^2}{\overline{OP}^2}, \quad k_2^2 = \frac{\overline{OM}^2}{\overline{OP}^2}$$

Hence

$$k_1^2 + k_2^2 = \frac{\overline{MP}^2}{\overline{OP}^2} + \frac{\overline{OM}^2}{\overline{OP}^2} = \frac{\overline{MP}^2 + \overline{OM}^2}{\overline{OP}^2}$$

But $\overline{MP}^2 + \overline{OM}^2 = \overline{OP}^2$

$$\therefore k_1^2 + k_2^2 = \frac{\overline{OP}^2}{\overline{OP}^2} = 1 \qquad \underline{Easy.}$$

2. k_1, k_2, k_3 are all connected thus:-

$$\frac{k_1}{k_2} = k_3$$

Thus $k_1 = \dfrac{MP}{OP}$, $\quad k_2 = \dfrac{OM}{OP}$, $\quad k_3 = \dfrac{MP}{OM}$

$$\frac{k_1}{k_2} = \frac{MP}{OP} \div \frac{OM}{OP} = \frac{MP}{OP} \times \frac{OP}{OM} = \frac{MP}{OM} = k_3$$

♦ As well as digesting the mathematics, Barnes's birthday caused Molly some problems. With her wishes for his happiness on 26th September came a little water-colour painting which she had done herself, but about which she was uncertain.

Molly: 20th Sept. 1923. Hampstead
I simply can't imagine me ever wanting to stop writing to you, Barnes. I promise you that if I ever do, I'll tell you; but I feel perfectly safe because I know that will never happen. It seems to me that it's the same with you – supposing you wanted to stop writing to me, you would let me know? You see I am only a very ordinary, every-day person, while you are – oh dear, I don't quite know what to say you are, that will include everything – while you are Barnes; and writing to you and having your letters is one of my great enjoyments and interests in life. I think it is about as likely that the sun won't rise as that I shall want to stop.

Molly: 25th Sept. 1923. Hampstead
Many many happy returns of your birthday, Barnes. I hope you will have a very happy day and a very happy year and heaps more happy birthdays – in fact I hope you will have all the happiness there is.

Barnes, I have spent ages and ages, wondering whether to give this thing to you or not; though I did it myself, I didn't make it all up. Firstly it is of no earthly use to anyone; and secondly I have spent a good time over it, but it is very badly done. You see it was like this – a long time ago I asked myself what I should give Barnes for his birthday. I have never given a present to a man except Daddy before, and I generally give him a book because I know what he is wanting at the time. But I didn't know what book you'd like because you are sure to have read all the books you like. Another time I shall ask you. Then I thought you might like something just to show how that how – oh, I always dash into a sentence and half way through find I can't finish it. Anyway, here it is with all my very best wishes and with many humble apologies because it is a pretty rotten birthday present.

Do you remember this time last year? It was the day you were to have come here for tea, and we were all downstairs, and the bell didn't ring and nobody heard your knock because the carpenters and men were making such a noise. And to-morrow you were going to Switzerland, and in six more days we would start our first term at College. Hasn't the year gone quickly – such an enjoyable, interesting year it has been for me, with different people and things and feelings. Has it gone quickly with you? You have had all sorts of new experiences – Switzerland and Germany, and Airships starting again. I wonder what will be happening this time next year; don't you feel excited when you look forward and think what will happen in a years time?

Barnes: 26th Sept. 1923. New Cross
My dear Molly,
    I've been stuck for a quarter of an hour since the last comma trying to make up a sentence which I shall be able to finish, and which will tell you how much I love and prize your sweetest of all birthday presents. This one – sentence I mean – has composed itself, and has got very near the tearing up point in consequence. But oh Molly, what could you have sent me better? I think its just the dearest little picture I've ever seen, and I do thank you very very much. I think its <u>wonderful</u>, and I love every bit of it. I dont know what I like the best, its all just perfect from the little frog striding so manfully thro' a forest of bullrushes to the excited little birds who are chattering on the topmost twig. I <u>do</u> think you are clever. Is that what you see Molly, when you and the Duck sit and look out of the window together? Molly, its me thats the ordinary everyday person, and you, with your sweet beauty of thought and mind, who are so wonderful to me. And you cant think how much I value being allowed to see the things you see, and hear the things you think, and I <u>do</u> thank you for it. It makes me feel very very humble. Thankyou again and again.

    I should think I did remember this time last year. Has the year gone quickly? I really dont know. No, on the whole I think not. It has strung itself out to the last possible minute, and as if I hadn't endured enough of it it kept me up till half past two this morning. But thats ungracious of me, because really its been the most wonderful of my life. Switzerland and Germany and airships were only incidents.

This letter might have been done left handed, for it has taken nearly an hour and a half! and now I must go and help clear supper and so on.

Goodbye, Molly for the present. All my thanks to you for everything; your letter, and wishes and your gift.

Barnes: 28th Sept. 1923. New Cross
Molly, I've been able, since writing to thank you for it, to look at my picture by day. I think its <u>exquisite</u> so delicate, just as if a real fairy had guided your hand. I'm going to have it framed, and hang it by my bed. I'm jolly glad you <u>did</u> send it. Thank you again Molly, so <u>very</u> much. My other letter was a poor appreciation, but you know I think all I feel about it.

Thank you so very much for your letter to Pulham. Yes, you did omit c/o Major Scott, and the servant at the Scott's thought it couldn't be me, and sent the letter on to the house of another officer on the airstation. When Scottie and I came in, I asked if there had been any letters for me, and she told us of this one. The people to whose house the letter had been sent were, we knew, all out, so there it was lying locked in their hall. Scottie knew a back way into their house, and said, in a casual way, oh yes he would walk over there after dinner and climb in and get it. After dinner, it was pouring with rain, and we had just had a hot bath and change, and he apparently forgot all about it. I made various shy remarks, about how I hoped no one would return unexpectedly, write Not Known on it, and repost it. Thats all right Wally, says Scott, you're not in a hurry for it are you? Seeing that we were in the depths of the country where the last out post goes at 5 in the afternoon, I couldn't scrape up any desperate reason regarding urgent matter, – immediate reply, and so on, and it was useless to say My good Scottie, its my letter from Molly!! for amongst a large circle of men friends I am I think regarded as a sort of cave-dwelling bachelor, with no girl friends whatever. Which is, I suppose in substance true, you being the first. I mean I <u>know</u> a good many of course, but being friends is a different thing. So there your much wanted letter stayed until nearly 11 pm. with me dancing a mental can-can of impatience all the evening. Scottie and his wife are the most delightful people. Scott is the pilot who flew R.34 [an airship built in Barrow under Barnes's design] across the Atlantic to New York and back. I made my first flight with him, when I was in the Air Service in 1915, and had my first accident with him too.

They've been married nearly 5 years, but seem just as much in love as ever – I quite felt in the way sometimes. He's the same age as me, and she was married when she was nineteen or twenty and they have two ripping children John who is 4 and Judy who is 4 months.

Scott and I came up to town on Wednesday, having been over the ships at Pulham, saw Commander Burney and went on to Bedford yesterday. I didn't get home till 8 in the evening and was rather tired, having been climbing about all the day. The shed roof is about 140 feet high and one climbs about to inspect various parts of the ships. Have you a good head for great heights? I hope so. Of course it makes a big difference whom you are with. If it is someone in whom you have complete confidence you can follow them almost anywhere. Ive been busy at the Air Ministry all day today. I'm in a very difficult and delicate position – its too long to write but I'd love to be able to tell you all about it.

Oh that dentist – I wish he <u>was</u> a ghost – I've been again today and had an awful time. How did you get on? I do hope there was nothing the matter. If things are not what they seem – as dentists for instance – why must they often be so ludicrously different? The man to whom I was recommended in Victoria Street, was called by the aristocratic name of Montgomery. For two visits I called him Mr Montgomery. At the end of the second, he said solemnly "I hope you are not under the impression that I am Mr MONTGOMERY. He has been dead for a year. My name is <u>PANK</u>!!! Loud gurgles from me. I love talking to my dentist, dont you? Thus:-

The dentist "Did you have a good holiday this year?

Me – (mouth full of pads and tubes and wide open) "Ya aah a waah a aah aaa"

Dentist "on the whole its been a fine summer, dont you think?" (why will the silly idiots always make remarks that require an answer).

Me "Yah aah aah"

Dentist "Im going away next month"

Me "ooh aah"

Dentist "Things in the Ruhr dont seem much better?"

Me "Yah aah aah ooo aaah eee eee oo ah ah eee aah ya ah"

Dentist "EXACTLY"

and so on.

Oh dear Molly, here am I with time so short, and wasting it all with silliness. I'm sorry.

Molly, to me <u>you</u> are the most wonderful person, and I keep on trying to make you see that I am ordinary and everyday. I dont think things are quite the same, because of course I <u>knew</u> that I should never want to stop writing a year ago or more (about 18 months).

Molly: 4th Oct. 1923. Hampstead
I am so sorry I forgot to put c/o Major Scott and therefore caused you so much bother. Barnes, you are <u>not</u> to call your letters rotten; they aren't rotten – quite the contrary, they are topping, the last one was, and so was the one before and so are they all.

I have never been anywhere very high, and nowhere that is the least bit dangerous, but I am always hoping I shall some day. The nearest I've been to it is near the top of Cader at the precipice above Llyn Cader. As you say, if you are with a person whom you could trust absolutely, I should think it would be quite all right. Anyway I'm quite sure I shouldn't mind great heights at all.

I would love above all things to come over an airship with you one day. I'm quite sure I shouldn't mind in the least climbing about in the roof – quite the contrary, I should love it, specially if you were there; then I should know it was absolutely all right.

I am so sorry Mr Pank gave you such a bad time. I hope you are all right now, and haven't got to go again. I too had an awful time. On Tuesday a tooth began to ache furiously so that I couldn't eat anything all that day.... On Wednesday I went to the dentist and he said a piece had broken off and the nerve was practically exposed. Of course he drilled right down to the nerve and then squeezed cold air in, at least it felt as if that was what he did.... Yes, aren't dentists silly the way they ask you questions while they are stopping your teeth. Mr Barrat – Mr Russel Barrat, very superior to your Mr Pank, though perhaps not equal to a Mr Montgomery – asked me what school I went to! I made an indignant noise in my throat, and I wanted to show him the back of my hair, [schoolgirls did not have hair that was "up"], only I couldn't because he was digging diligently at the tooth. Then he asked if I went to South Hampstead High School; I made a noise something like the way you talk to your dentist, which I suppose he thought was meant for a yes, for he went on discoursing for the rest of the time on school and how much nicer it was for me that it was for him when he was a boy. Dentists are queer people; but I do think he might have realised that I am heaps too old for school.

Are you really going to Bedford? I do wish you could tell me what the difficult and delicate position is, or was. That's the worst of only writing – there isn't half time to say all one wants to. Barnes, I know you didn't do other things instead of writing. I know you simply haven't time to write any more, though they are beautiful letters already, and what more I could expect I don't know.

I am so very glad you liked that present. You can't help seeing things like that sometimes, specially when there is a moon. Barnes, you shouldn't spend nearly an hour and a half writing to me. You are so ripping.

Molly: 6th Oct. 1923. [Unaddressed]
I was reading through your letters, and I hope you don't think I made up all that picture, because I didn't, I copied it from a book of Pam's. I believe I said so, but I can't leave you in any doubt.

♦ Barnes received this contrite little note when he returned from a weekend staying, as he had done since he was a small child, with an "adopted" aunt. 23 Pepys Road, which Barnes now shared with his father and Fanny was not the home in which he grew up. The vivid memories of childhood belonged to 241 New Cross Road, quarter of a mile down the hill. Dr Wallis had moved before his first wife Edie died, but some time after Barnes had left the house where he grew up.

Barnes: 6th–8th Oct. 1923. Brighton and New Cross
This house is Auntie Bell's house. She isn't an aunt at all really, but was my Mother's great schoolfriend....

They lived in the most beautiful houses, where John and I used to go and stay as kids. She is a brilliant pianist and my earliest recollections are of her and her music and her beautiful houses and carriages and servants – a sort of fairyland.

Just for this fortnight they have lent [this house] to Father and Fanny. I love coming here – everything is even more familiar to me than the things at home.

Harriet amused me yesterday – she has come down too – for I was talking to her in the kitchen and picked up a queer silver and oak tea caddy. "Thats a tea-caddy" says Harriet, thinking I suppose that I was wondering what it was. "Yes Harriet" says I "I've known it since I was six". The only thing thats changed is the grand piano, and that is the third that Auntie has had to the best of my recollection. The first that I can remember in the big house days, was one of those huge concert things with an enormous length of tail. And the floor of the drawing room was very highly polished and slippery, and it used to be a great game for us children to squat down like little gnomes, and propel ourselves about in and out amongst the legs and pedals of the piano on the slippery floor while Auntie Bell played appropriate music. Goodness, how many pairs of little knickers we must have worn out.

Another very clear memory to me is at their house on the North Downs. So big was it, that it had a dairy, and they used to churn their own butter. And the fat and important housekeeper was showing Mother and Father the dairy, and we two imps – I was 5 and John 6, were there too. And I, being even then of a mechanical turn of mind, began, while the others were talking, to investigate the churn – one like this:-[Barnes drew a little diagram of a rotating barrel churn]. Presently I came upon a bolt, like a door bolt, used to shoot into a slot to prevent the barrel from swinging round, when the churn was opened to remove the butter. Someone had taken out the butter, but had left the churn open, and full up with butter-milk, which was going to be given to the pigs. Of all these things I knew naught, being only 5, except that bolts would generally draw. So I quietly drew it.

To my horror the churn immediately swung over, and simply flooded the dairy floor with butter milk. I seem to have a vision of Mother, Father, the Housekeeper, John and myself all together trying to find standing room on the bracing bars at the bottom of the wooden stand to raise ourselves above the milky flood. And then dear Mother, intensely embarrassed, trying to explain me away to the Housekeeper, and after that a merciful oblivion has been drawn down by the intervening years. But I dont think I was even beaten.

That time is the last distinct recollection I have of Father before he was lame. How early can you remember I wonder?

All right, "orders is orders", and I wont call my letters rotten again, but I cannot often help feeling that they are. You see Molly when I do <u>anything</u> for you, I never feel as if it were good enough, and always want to apologise for it as it were. I cant help spending a long time sometimes – an hour and a half isn't long, most take longer than that – only that one was such a poor result (sorry – I wont disobey you again, – really) when it was done. But the more one feels things the harder it is to say anything somehow. I dont know if you will understand? Yes of course you will.

I'm sure you will be all right at a height if you feel as if you wanted to try – I dont mean for flying nobody minds that – its quite a different sensation; what I meant was climbing about the gangways and so on in the roof of a shed – as you'll perhaps do some day if you care to come over an airship with me. But thats looking very far ahead.

Molly I am so glad that the last year has been a happy one for you. the coming one will be even better, because coll. wont be so strange, and you will find the work easier this time. I have another lot of trig. nearly ready, and will send it on soon.

Mrs Blackman, a charlady whom you may remember, is looking after me. Oh dear, she reminds me of my days in lodgings. Harriet left a dish of stewed damsons behind when the family left home last Monday, and I was still having them on Friday! Regular landlady style. And the drawing room always seems so miserable without Fanny. If I go to Bedford, I really dont think I can face rooms again. I shall live in a hotel all the time. Tho' that is a miserable business, as I have tried it before. Still, houses in Bedford are not to be obtained I hear, and anyhow, a house for a bachelor is a very uneconomical business, and its an awful fluke if you get a decent housekeeper.

It felt so lonely at home without my old folk, that I have come away. I dont mind being alone so much in a place where one is accustomed to being alone. But to be alone in a place which one associates entirely with two very dear people gives me the absolute blues.

Talking of houses has made me think of furniture, and furniture makes me think of a very nice thing that has just happened. During the early days of the war, I served under, and was much associated with Air Commodore Masterman. I assisted him in carrying out experiments for mooring non-rigid Airships out in the open, and he invented and I worked out for him the final arrangement. Then we patented it – Abel and Imray did that for us – [the family firm of patent agents in which Molly's father was a partner] and I remember calling on Uncle Arthur long before he knew who I was. Now Commodore Masterman has made some money out of royalties on the patent, and wants me to share, and wrote the other day saying he had £42 in hand for me, and would send it along. This is the second time this has happened. For lots of reasons too long, as usual, to write, I dont feel I ought to accept the money and I have explained this to him. So he says all right, he wont make a fuss about it, but will I allow him to give me something as a memento of our work together. I couldn't very well refuse without hurting his feelings; and indeed I <u>should</u> value something from him most immensely. He suggested a fitted dressing bag, but they are after all rather useless things, and I have often thought how, if and when I do ever have a home of my own, I should so much like to have a Grandmother clock.

Grandmothers are the daintiest little things, with pretty silver faces; and the one I coveted chimes every quarter like Big Ben, with very deep, slow, sleepy chimes that make you feel most awfully homey and contented and peaceful. Not a bit loud – so quiet that you would hardly hear them, and yet quite penetrating, giving you quite a queer feeling of a sort of sympathetic personality pervading the house. Do you think I'm talking rubbish? I suppose it is absurd to attribute personality to a clock, but to me it has; and I think to you too? Its just like a dear old lady in cap and mittens, who sits in her chair, and says "there's plenty of time my dears; dont rush and worry; take time, take time". And then she folds her hands again and sits and waits for the next quarter.

My Grannie is going to be a very very sweet old lady. She's not really old in her physical being, as she has the very finest modern English works that you can buy. But she is old in that she embodies the spirit of hundreds that have gone before, and she herself they tell me, will live to a great age – say 200 years! Oh Molly you've only to see her to fall in love with her. I do hope you think I have chosen wisely? Please tell me if you think you'll like her. I thought it over for a long time.

Masterman met me and took me to lunch on Wednesday (how nice to be taken to lunch). And then we went to Harrods and chose my dear Grannie. And then we went up to the restaurant there, and over a cup of China tea, "Chaney my dear" he composed the following inscription which is to go on a silver plate just inside her door so as not to spoil her outside appearance.

to B.N.W. from E.A.M.

In happy recollection of a

successful experiment

1916 – 1918

Dont you think thats very cleverly written, and very sweetly put? Here is her picture – do not bother to return it, just tear it up.

I've been busy on Airship work all the week. Burney suggested that I should go up to Howden in Yorkshire to inspect one of the sheds there, but I dont much want to go, and think he has forgotten about it.

The new Airship is such an enormous step foward in size that the whole structure has to be designed anew. What I mean is we cant simply take existing forms and methods and multiply them by two as it were.

Can I give an example of what I mean? Bridges perhaps are the closest parallel. You have often seen the ordinary type of railway bridge with a girder like this [here he added a simple little diagram of an openwork structure resting on two supports]. One may perhaps use this type up to spans of 100 feet or slightly more. Much over that size however that particular method of construction becomes impossible, chiefly because the weight of the bridge itself becomes so great that it cannot hold itself up. And so a different kind altogether has been evolved for great spans – like the Forth Bridge for instance – one of the worlds great engineering marvels. We are in much the same position regarding Airships at the moment, only their structures are far more complex than even the greatest of bridges. I've had some rather good ideas this morning and am feeling rather pleased. The trouble is one can never tell whether they are really good, without days and days of elaborate mathematical calculations.

I am just home, and have found your little note waiting for me. Molly dear, you <u>did</u> tell me about the picture right from the very first. I do hope you haven't worried over thinking I didn't understand that? It doesn't detract from it in the very least. I took it that the picture appealed to you and you painted it and sent it to me because it represented so well all the charming things that please you and which sometimes you have told to me.

But I can so well appreciate your feeling if you thought you hadn't made it sufficiently clear, and it was <u>awfully</u> decent and brave of you. One sometimes gets in tangles quite unintentionally thro' someone else misunderstanding something, and it worries you more and more until you feel simply awful about it and you go, feeling like a criminal, and blurt out the whole truth. I've done it heaps of times myself, so I can sympathise. But isn't it a ripping feeling when you <u>have</u> done it, and you find the other person has been trusting you all the time like anything, and only thinks heaps more of you for what you have done, instead of thinking you a little liar as you half feared! Oh Molly, I've done it so often, and its one of the best sensations in the world, especially when its someone you admire very much and look up to. Dont you think so?

Molly: 18th Oct. 1923. Hampstead
You are the most wonderful person that ever was. How you could tell exactly what I was feeling on Saturday when I wrote the note [about her present] and all last week, is a perfect marvel to me. Oh Barnes, thank you so very very much for your letter, and specially that part of it. You can't think what a relief it was to get it. You see all through the week I was alternately hoping Barnes wouldn't think I was very dreadful, and wondering if he would. It wouldn't have mattered if it had been just anybody,

but when, as you say, it's somebody you admire and look up to, it matters more than anything in the world. Thank you for being so understanding and knowing how it feels. Yes, I should think it is ripping when you know the other person has been trusting you all the time; when I got your letter, I felt so jolly glad, I hardly knew what to do with myself. I could go on saying thank you for ever, but there is a limit to your patience.

Yes, I know just how hard it is to say things when one feels them very deeply. As a matter of fact I should imagine that it would be harder for a man than a woman, because I suppose he is naturally more reserved. Some people seem to find it fairly easy, but personally I find it most awfully hard to say things I feel very much.

Barnes, I think she is just the dearest sweetest Grannie that ever was. I know I shall love her directly I see her. What could you possibly have chosen better? I think it is a very wise and a very good choice. No, I don't see anything at all absurd in attributing personality to a clock, specially a dear Grannie clock like this one. I love to think of the old lady sitting calm and peaceful saying "take time, take time". It is a dear thought about a dear thing. She will make a home more homey and anything that does that is a good thing. That inscription is very sweetly put, I like "In happy recollection of a successful experiment" most awfully. Thank you so much for sending me her picture. Of course I shan't throw it away; I shall treasure it till the time when I can see and hear her.

I would love to see Auntie Bell. How you and John must have loved going to stay at her beautiful house. I can just imagine you two sliding about on the slippery floor while Auntie Bell played for you. I always think a great fascination to a child is the fact that everything is so much bigger than himself, so that the legs of a piano seem quite large and he can easily get under it and have it for a roof. After all, why should you be beaten? You weren't told not to touch the churn, and you couldn't possibly know the disasterous effect of simply drawing a bolt. You were only investigating a new thing, and I think it was very praiseworthy and very typical. But I can imagine how embarrassed your Mother felt when she was trying to "explain you away" to the Housekeeper. What a lovely expression! I wonder if she managed it; of course I suppose it all depended on whether the Housekeeper was an understanding sort of person or not. I wish I had been there to see you when the buttermilk flowed out! I guess you ran straight to your Mother as the likeliest person to help you out of the difficulty.

I can't remember anything very exciting happening when I was small. My chief recollections are of the night nursery where Baba, Betty and I slept with Nan. We used to be taken upstairs to go to sleep after dinner every day, and were each put in a cot and secured by having a net over us, which was attached to the four corners. I remember I used invariably to undo mine, climb out of bed, go to Baba and assist her, and then we got Betty out of bed, regardless of the fact that she might be wanting a nap after her lunch. Then we had a perfectly glorious time rushing about all over the top landing (the nursery was down below) until Nan, very irate, came up, spanked us and put us to bed again. In the end my net was tied right under the bed with string where I couldn't possibly get at the knot. But I found a place in the net where after much pushing day after day, I managed to get my head through. It wasn't much good however because the rest of me couldn't get through. How I hated that net; I believe if they had just put me to bed without one at all, I should have stayed there quite quietly, but to feel that I was caged in only made me want to get out all the more.

One night I remember, I decided that I ought to see what the garden looked like in the dark. I couldn't have been very old because I had only just moved into the third cot. You see, there were three cots, and as the newest baby came upstairs we all moved up one place. I suppose I must have been about $4\frac{1}{2}$. Anyway, I got out of bed and crept downstairs into the drawing room where there were French windows leading into the garden. I can remember peeping into the dining room where Daddy and Mother and some other people all in beautiful evening dresses were having dinner. The windows in the drawing room were open because it was a very warm night though there was a fire. I suppose it was in September, because it was quite dark outside, but it couldn't have been very late. I went outside and I can remember how horribly uncomfortable the gravel was to walk on in bare feet, and how delightfully cold and wet and soft the grass was. I sat down on the grass and waited – what for I don't know, but I remember expecting something to happen. It wasn't really dark because there was a bright moon and plenty of stars. I was just beginning to feel really excited because I was sure I saw something tiny moving beside a big red dahlia, and I suspected fairies, when to my dismay I saw Nan coming out of the house. She carried me ignominiously indoors and I was taken up to the nursery where my very wet nightress was taken off and dried while I sat on a chair in front of the fire very warm and comfortable with a blanket round me to keep off the draught and "just to be decent in case" as Nan said. She never got further than in case, or I never heard any more, and I have wondered ever since why I should be decent "in case". I don't remember anything more about that.

Another time I remember we all went to tea to the Pantins [cousins on the Bloxam side of the family] and played in a glorious hayfield at the end of their garden. We each had a present given to us, and mine was a box of tiny cooking utensils with tins of flour and currants and sugar – brown and white, and sultanas and other good things. We went home in a carriage of some sort and I offered there and then to divide the contents of each of the tins between Baba, Betty and me. They declined with thanks since we had all had a very large tea; so I struggled through all those tins by myself, eating the sultanas and currants and nicer things first, and ending up with the flour. I don't know what happened when we got home. I was evidently a very greedy child.

Do you know, I realised the other day for the first time that there really is some use in queer Physics things, like the triangle of forces. Apparently you need them for engineering. It makes it much more interesting when you think they are really useful in important things. And, by the way, in Physics the other day Mr Sutherland used a little Calculus as a simpler way of doing something, and I was awfully proud to think that I knew what he was doing (and I certainly didn't this time last year) while most of the others didn't understand it.

Chapter 10
THE END OF THE STALE BREAD

Barnes: 22nd Oct. 1923. New Cross

<u>Of course</u> all physics things are the everyday tools of the engineer. Didn't you know that? That's partly why I love to think you take them, so then I can talk much technical rubbish to you, and you will nod your head very wisely and say "yes, yes" at proper intervals, when all the time we both know perfectly well that being only a girl you can't understand a single word! Oh dear I feel so orangey, and theres that wretched cuckoo [the clock he brought back from Switzerland] just struck ten [p.m.], and the post goes at ten past. Wring its little neck I will. Good old Grannie, she's always a bit slow – says she hasn't got used to the place yet.

No, there is nothing very wonderful in being able to tell something of how you felt about the picture, for I think that you and I look at things in very much the same way.

You see I could tell from your little note that you were perhaps feeling a bit worried. And I have often and often noticed how you feel about many things just exactly as I have felt; so it was not difficult to put myself in your place. But I do wish I could persuade you never, never to feel one bit worried about <u>anything</u> you do in connection with me. Molly, when you hold the absolute trust of a person as you do mine, and as you always will do, dont you see that it doesn't matter what you do, or do not do, I shall understand – or if I dont understand, I shall know that there's an explanation. It reminds me of rather a neat little placard I saw hung somewhere, "If you trust you dont worry; if you worry you dont trust". I think it was intended to have a religious significance, but I dont see why it shouldn't apply to friends.

Oh Molly, I feel so mean about the maths. I've got another instalment well started, but never get a chance to go on. You see since I went to Pulham, I have been working entirely on Airships, and I have I think done an awfully exciting invention, which if successful will practically revolutionise Airship structures.

But before one can say definitely, one has to do weeks and weeks of calculations and drawings. And for many reasons it is most important that I should be in a position to put it forward <u>before</u> the new company starts. So I have been working day and night. And then Saturday and Sunday I played tennis, as I simply get rotten if I work too hard and cannot get any games.

Saturday was a day almost too good to be true. Firstly came your best of all letters, and then when I got home – there was Grannie. All in bits poor dear, and it took 3 hours to put her together. She <u>is</u> a darling – but you must come and see her for yourself. I told her all about you, and all the things you said at the end of your letter.

I simply loved the tales of when you were small. And <u>how</u> that gravel must have hurt your tiny feet. You were an awfully brave child. Molly, I dont wonder you are so perfectly ripping now, because Fanny has just unearthed a photo of you when you were about 6, and you were the most sweetly pretty child I have ever seen. She has another one too, when you were about $4\frac{1}{2}$ – just the age that you went and sat in the wet grass. I like the expression that you "suspected fairies". My dear Holmes!.

<u>Please</u> Molly write to me some more about yourself – I did so love your letter. Poor Auntie Fanny has been in bed with a chill. Can't you come to see her – I go away on Tuesday, to open the new Airship works at Bedford, but I will tell you my address. I dont want to leave home a bit. This does seem such a waste of my one chance to write to you because its such a r----- short (just in time) letter.

Barnes.

I'm sick of being your affectionate cousin its too awful for words, makes one feel like a model child.

Barnes: 1st Nov. 1923. In with Chapter VIII
I am very sorry that this chapter of maths should have been so long in hand. It is quite easy. In case you have no book of mathematical tables I am sending you an old one of mine which I got when working for Matric.

I'm jolly glad to say que je n'irai pas a Bedford after all for plusiers jolly semaines for which bien des remerciements or words to that effect, so I just wandered down there for the day on Wodensday to whom be praise. Never got a bite to eat from 8 a.m. till 3 p.m by which time j'avais beaucoup de faim et de soif. Je ne sais pas pourquoi je vous parle en francais. Wann Du die Deutsche sprache gelernt hast, dann will ich Dich in Deutsch schrieben which is probably pretty awful German, made up on the spur of the moment. Dont even know if a cousin is entitled to address you as "Du". P'raps not.

Sorry but I'm a bit orangey hier soir having persuaded an avaricious income tax man to reduce my income tax by over sixty pounds. Good work. Shall now be able to afford another pair of socks.

Wouldn't it be rummy if we went about like this:- Fare thee well fair coz, I bear thee ever in mind. Thou hastest a goodly number of pages of mathematics to digest.

Meantime I subscribe myself
   Yours truly
     Barnes

It simply had to be yours truly after all that!

Trigonometry. Chapter VIII.

There are many other relationships between k_1, k_2, and k_3, some of them quite simple, some of them extraordinarily complex. You will not need to master very many, but before we go any further let me introduce you to the real identity of k_1, k_2, and k_3. They are amongst the most important little fellows in the whole range of mathematics, physics, engineering, surveying, architecture, astronomy and a host of other dependent branches of applied maths too numerous to mention. You can juggle with them like a conjurer with a hat – turning rabbits into canaries; a paper streamer into the flags of all the allies. You, from your knowledge of trigonometry have recognised them long ago – I wish you hadn't – I have kept them in their clumsy form so long to try and cure you of the habit (if you have it) of thinking that modern trigonometry is derived from or concerned especially or solely with, triangles. I admit I cannot quote any textbook in support of my method, but I am convinced that to start right away by showing how so-called trig is founded on coordinate methods is the best and quickest.

k_1, k_2 and k_3 are then so important that they have been given special names, with regard to the angle with which they are connected.

If we are talking about the +ve angle θ, then

MP/OP or k_1 is called sine θ, written for short sin θ

OM/OP or k_2 " " cosine θ, " " " cos θ

MP/OM or k_3 " " tangent θ, " " " tan θ

The reciprocals of each of these, have also got distinguishing names. They are often useful and you must know them, but they are of only secondary importance, and you need not be so familiar with them as you must be with the primary three.

They are

OP/MP or $1/k_1$ called cosecant θ written for short cosec θ

OP/OM or $1/k_2$ " secant θ " " " sec θ

OM/MP or $1/k_3$ " cotangent θ " " " cot θ

Note that the reciprocals of the names without the prefix "co" in the primary series, are <u>given</u> this prefix in the secondary series thus:

the reciprocal of the sine   is the <u>co</u>secant

" " " " tangent " " <u>co</u>tangent

" " " " <u>co</u>sine " " secant

The cosine being the only primary which has the "co" prefix.

I advise you to get the habit of reading mentally the abbreviated names in full – thus read "sec θ" to yourself as "secant θ" – not "secktheta" and so with all of them. The reason is that there is a further variety of this curious family, known as "hyperbolic functions" which are designated "sinh θ", "cosh θ", "tanh θ" etc. If read in full these are "hyperbolic sine theta" "hyperbolic cosine theta" etc. This is really too long for anyone, even when done mentally, so here we <u>do</u> say either "shine theta" or "sink theta", "cosh theta" and "tank theta" etc. etc. Ugly names but no one seems able to invent anything better. And so closely resembling the others that you will readily appreciate my advice, but your prof. may prefer something different – there is no standard in this respect. Why not? Because maths is a wonderful shorthand appealing almost entirely to the eye, and not intended to be spoken I think. So that it is only the wretched teacher who has to contend with these difficulties [of pronunciation] and each adopts his own solution. However, until you become so familiar with trigonometry equations, that you can quick-read them, i.e. read them without repeating the words mentally it is best I think to adopt some regular method of emphasing to yourself the big difference which that little "h" on the end really makes.

In physics, we are quite as often as not, dealing with angles greater than one right angle, that is angles which if set down on our co-ordinate system would fall in the 2nd, 3rd, or 4th quadrants. Books of mathematical tables however, for shortness, only give the sine, cosine, tangent etc. for angles up to 90°. It is easy for us to calculate these ratios for <u>any</u> angle, when we are given the ratios up to 90°. By the way, the sine, cosine, tangent etc. are related to "Circular Functions", "Trigonometrical Ratios", or "Trigonometrical Functions", indifferently.

I. <u>Given an angle whose bounding line falls in the 2nd Quadrant</u>
to express its sine, cosine, and tangent in terms of its supplementary angle.

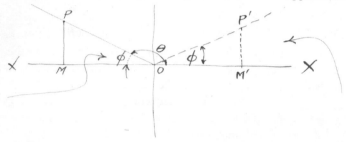

Here we are dealing with the angle θ which lies between 90° and 180° i.e. between $\pi/2$ and π. The angle ϕ is the supplement of θ, being the difference between 180° and θ.

Set off this angle in the 1st quadrant, (being necessarily an angle less than 90°, it will lie in the 1st). Mark off OP' = OP
and   OM' = OM
or simpler – drop perp. from P' on OX.

Now $\sin \phi = $ M'P'/OP'

and $\sin \theta = $ MP/OP

But is easily shown that the Δs OM'P' and OMP are identically equal (by geometry).

Hence M'P' = MP and OP' = OP $\therefore$ M'P'/OP' = MP/OP

$\therefore$ $\sin \theta = \sin \phi$

So that by looking up $\sin \phi$ in the tables, we shall get also $\sin \theta$.

But $\phi = (180° - \theta)$ $\therefore \sin \theta = \sin(180° - \theta)$

<u>General Note</u>. Hitherto I have been a little careless in the letters by which I have referred to angles. the following is a very generally accepted convention:-

When we wish to refer to an angle, which it is understood we wish to think of in degrees we give it a letter from the capitals of our own alphabet thus, angle A or B or C etc. and we should speak of sin A, tan B etc.

When we wish to refer to an angle, which it is understood we wish to think of in radians we give it a letter from the small Greek alphabet, i.e. α, β, γ, etc.

As I mentioned when talking about angular measurements, radians are used in all mathematical processes, and degrees in all surveying and practical work. Sometimes when we pass from the pure mathematical to the practical, it is difficult to adhere strictly to the above rule, but he is a poor mathematician who is bound fast by convention, and it doesn't really matter a hoot as long as one gets the right answer. Tho' there is "style" in maths, just as there is in literature.

Going back to our angle θ, we see that $OM/OP = \cos\theta$.

But from our Δs OM'P' and OMP, since they are identically equal, besides M'P' and OP' being equal to MP and OP respectively, we have also OM' = OM. Now as far as pure geometry is concerned the fact of OM being measured in the opposite direction to OM' would not make any difference to the identical equality of the triangles. But by our convention of signs we must say

OM' = –OM, or if you like

OM = OM'

it doesn't matter which way round it is written.

So that $\dfrac{OM}{OP} = \dfrac{-OM'}{OP} = -\dfrac{OM'}{OP} = -\cos\phi$

So we see that $\cos\theta = -\cos\phi$ when θ falls in 2nd quadrant.
But $\theta + \phi = \pi^c$ (= 180° of course)

$\therefore \quad \theta = \pi - \phi$ or 180° $-\phi$

and $\phi = \pi - \theta$ or 180° $-\theta$

So that we can say, as a general formula when θ lies in 2nd quadrant:-

$\cos\theta = -\cos\phi = -\cos(\pi-\theta) = -\cos(180°-\theta)$

Thus, by looking up $\cos\phi$ in the tables (ϕ being found by simply subtracting the given value of θ (in degrees) from 180°) and putting – in front of it, we get $\cos\theta$.

Now for the tangent:-

$\tan\theta = MP/OM$ but $MP = MP'$ and $OM = OM'$

$$\therefore \frac{MP}{OM} = \frac{MP'}{-OM'} = -\frac{MP'}{OM'} = -\tan\phi$$

$$\therefore \tan\theta = -\tan\phi = -\tan(180° - \theta)$$

I hope I haven't confused you by marking both angle MOP and M'OP' as ϕ in the figure. The two ϕ's are of course equal because we made M'OP' = MOP, but in all the above deductions, the ϕ to which I refer is of course the ϕ in the 1st quad. (i.e. M'OP').

You, having learnt your trig. from a book may say to me, "we have

already got a neat little ϕ in the angle MOP, – why can you not just take the ready made triangle MOP and derive the functions, i.e. sine, cosine, tangent, from that. The book tells me that $\sin\phi$ = MP/ OP, whether my triangle is standing on its head, or to the left, or to the right."

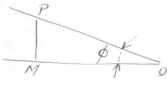

What you say is quite correct if we regard MOP as an independant figure, with no relation to the XX and YY ordinates. But placed where it is on [p. 184] – the line OM is a negative distance, and in order to derive the functions of an angle we must think of it as being placed in its correct quadrant, or some of the lines will be of the wrong sign (i.e. neg. instead of pos. or vice versa, as the case may be).

So that before co-relating the functions of θ and ϕ, we must redraw ϕ in its correct quadrant in order that the lines OM', OP', M'P' may have their correct signs for the angle ϕ.

This is clearly illustrated, if you think for a moment, by the results we have obtained.

See. <u>Numerically</u> the sine, cosine and tangent of θ are equal to the sine, cosine and tangent of ϕ resp[ly], but actually one of the lines from which we derive the functions of θ is of opposite sign to one of the lines from which we derive the functions of ϕ, when ϕ is placed in its proper quadrant (i.e. OM = –OM')

Hence, the functions of θ in which that particular line plays a part (i.e. cosine and tangent) are of opposite sign to the corresponding functions for ϕ.

All this shows clearly why I have tried to get away from the triangle idea. If one thinks in triangles, it is difficult to see how θ can have any sine etc at all!! because ϕ isn't <u>in</u> a [right angle] triangle!

Just remember that the sine, cosine, tangent etc are simply the ratios of the x ordinate and the y ordinate of the point P, wherever we may chose P on the radius vector, combined with the length OP of the radius vector itself.

II. <u>Given an angle whose bounding line falls in the 3rd Quadrant</u> to find expressions for its sine, cosine and tangent in terms of an angle less than 90° (i.e. $\pi^c/2$)

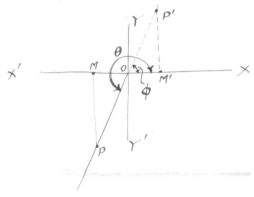

This time θ lies between 180° and 270° (π and $3\pi/2$)

If we produce PO backwards, making OP′ = OP we get, by geometry, the angle MOP = the vertically opposite angle M′OP′. Also since MP and M′P′ are parallel (both being perpendicular to X′X) the alternate angles MPO and M′P′O are also equal.

The remaining angle PMO is of course equal to P′M′O, both being right angles. We have already made one side of one Δ = one side of the other, hence they are identically equal, – geometrically.

But as before OM is a negative quantity, and this time MP is a negative quantity also, so that we must write

OM = –OM′ and MP = –M′P′

then $\sin \theta$ = MP/OP (remember wherever OP lies it is always considered positive – convention again)

Substituting for MP and OP we get

$$\sin \theta = \frac{-M'P'}{OP'} = -\frac{M'P'}{OP'}$$

but if we call the angle M′OP′ ϕ as before we have

$$\sin \phi = M'P'/OP' \qquad \therefore \sin \theta = -\sin \phi$$

Now θ may be considered as made up of two bits, for imagine the radius vector to be starting on its journey, and starting off from the line OX where it always lies when we are not using it, it rotates about 0. When it has rotated as far as OX′ it has described 180° (= π^c) and has only the angle X′OP (or MOP if you like) to turn thro' to reach its finishing position along OP.

So that angle	θ	$= 180° + \angle$MOP
But	$\angle$MOP	$=$ M′OP′ $= \phi$
$\therefore$	θ	$= 180° + \phi$
$\therefore$	ϕ	$= \theta - 180°$ (by changing 180 to the other side and then turning the whole equation round)

Substituting this value for ϕ in our equation $\sin \theta = -\sin \phi$ we get

$$\underline{\sin \theta = - \sin(\theta - 180°)}$$

Now here again ϕ or $(\theta - 180°)$ must always be less than 90°, for as OP falls in the 3rd quadrant it is obvious that its prolongation backwards (i.e. OP′) must always come in the 1st quadrant.

Remember that we <u>know</u> the value of θ – I only dont give it a particular value in order that our investigation may be perfectly general. So that by looking up the value of $\sin(\theta - 180°)$, whatever the subtraction may come to, in our tables, and putting – before it we get the value of $\sin \theta$ itself.

Sin θ then is numerically equal to $\sin(\theta - 180)$ but is of opposite sign, i.e. sin θ when θ lies between 180° and 270° (or in circular measure between π^c and $\frac{3}{2}\pi^c$) is a negative quantity.

We need not spend so long over $\cos\theta$ and $\tan\theta$ as you will be able to see them quite readily for yourself.

$$\cos \theta = \frac{\text{OM}}{\text{OP}} = \frac{-\text{OM}'}{\text{OP}'} = -\frac{\text{OM}'}{\text{OP}'} = -\cos \phi$$

and as before $\cos \phi = \cos(\theta - 180°)$.

$$\therefore \underline{\cos \theta = - \cos(\theta - 180°)}$$

The tangent:-

$$\tan \theta = \frac{\text{MP}}{\text{OM}} = \frac{-\text{M}'\text{P}'}{-\text{OM}'}$$

Remembering our rule of signs in algebra that minus × minus makes + and minus ÷ minus makes +

we get $\dfrac{-M'P'}{-OM'} = +\dfrac{M'P'}{OM'}$

(I have only written in the + to emphasise it – it isn't necessary)

But $\dfrac{M'P'}{OM'} = \tan\theta$

$\qquad = \tan(\theta - 180°)$

so that $\tan\theta = \tan(\theta - 180°)$

Notice then that in the 3rd quadrant $\sin\theta$ and $\cos\theta$ are negative quantities, while $\tan\theta$ is a positive quantity.

We might have deduced this result for the tangent another way. You will remember that while we were in the k_1, k_2, k_3 stage we found that

$k_3 = \dfrac{k_1}{k_2}$

or $\tan\theta = \dfrac{\sin\theta}{\cos\theta}$

since $\sin\theta$ and $\cos\theta$ are both negative quantities, $\tan\theta$ must must [sic] be a positive quantity, being equal to the <u>quotient</u> of two negs. Thus

$\tan\theta = \dfrac{-\sin(180° - \theta)}{-\cos(180° - \theta)} = +\dfrac{\sin(180° - \theta)}{\cos(180° - \theta)}$

$\qquad\qquad = \tan(180° - \theta)$

III. Given an angle whose bounding line falls in the 4th quadrant, to express its sine, cosine and tangent in terms of those of an angle less than 90°.

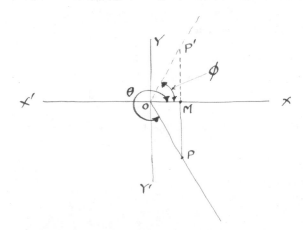

With OP lying in the 4th quad we are considering an angle θ between 270° and 360° ($3\pi/2$ and 2π).

Chose P and drop PM onto XX′ perpendicular as usual. Produce PM upwards making MP′=MP. Join OP′.

Then in Δ's MOP and MOP′

$$\therefore \begin{cases} MP = MP' \\ OM \text{ is common} \\ O\hat{M}P = O\hat{M}P' \quad \text{which is the included angle} \end{cases}$$

∴ Triangles are identically equal and in particular

OP = OP′

MOP = MOP′

As before we will call MOP′ the angle ϕ. Notice that in the geometrical proof of the identical equality of the two triangles which I have used I have written MP = MP′. So it does – in length because we made it so. As I have said before that is all that geometry worries about. But for our further purpose we must again take note that to us, MP′ is a pos. and MP a neg. quantity. Hence MP′ = –MP or MP = –MP′ as you like.

I beg your indulgence for reiterating this negative quantity business, if you feel you have fully grasped it. Note that the mere act of prefixing – in front of MP′ above, does not make MP′ a negative quantity. In the equation

MP = –MP′

MP is still the negative quantity and MP′ is still the positive quantity. The equation merely tells us that, of the two quantities MP and MP′, one is pos. and one neg. because + the one = – the other. The equation does <u>not</u> tell us which is which. That is why we can write MP′ = –MP or MP = –MP′ indifferently. If we had not the additional information which the diagram on [p. 189] gives us, as to which is the neg. and which the pos. quantity we could not get any further. I write this to make quite certain of the point once more, that the negativity is a <u>property</u> of the line MP. Thus, suppose for a moment that we made MP exactly 2 ins long. Then writing it in equation form we get

MP = –2″

If we put in the understood +ve sign we get

+MP = –2″

Now OP has nearly completed a circle. If it went on until it lay along OX again, it would have turned thro' 360° or 2π radians. It is short of 360° by the angle MOP

$$\therefore \theta = 360° - M\hat{O}P \qquad \text{But } M\hat{O}P = \phi$$

$$\therefore \theta = 360° - \phi$$

$$\text{or } \phi = 360° - \theta$$

which is the way round we want to have it.

Then

$$\sin\theta = \frac{MP}{OP} = \frac{-MP'}{OP'} = -\frac{MP'}{OP'} = -\sin\phi$$

$$\therefore \underline{\sin\theta = -\sin(360° - \theta)}$$

$$\cos\theta = \frac{OM}{OP} = \frac{OM}{OP'} = \cos\phi$$

$$\therefore \underline{\cos\theta = \cos(360° - \theta)}$$

$$\tan\theta = \frac{MP}{OM} = \frac{-MP'}{OM} = -\tan\phi$$

$$\therefore \underline{\tan\theta = -\tan(360° - \theta)}$$

So now we are in a position, given a book of tables covering angles up to 90°, to find the functions of any angles whatever.

Four special cases I have omitted. These are when OP happens to lie along any one of the four lines OX, OY, OX', OY'. If OP stopped at any of these positions it would have described 90°, 180°, 270° and 360° or (0°) respectively.

I will deal with these 4 special cases in the next chapter.
In the meantime let us just summarise what we have learnt.

Do not bother to memorise any of these values. It is so easy, when you do want to use them, just to draw a little figure roughly like this, when you can see them at a glance.

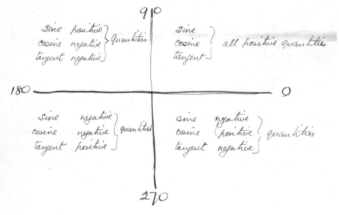

	When Value of θ lies between			
	0° & 90°	90° & 180°	180° & 270°	270° & 360°
sin θ =	As given in the table for the value of θ required	sin (180°−θ)	−sin (θ−180°)	−sin (360°−θ)
cos θ =		−cos (180°−θ)	−cos (θ−180°)	cos (360°−θ)
tan θ =		−tan (180°−θ)	tan (θ−180°)	−tan (360°−θ)

An interesting way of showing whether the functions are positive or negative quantities in the various quadrants is to draw a little diagram so

Molly's next letter was sent after a weekend with her old school friend Mary. The following weekend Barnes also went to an old friend, Air Commodore Masterman, from whom the beloved Granny clock had come. Both found their fortnightly letter had to be delayed; both scribbled hasty interim notes.

Molly: 2nd–5th Nov. 1923. *Princes Risborough*
<u>Sunday morning in bed before breakfast</u>. I had a glorious day yesterday. In the morning we went for a long walk, and of course the woods were wonderful, and it was so windy we stood on the top of Whiteleaf Cross, and

the beautiful wind blew right through us, and then we ran all the way down to the bottom over dead brown beech leaves. We had no hats and only blazers instead of thick coats, and Mary shed hairpins right and left all the way down, so I had to lend her as many as I could out of my hair. On the top of Whiteleaf – at least in the woods near it – I caught two leaves, and therefore had two wishes. I do love wind, and the smell of damp earth and leaves, and coloured woods and clouds racing across the sky, and dead leaves under your feet; in fact I love the whole of Autumn, don't you? As each season comes, I think I like it best, and I never can decide which is my favourite. Have you got one?...

My dear orangey Barnes, you may talk in French as much as you like, but German I don't think is fair even if you have overcome the arts of an avaricious income tax man, and are therefore feeling orangey. What does "Du" mean? thou, I suppose. And why shouldn't you address a cousin as "Du"? Will you please translate what you wrote? I guess it's something like "when you have learnt the German language, then I will talk to you in German", though I can't find "gelernt" anywhere in the dictionary, I suppose the infinitive is quite different from that part of the verb. I admire your French most awfully; one can see the effect of a winter in Switzerland. What a relief about the socks; it sounds as if you were very hard up for a new pair. I fear the goodly number of pages thou sentest me have not all been digested as yet, but I trust they may be so ere many moons have waxed and waned (I really mean suns, only it doesn't sound right). Oh, and thank you very much for the book of mathematical tables, which I also haven't yet had time to examine thoroughly.

I don't know what there is to write and tell you about me, besides there are heaps of very much more interesting things to talk about. I don't think I was particularly brave when I was small. I wanted to tread on the nice wet grass, and didn't mind undergoing a little discomfort to attain my end. Also I don't think I ever was afraid of the Dark – (I always thought of him with a capital letter) – he seemed so great and loving, covering everything with soft blackness so that the birds and flowers could go to sleep comfortably, and so that the Tiny Ones could dance and play without fear of being seen by grown-up people. I wonder why so many children fear the Dark. There must be some way of showing them that there is nothing at all to be afraid of; I would like to try with a child.

Barnes: 9th Nov. 1923. New Cross and Andover
... here is Friday night, and I cannot get [my letter] done. I do hope you wont mind? I've had a tremendously busy week, as I have started work on

the new airship in earnest, and have been working late every night. Burney wanted to see what my proposals were, and so I've had a rush.

Thank you very much indeed for your lovely long letter and for your note. It <u>was</u> decent of you to think of sending it Molly. Of course I didn't mind waiting. Having had your note, turned it from what would have been a time (in spite of my saying I never expect an answer) of supreme anxiety, to a time of very pleasant anticipation. I do think you are a thoughtful person for other people Molly. You jolly well saved me a miserable weekend. There – there go all my good resolutions. But how <u>can</u> I pretend I should not mind not receiving your letter, when its the one thing I <u>long</u> for all the week?

I'm so glad you had such a good time with Mary, and I loved your description of everything. Your train writing is jolly good. Mine always looks as if I were intoxicated, or else the trains I go in are much more jiggley than yours. Fancy keeping on writing right up to the time that you saw Mary – you <u>are</u> ripping. Did you fling everything – pens ink and paper, out onto the platform? I've never heard of catching leaves before. Talking of wishes reminds me of the last time Auntie Nellie was here. She had a wish bone for supper, and told me to pull it with her. I said jolly well no, because I <u>never</u> get the right half in that sort of thing, and wasn't going to run the risk of not having my wish come true now-a-days. So she said she was quite certain we should both wish for the same thing, so that it wouldn't matter which of us got the half. Jolly ingenious of her wasn't it. So I pulled – wished – and won!

You are funny about your hairpins. What I cannot understand is, if you are prepared to part with, say fifty percent of your pins to help Mary, without any serious consequences to yourself, why put the extra ones in! Just feminine inconsequence I suppose, or is it "in case"? It must be rum to have hair. I do think yours is pretty.

Seasons – oh, Spring and Summer are my favourites, one gets out of doors so much more. And I hate thick undies and overcoats. Autumn is not so bad especially in the country, when you can go and kick your feet thro' rustly leaves. But I simply cant get up any enthusiasm for winter – except in Switzerland. What can one <u>do</u>? Of course at Bedford I shall get golf, but in town its perfectly putrid. Now tell me truly, what can you see in a winter in London? By the way is your bath water c-c-c-c-old enough these days?

How frightfully energetic of you to walk 3 miles to Church. Are you <u>very</u> particular about going to Church Molly? Because I played tennis all that morning. Would you be shocked at my doing that?

You need not be afraid of my writing German to you, for I have forgotten all I ever knew, which wasn't much. Nor French either. Yes Du means "thou" and you have translated quite correctly. All regular verbs have the prefix ge- in the past participle, so lernen to learn, becomes gelernt. The 2nd person singular is very restricted in its use. When I was out there, they were telling me something about it, but as usual I have forgotten. But I think they said that even when you were engaged, you might not say "du" to your fiancee, until you were actually married. So I should think stepcousins would not be permitted, but I didn't think of it until after it was written. By the way the u is pronounced something like that in our "true". German pronunciation is harder than French I think.

How funny – I see my dictionary says "mit einem auf du und du stehen (lit. with one on thou and thou to stand) means to be on intimate terms with some one". May I consider that I "mit dir auf du und du stehen" Molly? If so du's all right. Queer that the French should have the similar "tutoyer". I think it was jolly clever of you to translate it.

Dear Herbert [his American pupil from Chillon] is in Spain somewhere, studying the language and history of the country. They were recently in Paris for some time whilst dear Herberts tutor-of-the-moment got married. It seems a great joke to them that every single tutor dear Herbert has had has got married shortly after. He wrote me a card commenting on this latest proof of the rule, and ended "to be near us is sure death". You have to say this thro' your nose with a pronounced American accent to get the proper effect. Mrs Thomas never could make out why the charm did not work in my case. I never enlightened her, and in consequence she has threatened to run off with me herself to a registry office if I am not married by the time she comes back to England. Lets hope an all-comprehending Providence will protect me, for I might almost be one of her own children. You have no idea of the risks a poor bachelor has to undergo!!

Yesterday – Sunday, the children took me to the little village church. Commodore and Mrs Masterman had to go to the Aerodrome Service at half past nine. I was writing to you soon after ten, when a shy little Peter [female] came into the drawing room and said "Are you coming to Church". Even a letter to you Molly had to give way to a demand from Peter. I'm sorry, but little girls are perfect tyrants – all my children shall be little boys, or I shan't be able to call my soul my own.

This is the last piece of my lovely paper, [from Montreux] it has served me well, has it not? I have a feeling that nothing else is good enough to write to you on, so must scour London to see if I can get some more like it. I have never regretted that 7 francs.

Molly: 14th Nov. 1923. Hampstead
It is funny how they use thee and thou in other languages but not in English. Yes, you "mit mir auf du und mir stehen" (I don't know if mir is right, but since thee is dich and me mich, mir might be with me) and du is all right and I like it.

Thank you so much for your real letter and the other one. It would have been just the same with me, if I hadn't had the extra one I should have wondered and wondered what had happened to you or the letter.

Yes, your paper has indeed served you well. I am sorry it is all used up because I do love it so. I can so well remember the day you bought it – at least not exactly that, but the first letter you wrote to me on it. It was such precious paper, that you had to be very economical, and so you pushed the date right up into the left corner, and you wrote very small and neat. And you told me how you had been into Montreux to get it, and how you got excited over your French and therefore paid more than you meant to for it. It has lasted just about a year; I believe you wrote your first letter on it on November 13th, but I'm not quite sure.

<u>Thursday</u> It was the 13th; I looked it up last night. Also I tried to count the number of pages you have used, but it's no good; I kept on making little excursions into the letters because something interesting caught my eye, and then I forgot where I had got to. After two or three beginnings I got as far as 31 or 41, I couldn't remember which, so I gave it up as a bad job.

It was ingenious of Auntie Nellie to get out of the wish bone difficulty like that, though I don't know how she could tell for sure that you would both wish the same thing. Oh yes, you always wish when you catch a leaf; it is a jolly difficult thing to do (catching a leaf, I mean) and I was rather proud of two in one morning. I wish one could tell one's wishes before they come true; I'm always so curious to know what people wish. Will you tell me yours one day, after it has come true of course. When I have wishes I always go on wishing the same thing until it comes true.

That is just exactly it – I put in a lot of extra hairpins in case, specially when walking on a windy day without a hat; then, in moments of stress I can lend some to Mary, though the result is that my hair gets most dreadfully untidy. It must be rummier to have to shave every morning.

Dear no, I am certainly not very particular about going to Church. I like going to little churches when I am staying with Mary, specially if there is a nice long walk beforehand; and Kimble [where the two girls went] has such a dear little Church and I had never been there before. But if I go down there on Saturday and we haven't another chance of a proper long walk, we always go for one on Sunday morning. Of course I shouldn't be shocked at you playing tennis on Sunday morning. To begin with I shouldn't be shocked at anything you did; to – (I don't know what to say, I mean – to middle with) I'm not half good enough myself to be shocked by anything like that; and to end with I don't see why a person should go to Church every Sunday, if they don't feel like it, and specially if they get very little exercise during the week and Sunday is practically their only chance. Daddy is most awfully particular about going to Church every Sunday, so we always do, but I don't see why you should. It seems to me it is just as good to take exercise and keep oneself fit, or to go for a walk and see everything that is beautiful and wonderful, and one can be just as praising and worshipping that way. Personally, I'd rather say my prayers with my eyes open and something beautiful in front of me, than with them shut. Only I can quite understand that anybody wouldn't be like that; I know Daddy wouldn't.

Are little girls more tyrannical than little boys? I like little girls. I think it would be more fun to have a family consisting of half of each.

It is bed time and I must stop, though there is heaps more to say.

Goodnight, Barnes. Molly.

Barnes: 19th Nov. 1923. New Cross
This is the new paper. I went to the Army and Navy Stores, who are supposed to be very good for this sort of thing, but I dont like it nearly so well as my old. And I cant write without lines half as well as you can. But perhaps I shall improve with practice. I think the man was rather amused at me, for he was going to serve out tons of envelopes, and I said hastily, "Oh no, I only want a very few". There are 100 sheets in the pad, and I had to get 25 envies. I've still got ten of the Swiss envies left. Molly, how on earth did you remember about Nov 13th? Why, I hadn't the remotest recollection of it myself. You really are very wonderful. You are amusing too, not being able to count the sheets. Why didn't you turn the letters upside down? When I counted your last paper, I solved the problem by reading them all first. I think there ought to be a sort of home for envelopes that haven't had a chance to get used. Just think of

those poor ten, lying side by side in the drawer, watching the rapidly dwindling pad, and crying out "We shall <u>never</u> get sent to Molly". How they must have counted the numbers before their turn could come. A million times worse than standing in a queue – (right first time) dozens long when you only had a minute to catch a train, and an old lady is having a prolonged and foolish argument with the booking clerk. Tho' somewhat the same sensation, impotent impatience; but after all losing a train is only losing a train, whereas not being sent to you must be worse than anything for a poor little envie.

They are simply rustling with jealousy on seeing their newer and more fortunate supplanters. I tell them they hold a special place in <u>my</u> heart, but they dont seem to care a hoot about that. If they are not quiet I shall lick all their gums and stick them down for good.

Now I must stop to do some maths. I am taking some private lessons from a very clever coach, in rigid dynamics. Its the most appalling stuff, and quite beyond me to work up by myself, but one needs it for stability calculations for Airships. You have no idea of the delicious sensation it is to be taught.

Yes, I shall tell you my wish some day. Auntie Nellie is a wizard and reads my thoughts. I have been making desperate efforts to catch a leaf ever since, without success, in the neighbourhood of Westminster Abbey and Victoria Street where there are trees. But one simply can't go careering all over the road when dressed up with walking stick and gloves. I did run a little way after one, and people looked at me as if I were an escaped lunatic. Poor folk, perhaps they have no Molly to tell them about such things. If you were a really charitable child you would organise a society for telling busy people fairy things – tho' perhaps you do.

Little girls are only tyrannical to me, because they are so perfectly irresistible. If I have any in my family I shall spoil them most completely, and their Mother will have to do the stern business, whereas with little boys I can be very stern myself. Yes I think half and half is best of all if only one could arrange it.

I think I can say that my invention will be successful now. Everyone likes it very much. I say, could you come early tomorrow, so that we could go for a walk? Molly, I simply haven't had a moment to write any maths – I'm most awfully sorry. Shall I give you a lesson tomorrow? But I suppose you will be whisked away at 6 punctually. Do you realise that I have not seen you since September 8th?

This wretched election is upsetting all our Airship plans, because the Bill, authorising the subsidy from the Government, was to have been passed this session; and now of course, it can scarcely get thro' the new Parliament until say March. And if the Liberals come back that awful man Winston Churchill is a bitter opponent of Rigid Airships, and very likely he may try to stop it altogether. It puts me in a very difficult position as I had started engaging a new staff of engineers, and it is awkward if a man is already in a good post, and perhaps married, to ask him to give it up to join me, and then perhaps have to wait for ages.

The bottom envelope in the new pile is already turning a little blue after doing some fierce brain work on the subject of averages. "Twenty fives into a hundred", he muttered hoarsely "goes 4, and the wretched man has exceeded the average by one on his first letter"!

Molly: 28th Nov. 1923. Hampstead
I like the new paper very much; though I love the old so, I don't think anything could quite come up to it. Why don't you use those ten poor little envies up with ordinary paper? I do hate to think of them lying in the drawer never getting used. There's nothing a bit wonderful in remembering about November 13th. I remember noticing when I read it last year that you started it two days after Armistice Day, though I didn't get it till some time after because you had a very bad cold soon after you began it, and couldn't finish it for a little while. And I don't see how I could forget it seeing what a lot of things there are to remind me of it – it was the first letter on that paper, it contained a very interesting chapter of Maths about minutes and seconds and small quantities, and the letter itself is quite unforgettable.

I <u>am</u> glad about your invention. It makes me very proud to think that I should have such a clever, ripping friend as you, Barnes. Then I hope Winston Churchill doesn't have a chance to stop the building of Rigid Airships. Whatever could anyone have against them?

I am so sorry we were so late on Saturday, it was the cook's fault, and she isn't coming after all. By superhuman efforts we managed to get George cleaned by about quarter past two, and Mother wasn't ready till after three, by which time George was looking decidedly shady again, so that we didn't get off till about quarter past three.... I did enjoy myself so much, and Pam and George, specially the latter, are still marvelling at your extraordinary braininess.

Oh Barnes, I <u>do</u> love your Grannie, she is even dearer than I had expected she would be. I could tell more or less what she would look like, but I had no idea what a beautiful chime she had nor how wonderfully perfect her works are not what a quiet soothing tick she has – you are quite right, she does say "Take time, take time". I could stand and watch her and listen to her for ages. Isn't it wonderful to think that she will be alive after 200 years, and for all that time she will go on quietly ticking away, and other children will want to wind her up, and other people will watch her hammers go when she chimes. I love to think of her being there, watching all the generations of children coming and growing up and getting old, and perhaps going away for a while, and when they return she will welcome them back as only an old Grannie can. She was certainly the very best thing you could have chosen Barnes.

Barnes: 4th Dec. 1923. New Cross
There's nothing amusing in seeing me run down Victoria Street. If you fix your eyes on a far distant bus – which there is no possible chance of catching, and run fast enough no one will look at you twice. It's when you gyrate and caracol (whatever that may be) while gazing expectantly into the skies, right in front of an oncoming taxi, that people begin to stare. I haven't caught a leaf yet. Certainly I do think you were to be congratulated on two in one morning.

In the train. Friday. I am on my way to Shoreham-by-sea near Brighton to have a consultation about some new engines for the Airship. This is a most luxurious Pullman car, but the train goes so fast that it shakes like anything as you can see.

Molly, I am so glad that you approve of Grannie. I do think its ripping of you to be so interested in things. Since you were here I have changed the wireless – in fact that Saturday evening, as I could not work after you left; and it is now nearly perfect, and so loud that I have to reduce the power to make it comfortable. Otherwise I have done practically nothing but work. We had the 1st meeting of the new airship design committee last Friday and I submitted my new design, which they decided to adopt. Molly I do wish you wouldn't think of me as clever, because I'm distinctly mediocre, which is one reason why I have to work so fearfully hard in my attempts to keep pace with the really clever people. And I've never tried to make you think I was brainey. And I only wrote the maths because I delight in trying to serve you, not to show off. You see I found that, in teaching the boys at any rate, I found things so difficult myself that I knew their difficulties, which is what a really clever man never can do, because to him the things aren't difficult. And because you think a man clever is a poor reason for liking him, because you are sure to find him out sooner or later.

(Home again) No, that's a nasty, horrid, ungracious remark, and I unreservedly withdraw it Molly. I think that what I meant was that cleverness is no indication of character – clever people are often jolly unpleasant – and tho' I think for a firm and lasting companionship it is necessary to have some equality of intellectual power yet absolute braininess is not by any means essential. Anyhow lots of perfectly stupid men seem to succeed without difficulty in getting accepted by very charming women, so it seems to pay to be moderately dense.

♦ The arrival of a beautifully written invitation card from Molly interrupted Barnes's letter-writing at this point.

Molly: 5th Dec. 1923. Hampstead

Barbara and Molly Bloxam
request the pleasure of
Barnes'
company at a small Dance
on Friday the 28th December from
8 o'clock to 12 o'clock

Dancing   R.S.V.P.

Barnes: 4th Dec. 1923. [continued 5th]

Molly, thank you most awfully for the invitation you sent me for your dance. And I am very very sorry I can't come, but the reason is this. For some 2 months now, I've got so sick of having a cold in the head and using about 3 dozen hanks a week, that I have been to see a specialist, and he says it is caused by the bash on the nose which I got when boxing years ago, and things inside are all out of place. And although I have been doing a treatment by pushing various evil chemicals up and sniffing all sorts of things, he says he cannot do any permanent good unless I have an operation as one side is practically closed up at the back inside, and apparently germs and microbes sit in there and laugh at him. So he wants to chop it all straight again. (Why do you want to be a doctor?) As I cannot spare any other time, being so very busy, I had already arranged with him to go into a nursing home at Christmas and he will do the op. for me. I go in on the Thursday evening before Christmas, and he will do it on the Friday morning. I dont suppose I shall be fit to come out till after Christmas, and certainly, even if I get out by the 28th I should not be presentable at a dance I fear.

I gather it is no worse than having adenoids removed, but takes much longer to heal up and gives a good deal of discomfort for a week or two. Still, it will be a jolly good thing, as I simply hate myself now, and it must be awfully unpleasant to others to be always sniffing! So thats the reason Molly. I cant tell you how disappointed I am. I do, do hope you will have a very happy time. Being busy has its drawbacks!

Molly: 13th Dec. 1923. Hampstead
No, one wouldn't like a person simply because they were clever. In fact, if they were lovable and friendly, one wouldn't care a rap if they were clever or not. And I won't think you clever Barnes, if you don't want. I know you've never tried to make me think you were brainy – quite the contrary, you've always done just the opposite. Really, I believe I'd admire a person more, if he were not so awfully brainy but had to work hard to keep up with the very clever people, than I would one who was just so clever that he wouldn't have to bother. Anyway, I can't help thinking that you are just a little bit clever, not awfully brainy, but just a little – see? You won't mind that, will you?

We have been having end of term exams all this week and last. We have heard some of our results and I have managed to get through three of them – we haven't heard about Chemistry yet. I don't know if you will be interested to hear about them; they aren't a bit good considering. Botany was 70 and I don't know where I was; Physics he only told us what class we were in, and I was in the 2nd, and Zoology 2nd with 80, only Dr Fraser said it was really first because my practical drawing book had A+ and no one else had even A, the next was B+ , and the top person who had 85 had B for her book. Oh dear, when I read all that through, it sounds horribly cocky; but I'm not, Barnes, in the least. To begin with there's nothing at all to be cocky about, because I ought to have done much better considering its my second year, and I feel most exceedingly un-proud of myself. You see, I like to tell you about my exams however badly I've done, because I guess you'll be interested to hear about them. But please, you won't think I'm too awful when I do badly, will you? Because I shan't know what to do if you do. The physics are going very much better this term, though I still find them the hardest subject of all, as you perceive. But, thanks to you, I can understand what he is talking about when he speaks of sines and cosines and tangents and the limit of a fraction, and I can resolve forces and easy things like that which need a little trig., but it still takes ages and ages to understand things. I do envy those lucky people who can understand everything without having to work at it for ever so long.

Barnes, I am so awfully sorry about the operation. What shall I do? I can't bear to think of you being in a nursing home on Christmas day. All the time I shall be wondering what you are doing and if it is hurting. Will it hurt very much? I do hope it won't. Isn't it much worse to think of somebody else having an operation than of you having it yourself, how I wish it was my nose.

I too am so disappointed that you can't come to the dance. Where is the nursing home? Could one write letters there? And will you really and truly be there on Christmas day? If its no worse than having adenoids removed, it can't be dangerous or anything, can it. I keep on telling myself that it is all right, but I keep on worrying all the same. I shan't expect a letter next Saturday, but please will you just let me know if one could write there.

Please, Barnes, it will be all right won't it? I'd give anything if it could be me and not you. Goodbye, Barnes. Molly.

♦ Many inventions were involved in the design of the new airship. Not surprisingly, Barnes did not describe the particular one referred to in his letter of 19th November. He was sadly disappointed to lose his first opportunity to dance with Molly, and that under Mr Bloxam's own roof; but at last, on a visit which Molly made to New Cross, he had introduced her to Grannie clock. She took Pam and George with her and, as he was to do with many children over the next decades, he explained the mechanism to them, showing them how to wind her. Granny clock still says 'take time, take time' with her measured and peaceful stroke.

Chapter 11
FROM THE SICK BED

Barnes spent the weekend before his operation in Sussex with his kind Auntie May, who had guessed all about Molly. He worked on the next chapter of trigonometry, anxious that it should reach Molly before he disappeared into his hospital bed.

Barnes: 16th Dec. 1923. Rudgwick
I quite know what exam time is like from sad experience. I think you have done splendidly. If the marks you quote are out of 100 total then they are very good indeed. And whatever happened, I should know you had not done badly – exams are not a really good test of either knowledge or intellectual power. No, if ever you <u>do</u> have a stroke of bad luck I shall understand and comfort you, and think all the more of you. I think it is very unfair that the class should be decided by exam marks only, and not by the practical books as well. You must have done jolly good work during the term to get A+ on your book. I should have thought that 80 and A+ represented a much higher standard than 85 and B.

You must not be at all worried about the op. Its nothing at all, and I wouldnt have told you, only I wanted you to know the real reason why I could not come to your dance. You are quite right when you say that it is far worse to think of someone else, than of oneself. But you really need not think twice about it, and just think how good to be cured for ever of my wretched catarrh.

And as for Christmas Day in a Nursing home – I daresay its as cheery as anywhere else under existing circumstances. I'm awfully sorry Molly that I should have worried you. <u>Please</u> dont think any more about it, and I'm jolly glad it <u>isn't</u> you instead of me.

My dear Molly, as a catcher of leaves, you are not in the same class with me. At last I have achieved my ambition. During a walk in the country this (Sunday) afternoon, I have caught no less than FIVE, one for every letter of your name. And I wished the same wish for each one. Four I caught in one place, halfway down a hill in a lonely country road, holding a large bunch of holly in one hand, together with a walking stick. Leaping and capering like a lunatic, with a deep and clayey ditch on either hand, I couldn't help laughing at myself. I couldn't go on to catch any more for Mary Frances as well, as I was jolly hot, and it takes some time to do. I kept the last three – the -LLY ones. I wish I had thought to keep the others, but I never guessed I should have such luck. Two were most beautiful catches, straight into my hand at the first swoop. The other three I fumbled rather, but got them all before touching the ground. Dont you think the one for each letter is a jolly good idea? All my own too. You know I think quite often of you now as "Mary". I do think its such a beautiful name. Dont you like it?

Would you mind telling me whether, in physics you do pendulums, sound waves, etc etc?

Trigonometry. Chapter IX.
It is only left to consider the 4 special cases when θ = 90°, 180°, 270° and 0 or 360° to complete our work on the circular functions of any angle. Before you do this, it would be a good thing to re-read the chapter on Zero and Infinity, because it was really written as a preparation for this chapter.

Let us start with the sine, cosine and tangent of 90° first.

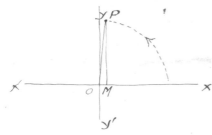

Imagine the radius vector starting from OX and rotating anti-clockwise, until it has nearly reached OY.

Angle MOP is nearly 90°, and as you see OM is getting very small, while MP is very nearly equal to OP itself, in length.

As P draws nearer and nearer to the OY line, OM gets smaller and smaller – Note – in this case I have taken OP of fixed length – any length will do – say 1" so that OP is like the radius of a circle sweeping round. Draw OP still closer – OM is smaller still.

Suppose we measure them, and find OP = 1" OM = 1/100" (my diagram is not to scale)

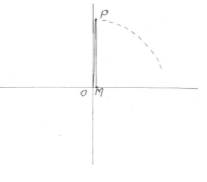

Then by calculation

$$PM^2 = OP^2 - OM^2$$
$$= 1^2 - 1/100^2$$
$$= 1 - 1/10,000$$
$$= .99999$$
$$\therefore PM = \sqrt{.99999}$$
$$= .99999999 \text{ (say)}$$

(I haven't worked it out, it isn't worth it).

So for this particular θ which is nearly 90° we get

$$\sin \theta = MP/OP = .9999999/1$$
$$= \underline{.99999999} \qquad (1)$$
$$\cos \theta = OM/OP = (1/100)/1$$
$$= \underline{1/100} = .001 \qquad (2)$$
$$\tan \dot\theta = MP/OM = .9999999/(1/100)$$
$$= \underline{99.9999} \qquad (3)$$

Examining these results we see that $\sin \theta$ is very nearly 1, $\cos \theta$ is very nearly 0 and $\tan \theta$ is a very large (relatively) quantity.

This is just our old friends zero and infinity popping up again, for when P actually reaches the OY line, OM has vanished – or in another way, M coincides with O the origin, so that the distance OM = 0 (nought) while MP coincides with, and is therefore equal to OP.

So we may write, when $\theta = 90°$

$$\sin 90° = MP/OP = OP/OP = 1/1$$
$$= 1$$
$$\cos 90° = OM/OP = 0/1 = 0 \qquad \text{(zero)}$$
$$\tan 90° = MP/OM = OP/OM = 1/0 = \infty \quad \text{(infinity)}$$

Is there any difficulty about that?

As P swings further round, and comes out on the other side of OY line, OM becomes negative, as we saw when examining the ratios of angles lying between 90° and 180°.

Now let OP go on till it approaches OX′ and θ is nearly 180°

You will see that this time it is the turn of MP to become zero as P finally reaches the OX′ line, while OM becomes numerically equal in length to OP, but of opposite sign, because OM is negative while OP the radius vector is <u>always</u> positive (convention).

So that when $\theta = 180°$

$\sin 180° = MP / OP = 0 / 1 = 0$

$\cos 180° = OM / OP = -OP / OP = -1$

$\tan 180° = MP / OM = 0 / -1 = 0$ (+ or − 0 is the same thing)

Pass on till OP approaches OY′ and θ is nearly 270°,

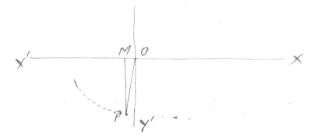

Here, OM, having reached a maximum negative value as OP passed thro' 180°, is again diminishing and dwindles to 0 as P reaches the 270° position, while MP is negative and equal to OP in length.

So that when $\theta = 270°$

$\sin \theta = MP / OP = -OP / OP = -1$

$\cos \theta = OM / OP = 0 / 1 = 0$

$\tan \theta = MP / OM = -OP / OM = -1 / 0 = -\infty$ (minus infinity)

(+ infinity and − infinity are <u>not</u> the same thing).

♦ Barnes made an error at this point in his analysis – exhaustion and lack of time may have cut him short. In fact tan 90° and tan 270° are <u>both</u> just as much +∞ as they are −∞. He neglected the fact that while OP is in the 3rd quadrant, before it reaches 270°, OM is negative. Tan θ then is the ratio of two quantities <u>both</u> of which are <u>negative</u>: one divided by the other is therefore positive and very large. As the <u>very</u> small denominator

OM tends to 0, tan θ tends to <u>plus</u> ∞. But if OP approaches from the <u>other</u> side of 270°, then OM is positive but OP is negative, and tan θ tends to −∞. These ideas had appeared previously in Trig chapter III [p. 106], when Barnes had discussed "poor old y, shooting out to infinity and back the other side whether he likes it or not". There he had explained that as x approached the "critical value", y had to prepare himself though his bag was not packed and his laundry not yet returned! But off he had to go, round the heavenly spheres to infinity and back again punctually, having described the "discontinuous curve". Lying in bed suffering considerable post-operative discomfort, Barnes thought over what he had written and realised his mistake. "What I meant to say was, as I was referring to the <u>sine</u> that MP cd. not be greater. I hope this did not muddle you. It came into my head afterwards when I was thinking". To Molly's credit, and to her teacher's, she had spotted the mistake and commented on it in answer. "Yes, you did say that OM could not be longer than OP when you were referring to the sine, but I concluded that what you meant was that MP could not be longer".

Going on, till OP is approaching OX having nearly completed the circle of 360°, we see that it is again MP's turn to become zero, while OM is equal to OP in both length and sign.

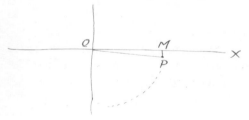

So, when θ = 360° or 0°, (the position being coincident for both)
$$\sin 360° = \sin 0° = MP / OP = 0 / 1 = 0$$
$$\cos 360° = \cos 0° = OM / OP = OP / OP = 1$$
$$\tan 360° = \tan 0° = MP / OM = 0 / OP = 1$$

Let us collect our results.

	0°	90°	180°	270°	360°
sine	0	1	0	1	0
cosine	1	0	1	0	1
tangent	0	∞	0	−∞	0

Notice:- Sine is alternately 0 and 1
          cosine   "   "   1 and 0
          tangent "   "   0 ∞ 0 -∞

Now we can examine both these results, and those obtained in the last chapter, and we come to some very interesting and important conclusions. Just imagine that the Radius Vector is sweeping slowly round starting from zero. (Like the big hand of a clock going backwards). What is happening to the various functions? Let us see.

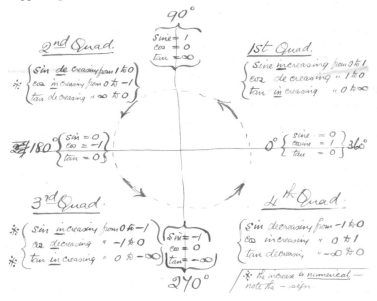

The diagram that I have drawn is not sufficient to show us exactly <u>how</u> or at what rate the sine, etc changes between the 4 cardinal points which we have accurately determined. Speaking offhand, one might say that the sine probably varied directly as the angle – that is, if one doubled the angle, the sine would also be double. You probably know already that this is not so, but it is interesting to think of it a little longer.

Supposing we try to represent the way the sine varies as OP goes round, in a graph. We will plot (to any scale) the position of OP along the base, and the size of the sine up the side i.e. as ordinate).

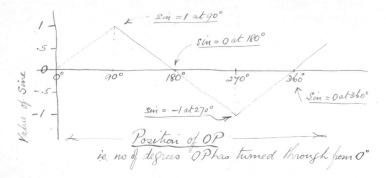

I have just filled in the values which we definitely know, and joined up the points by straight lines. This is the graph as it would be if the sine varied directly as the angle.

In marking off the scale of angles – 90° – 180° etc. we just imagine that the circumference of a protractor has been unrolled as it were so that all

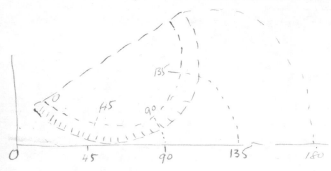

the angles are represented by equal spaces along our base line – e.g. 45° is exactly $\frac{1}{2}$ way between 0 and 90° etc, etc. or another way – suppose it suits me to have a base line 6 inches long – then 360° = 6" and 1" = 1/6 of 360° = 60° and we can fill in as many of the divisions as we please, and of

course –60° would be represented by –1", that is 1" measured to the <u>left</u> of the zero point. And so on.

How can we find out whether the curve we have indicated on [p. 210] is the correct one, or not? Well, at present, of our own knowledge, we could only do it approximately, by drawing a number of triangles with varying angle, and measuring the lengths of the sides. But such an exercise is excellent practice, in bringing home to us the physical (or better I suppose, the geometrical) meaning of the names sine, cosine and tangent, so let us try.

Look at the diagram [below]. (Oh Molly, I <u>do</u> hope this isn't boring you). I have drawn a circle 2 inches radius; and by the simple process of stepping round with a pair of dividers, I have marked off firstly the series of angles 30°, 60°, 90°, 120° etc., and then the series 45°, 135° etc. These points represent successive halting places of OP. By dividing the circle more closely, we could get almost any number of angles, but the ones I have chosen are quite enough to give us sufficient points on our curve to see its general shape. I have only lettered the 1st triangle OMP, but in every case, MP represents the vertical distance of P from XX′.

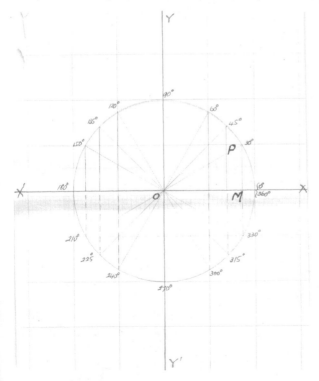

The upward lines in the 3rd and 4th quadrants I have drawn dotted, so that they will not confuse you where they join the downward lines on the XX′ line.

Now take a rule and measure the MP and OM for every angle:-

angle	MP	OM
0	0 ~~1·0~~	2·0″
30	1·0 ~~1·42~~	1·73″
45	1·42	1·42″
60	1·74	1·0″
90	2·0	0
120	1·74	−1·0
135	1·42	−1·42″
150	1·0″	−1·73″
180	0	−2″
210	−1·0″	−1·73
225	−1·42″	−1·42
240	−1·74″	−1·0
270	−2″	0
300	−1·74	1·0
315	−1·42	1·42
330	−1·0	1·73
360	0	2″

OP is of course a constant 2″ and so we get the three following columns [see p. 213].

I have shamelessly left the cosine to you as an exercise. All the necessary figures are given. If you have time do try it – unless of course all this is old work to you, in which case dont bother. That is where I am so handicapped. I often think I may be writing things that you are already perfectly familiar with. It isn't that I think you are very ignorant Molly – its only that I feel it is better for me to spend the time than possibly scamp some point that may really puzzle you.

Now let us plot these values for sine on another piece of squared paper. I have taken 6″ = 360° so that 1″ = 60°. Vertically, as a convenient scale let us say that 1″ represents 1 unit. Then marking off the values for the

Angle.	$\dfrac{MP}{OP}$ = Sine	$\dfrac{OM}{OP}$ = cosine	$\dfrac{MP}{OM}$ = tangent
0	$\frac{0}{2} = 0$		$\frac{0}{2} = 0$
30	$\frac{1}{2} = .5$		$\frac{1.0}{1.73} = .578$
45	$\frac{1.42}{2} = .71$		$\frac{1.42}{1.42} = 1.0$
60	$\frac{1.74}{2} = .87$		$\frac{1.74}{1.0} = 1.74$
90	$\frac{2}{2} = 1.0$		$\frac{2}{0} = \infty$
120	$\frac{1.74}{2} = .87$		$\frac{1.74}{-1.0} = -1.74$
135	$\frac{1.42}{2} = .71$		$\frac{1.42}{-1.42} = -1$
150	$\frac{1.0}{2} = .5$		$\frac{1}{-1.73} = -.578$
180	$\frac{0}{2} = 0$		$\frac{0}{-2} = 0$
210	$\frac{-1.0}{2} = -.5$		$\frac{-1}{-1.73} = +.578$
225	$\frac{-1.42}{2} = -.71$		$\frac{-1.42}{-1.42} = +1$
240	$\frac{-1.73}{2} = -.87$		$\frac{-1.74}{-1} = +1.74$
270	$\frac{-2}{2} = -1$		$\frac{-2}{0} = \infty$
300	$\frac{-1.73}{2} = -.87$		$\frac{-1.74}{1.0} = -1.74$
315	$\frac{-1.42}{2} = -.71$		$\frac{-1.42}{1.42} = -1$
330	$\frac{-1.0}{2} = -.5$		$\frac{-1}{1.73} = -.578$
360	$\frac{0}{2} = 0$		$\frac{0}{2} = 0$

sine as ordinates at the angles from which they were obtained we get the beautiful curve which you see.

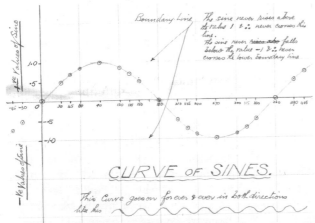

Boundary Line The sine never rises above the value 1 & ∴ never crosses his line.
The sine never falls below the value −1 & ∴ never crosses the lower boundary line

CURVE OF SINES.

This Curve goes on for ever & ever in both directions like this

Note the following points:-

1) The sine of an angle never exceeds ±1; obvious because OM <u>cannot</u> be longer than OP.

2) The curve simply repeats after 360°; and if OP rotated negatively, we should get a similar result on the negative side; and as OP can fly round and round as often as we like to make it, we get a curve which goes on for ever in both directions.

Since 0°–360° shows the complete variation of the curve after which it repeats, we call 360° the "period" of the curve.

The total distance thro' which a point moving along on the curve, moves vertically up and down – i.e. from + 1 to –1 – is called the "amplitude" of the curve. In this case the amplitude is 2, i.e., the total vertical distance moved thro' by the point is 2 units.

The points where the curve crosses the XX line i.e. at 0, 180°, 360°, etc., are called "nodes".

The points where the curve becomes a maximum or a minimum are called "anodes" i.e. at 90°, 270° etc.

The curve of sines is a very useful and important one in all branches of maths, physics, and engineering. For instance it shows us

1) The way in which a pendulum vibrates.

2) The way a sound wave travels thro' air.

3) The position of the piston of an engine in its cylinder, for any angle of the crank.

4) The way in which a horizontal magnet vibrates when suspended in a constant magnetic field.

Molly: 20th Dec. 1923. Hampstead

I couldn't wait to thank you for the maths till next week, but this isn't really a letter. It is all beautifully clear, and you know it couldn't bore me. I like that curve of the sines, I suppose you could draw one for cosines and tangents as well, only for the tangents you'd have to have one of those ∞ curves that goes right off the page and comes up the back way, I suppose. Of course I shall have heaps of time to work out the values of the cosine. You needn't worry – it is all absolutely new to me; all the trig I had ever done was the stuff I started to learn at the beginning of the summer term before you taught me any, and something at school which I have lately discovered must have been meant to be trig. All we did was to draw three graphs, one of which I now recognise as the curve of sines which you have drawn. I don't think any of us knew what we were trying to do; I know I didn't. We were given certain values of some things called sines, cosines and tangents and told to draw graphs. It seems to me that was rather putting the cart before the horse.

Goodbye Barnes. I hope you'll be all right to-morrow. I'm awfully glad you are having it done, and I think you are jolly sensible, and also awfully brave to have it done just before Christmas.

Barnes: 21st Dec. 1923. Park Lodge Nursing Home.
<u>Friday evening</u> Just a line to tell you that everything went off awfully well. I felt fearfully important as I had 4 doctors altogether.

Goodbye Molly, please excuse the writing, as I am lying down.

Barnes: 23rd Dec. 1923. Park Lodge Nursing Home.
Just a line to wish you every happiness this Christmas, and the best and most prosperous of New Years. I hope you will like the little fan which I am venturing to send you. If you dont, – or already have one, do not hesitate to return it, and I will get it changed. You can get coloured ones, but, alas, I have never seen any of your dance frocks – is it frocks or dresses, I never know, – so did not choose a colour.

And I hope you will have the most delightful dance on Friday, and lots more too throughout the year.

I am sorry my letter was cut off so short. I did not tell you half the things. But the last week was such a rush.

Goodbye Molly. All the best wishes in the world are yours.

Molly: 23rd Dec. 1923. Hampstead
A very happy Christmas to you. I do so hope you will have a good time even though you are in the nursing home. I do <u>do</u> wish I could see you; it won't be proper, complete Christmas without.

I tried to get a nice respectable copy of "The Prisoner of Chillon", but not one of the book shops I went to – (seven altogether) could get one, so in desperation I had to get that one. I remember when you first went to Switzerland, you mentioned it and said you had never read it.

Barnes, you really are clever – five leaves in one afternoon is absolutely wonderful. However you managed to do it, I can't think. I have never in all my life caught more than two at once, but next Autumn I shall go seriously to work, I'm afraid it is too late in the year to do much now. Of course Barnes has six letters, so if I caught six, I should feel really superior and very proud of myself. Oh, I do wish I could have been there to see you; I should have held your stick and the holly and then it would have been easier for you – no I wouldn't, though because I should have wanted to catch some myself. I am afraid my two only do for B.W. Yes it is a jolly good idea having one for each letter. Never mind, when I have caught my six, I shall be jolly triumphant. I suppose they must all be caught on one walk.

I love the name Mary, why on earth they wanted to change it to Molly I can't think. I always think it is such a pity that your name would have to be decided for you before you are old enough to give your opinion on the matter. After all it is you who have to be called by it all the rest of your life, so you are the chief person concerned. To go back to Mary, don't you like it heaps better than Molly? I do. Would you rather I were called Mary? I wonder if I made a great effort, I could even get people to call me Mary. If you preferred it, I'd have a jolly good try. It would be funny to think of and talk to myself as Mary, but I'd very much rather that than Molly. As a matter of fact I believe you do prefer Mary, because I remember when we were staying at New Cross the first time, you asked me if I had been christened Mary and then you wondered why I wasn't called by it because it was such a nice name.

It was so sweet of Uncle on Friday – you see all the morning I had been wondering and wondering how you had been getting on, and if you were all right, and I had decided that I couldn't telephone till the evening in case I should be disturbing either of them. Then at about two o'clock he 'phoned and told me that he had just left you and that you were getting on well. I felt so happy and relieved that I got George to come into the garden and we had a most glorious snowball fight then and there. But I do think it was ripping of Uncle Charlie to 'phone because I couldn't have enjoyed the snowballing a bit if I hadn't known, in fact I don't suppose we should have snowballed at all.

My poor Barnes, I am so sorry for you; it must be so dreadfully uncomfortable; I do hope it will be better on Christmas day. I'd give anything if I could be with you on Christmas Day, it seems so mean of me to be here enjoying myself, but I should enjoy myself ever so much more if I could go and see you – that would make a Christmas quite perfect.

Goodbye. You are so dear and cheerful Barnes.

Please have a very happy Christmas. Molly.

Molly: 26th Dec. 1923. Hampstead
It is the daintiest, prettiest, delicatest little fan of all the fans that were ever made, and I love it. Thank you ever and ever so much for it. I shall use it on Friday [the dance], and it is just right because the <u>frock</u> I am

going to wear is white, at least it is practically all white but it has some bluish sort of flowers round the waist, so that the fan will look ripping with it. I am awfully glad you chose a white one. I like the way you say you have never seen <u>any</u> of my dresses, no, frocks (I always call them dresses, though I know they ought to be frocks); sounds as if I were a lady of fashion and had at least 7 or 8!

Thank you very much too for your letter which arrived on Christmas day. I didn't realise before how difficult it must have been for you to write that little extra bit of letter on Friday; as you had only just finished with the anaesthetic you must have been feeling absolutely rotten and I think it was awfully ripping of you to write. Of course it was quite coherent though how you managed it I don't know.

I do hope you are better and thank you very very much for the dear little fan (as a matter of fact, it isn't really little at all).

Barnes: 27th Dec. 1923. Nursing Home
Thank you very very much indeed for your delightful presents.... I love "The Prisoner of Chillon" too, and am so glad at last to have a copy. I like that edition very much, for the notes are jolly interesting. But tell me oh learned one, what is a "hypallage", for the notes say that a splendid example of that figure occurs in lines 142 and 143, but omit to state what it is. Thank you ever so much for that too.

I <u>am</u> so glad you love Mary too. No, I wouldn't get people to change now. I can think of a much more beautiful way of making use of it than that. And after all Molly is a ripping name too. To me they stand for two different personalities almost. Molly is a sturdy, self reliant little person who trudges along with her hands in her pockets, – two of which she is careful to insist on having in all her work-a-day clothes – Molly often has a puzzled little frown, because some of lifes problems <u>are</u> puzzling; but Mary comes and smoothes it out again, with a soft and gentle hand. So because of the Molly side, I like Molly; and because of the Mary side I like Mary.

Barnes: 31st Dec. 1923. New Cross
A very, very happy and successful New Year to you. This isn't a letter, but just to send you all my wishes.

I left the home on Friday, and today have been out for the first time, to see my darling specialist. He's awfully pleased with me; says I'm a great success in view of the difficulty (in my case) of the operation, as apparently the septum – whatever that may be, was split, due I suppose to the injury; and he had to dissect it with a knife and forceps – p'raps you may understand some of this. Its quite an uncanny feeling being able to breathe thro' the right side of my nose – a thing I haven't done for years – it feels as tho' there was a sort of extra hole in my head. Hurrah, hurrah, no more <u>beastly</u> colds and sniffs.

Of course, I'm worse than ever just at the moment, but that's only while it is healing. I am so glad that I had it done, tho' its cost me a fearful lot – over £30! I know it seems an extravagance but it will save so much trouble in years to come, and lots of men have it done – three of my friends have.

Molly: 31st Dec. 1923. Hampstead
I think the dance was a success; anyway people seemed to enjoy themselves. The last person went at about half past twelve and we went up to bed at a quarter to one. I think I enjoyed the going to bed as much as anything.

Mary slept with Barbara and me and we had a fire in our room because people took off their coats there. You can imagine how cosy it was undressing by the fire, and Mary and I talked for a long time after we were in bed. I tied a piece of white ribbon onto the handle of the fan and carried it on my wrist. You can't think how proud I was of it and how pretty it looked and how everybody admired it. I fanned myself very ostentatiously (or allowed my partner to do so) between every dance and whenever else I had an opportunity. I'm awfully glad it is white, somehow I don't think a coloured one could look so fresh and dainty. I love it so. I don't want to wait for the next dance to use it.

You asked me if I do pendulums and sound waves in physics. All we have done about pendulums is to find the time of swing when there are different lengths of thread and also with different bobs – wood and metal.

We haven't come to sound this year yet, but last year we did an experiment with a thing called Kundt's Tube. You have a lot of cork dust in a tube and a rod is put inside it. You rub the rod and the cork in the tube arranges itself into a curve which I believe looked like the sine curve. I don't know if it is only a coincidence or if sound waves are always like that.

I am glad the book was all right. I believe "hypallage" (I'm so jolly thankful that I haven't got to say it, I never know how to pronounce the wretched word, but I know everything is accented which in respectable words isn't sounded at all) – it is when an epithet is applied to the wrong word – like – oh, I can't think of one now. But I thought it was supposed to be rather a bad thing to use, but I suppose it can be beautiful, and anyway Byron would be allowed to use it. Oh yes, I know one, I believe we had it as an example at school and it comes in the Prayer Book – "O ye holy and humble men of heart", when it really means Men, holy and humble of heart.

Tuesday This is New Year's Day. I hope you will have a very very happy New Year Barnes.

Were you awake last night to see the New Year in? I didn't mean to be because I was much too sleepy, but I woke up just before 12 and I heard the bells ringing, so I stayed awake for some time. I like being awake when it is all dark and quiet and you know it is the beginning of the year; it is all so still and solemn.

I do like your way of thinking of Mary and Molly; it is very quaint and very dear, and just like you to think of it.

♦ 1923 faded into 1924 as the bells rang, and the recently founded BBC broadcast the chimes of Big Ben to mark the passing of the year. Barnes and Molly had known each other for nearly two years and had exchanged over a hundred letters. Their meetings had been brief and infrequent, but their mutual interest had not waned. Their conversation on paper was more than ever affectionate, detailed, vivid and humorous. Barnes's work was increasingly demanding, and Molly faced her first year final exams. More serious was Mr Bloxam's stubborn stance; would he change in 1924?

Chapter 12
THE FLOODGATES OPENED

Barnes: 3rd Jan. 1924. New Cross

<u>Friday</u> I have been up to Harley Street again to see my specialist. He is most awfully pleased with me, and says everything has gone perfectly from start to finish. Of course it isn't healed up yet, so I have to put a light plug of cotton wool in the worst side, when I go out. He says the jolly old nose will be rather more aristocratic in shape, a little thinner, when all the swelling and irritation has gone down. I went to my office today also. Of course I've been doing a certain amount of work all the time, and have had the good fortune to have a most priceless brain wave – while in bed one night, I was worrying over a certain part of my new structure for Airships, and I got this idea, dimly, in bed.

Do you ever get a sort of instinctive feeling that something <u>can</u> be done, if only one could find the way? I've not been satisfied with this particular part for weeks, and my mind has been groping for a better solution. In fact I have a sort of enemy on the design Committee – a man who has never had anything to do with the design or construction of airships, but who thinks he knows a great deal (certainly he knows a lot of maths – more than I do), and who occupies an influential position at the Air Ministry and has managed to squeeze himself on to the Committee; in fact he asked Commander Burney (who is acting as Managing Director of the new Company) for the post of Chief Designer, which is virtually the post I hold; anyhow I think his nose is rather out of joint at the moment – he rather tries to prove that anything I do is wrong, and over this particular part he has gone to the trouble of doing endless complicated maths to show it wont work. I told him plainly that he was wasting his time, as the premises on which he worked were false, but at the same time, I was rather worried, because a very great deal hangs on this point, and although I <u>guessed</u> I was right the actual conditions were so complex as to prevent one's putting forward a definite mathematical proof. So I've been groping away for weeks, all the time I was in the home, and at last two or three nights ago, the solution came to me. I believe it will make a master patent; and, as usual, everyone to whom I have shown it says "How simple". That remark always makes me feel a perfect fool. If the thing is so jolly simple why on earth did it take several weeks bungling on my part to drop on it? And yet, when its done it <u>is</u> simple. I think I must be a very clumsy and inaccurate thinker. My mind has to wander down so many by-paths before I strike the right one.

I will try to explain the pendulum and Kundts tube in my next maths. No, the dust doesn't really form a sine curve – it is blown away from the antinodes and collects at the nodes, or should do.

Please Molly forgive this letter. Goodbye.

Barnes.

Do you realise that in 21 months you will be 21!

Molly: 9th Jan. 1924. Hampstead
I am so glad you were able to come after all on Saturday; it made it so much nicer. But I see you so awfully seldom now. You know, I have only seen you four times in over six months – once soon after you came back from Germany, once in September; once at the end of November when we went over to New Cross, and then on Saturday.

No, I don't think I have ever had that feeling that something can be done if you could only find the way, unless you count when I was young and trying to make a theatre once; only they weren't brain waves, but just thinking out how to do things. What a horrid man that is – the idea of trying to prove that what you do is wrong. Anyway, he will waste a lot of time and trouble and he will be pretty well squashed. I don't think it is so surprising that you do not think of the simple thing at once; all the time you are looking for something difficult and complicated and you don't think the solution could possibly be so simple as it really is. Anyway, I'm awfully glad you've got the solution now.

I say, you must have been a very superior patient for everything to have gone so well and for your specialist to be so pleased with you. I suppose now there is no place for the poor little germs to sit and laugh at him. I think it always was a jolly nice nose, and it still is.

Will I be 21 in 21 months? Yes, so I will. I wish it were days instead of months.

♦ At 21, Molly would be of age and able to decide the future for herself. If Barnes were to be included in her life thereafter, he needed to hold her affection until then. The maths coaching would probably not be needed after the end of the academic year; and he could not be certain how much the news of his own work, his inspirations, efforts, successes and failures really interested her. He decided to write to Mr Bloxam; a very rough draft of his letter. with many alterations and excisions, has survived.

Barnes to Mr Bloxam. No date.
Dear Arthur,

I am writing to ask whether you think the time has come when I might be allowed greater freedom in my intercourse with Molly.

Out of regard for the wish you expressed that Molly should have time to look round I have refrained for over a year from worrying you. This has not been easy.

Of course I do not know for certain the state of Molly's feelings, for no word referring to the matter has passed between us since Dec. 19th 1922, but I cannot help feeling that a state of considerable strain exists on both sides.

Surely if Molly has made up her mind, the present restrictions will do nothing to make her change it, while on the other hand they are depriving both of us of a time that would otherwise be one of the happiest of our lives.

Whether Molly has or has not made up her mind, I do not wish to press for an engagement or committal of any sort on her part. I only ask for unrestricted intercourse and friendship, and I would undertake not to attempt to urge her or hurry her into a decision in any way, without your further approval.

I should like you to know that I hold an assured position in Vickers with every prospect of doing well and am insured for £1650 as a 1st class life in 3 companies and also come under a pension scheme maintained by the firm.

Should you wish to see me before giving a decision I would gladly call on you in town at any time.

Yours sincerely,
   B.N.Wallis.

♦ On the 5th January Barnes visited Hampstead, and at some point during this visit, Uncle Arthur summoned him. Influenced by Barnes's patience and Molly's persistence, Mr Bloxam's attitude had shifted. He had waged his battle without support from any of his family; he had begun to respect and to like Barnes; and knowing Molly to be stubborn, had no wish to alienate her. He relaxed the rules governing the relationship between the two, on one side at least, although they were not freed from regulation. Barnes could, within reason, write as frequently as he wanted and, again within reason, express himself openly; Molly must restrict her letters and entirely exclude any reference to her feelings or any form of commitment as to the future. In the circumstances in which the two grew up a promise was binding; Mr Bloxam's purpose was to guard her from the danger of such a commitment. Barnes perforce concurred, binding himself to expect nothing, take nothing for granted and exert no emotional pressure. Amidst the shower of letters and emotion which now swirled about her, Molly did her best to curb her responses.

Barnes: 14th Jan. 1924. New Cross
My dearest Molly,

This is not an answer to your letter [of Jan.9th], for which thank you very very much, but to tell you the result of my talk with Uncle Arthur. He was most <u>awfully</u> kind to me Molly, and has consented to remove practically all restriction on our intercourse. Only two stipulations were made. The first was that (presupposing that you care for me in any way) he could not sanction an engagement, or any understanding on your part between us for a little while yet. The second was that I must be "reasonable" in my behaviour towards you.

On the other hand, I may make frank reference to my feelings for you, and in other respects may act practically as an elder brother to you and Barbara.

I think Uncle Arthur's great wish now is that you should not only abstain from giving any pledge or understanding to me; but that you should be guarded from any <u>feeling</u> in your own mind that our friendship in the past or in the future had committed or would commit you in the very slightest degree. Until he considers that you have sufficient experience of men and matters to make a decision for yourself I must loyally support him in the course he desires.

I do not think he quite realised how completely we have tried to keep the promise which he exacted from me that without further permission I would never again refer to my feelings for you. As a result of that promise I have today, and rightly, no more knowledge of what you may feel for me than I had when I proposed to you. The situation, as I accepted it then, and as it stands to me today is that you are absolutely and entirely free; that the whole responsibility for our continued intercourse rests on my shoulders; and that whether you ultimately accept or reject me, I am for ever indebted to you for the very sweetest and most gracious comradeship that a man could be privileged to enjoy.

I have never had a chance to explain to you all that has happened. I would like to do so now. I fell in love with you Molly at first sight. The first conscious thought that you were the one girl whom I must love and cherish with all my might came to me in the bus on our way to Southwark Cathedral [23rd April 1922]. Molly, it just <u>happened</u> – there was no question of "choice" or deliberate seeking. From that first moment, I simply loved you. At first, realising that you were very young still, I tried not to encourage it. I didn't try to win your affection and I even tried to keep away from High Wycombe when I was asked down there.

But it grew and grew and I was selfish Molly, and couldn't let you go out of my life. So I started writing. When I had to go to Switzerland at a moments notice I felt I couldn't go longer without definite recognition of my position by Uncle Arthur, but as there was no time to see him, Fanny saw him on my behalf. She got his permission for me to write freely to you "but not love letters, as yet". I came to England at Christmas on purpose to see you, and on my arrival asked Auntie Fanny whether it was all right for me to see you and propose to you, as she had seen your Father and knew his wishes. She said yes, certainly, so, as you know I did. I will not refer to what passed on Christmas day, except to say that some awful misunderstanding must have occurred between Fanny and Uncle Arthur. <u>Of course</u> I expected you to tell your Mother and Father. I'll never forget, to the end of my life how you trusted me when everything must have seemed against me. Oh Molly, you can't think what a comfort your sweet message was in the time that followed. And then came Whitsun, when Uncle Arthur wrote asking me not to come to the house so frequently. And again your little note of consolation. You <u>have</u> been a brick to me Molly.

This year of waiting has very much deepened and refined my love Molly. My first thought now is not my happiness but yours. Oh Child if I have ever brought you a moment of unhappiness I do most humbly ask you to forgive me. I <u>know</u> I've been selfish, but it is so hard, just at first. Now I have schooled myself to wait patiently. In due course I shall come to you for your decision. I will wait until I die if necessary. In the meantime I do not want you to feel the least concern or worry. Remember, <u>you</u> are not responsible because a man falls in love with you and it is better to make him unhappy for some time by refusing him, than to accept him out of pity, only for him ultimately to realise that by his selfishness he had led you to ruin your life. Such a thought would <u>break my heart</u>. I couldn't <u>bear</u> you to be unhappy Molly dear. I love you so dearly that even if in the end you say no, I should feel some consolation in the thought that even then I was serving you by decently and quietly going out of your life.

God bless you Molly dear. You mustn't answer this.

Barnes.

Molly: 16th Jan. 1924. Hampstead
Barnes, I am not going to answer your letter – not now at any rate. I can't thank you for it in ordinary words because they seem so poor and inadequate, but all the time I'm thanking and thanking you inside me. You have made me feel so proud and honoured that it makes me most awfully humble.

Shall you be at New Cross next week-end? It is ripping of Auntie Fanny to ask us. If it is fine and you are there, do you think we could go for a walk, because we couldn't last time we were at New Cross. Molly.

Barnes: 17th Jan. 1924. New Cross.
<u>Of course</u> I shall be at home this weekend. Yes, I shall just love a walk with you. I dont mind what it is – I should be happy to spend the afternoon in the waiting room of a railway station without a fire, provided you were there. Oh Molly, how perfectly indescribable to think that I can take you out anywhere.

Thank you ever and ever so much for both your letters. Molly, you <u>mustn't</u> feel like that. Its <u>my</u> prerogative. You'll have to put up with being simply <u>worshipped</u>, so you may as well try to get used to it now, and be very Queenly and crushing. I shall love you like that.

Ive been fearfully busy all the last two weeks. I am trying to make a model of one of my inventions at home, to see how it works, and Fanny is helping me. Its a most uninteresting model really to look at. But you shall see it when you come.

I've left myself so little time that this can only be a short letter. But as I can write to you every day in the week I don't feel as if it were an opportunity wasted like I used to (lovely grammar). I keep on sort of hugging myself to see if I really am awake or only dreaming. And I can use my dear Jane Austen-y ending again. I felt I had to give it up Molly – did you notice? – because it was <u>true</u> – thats why I loved it so – and therefore it might seem like a reference to my poor old feelings. Oh dear oh dear, how miserable I've been. Oh miserable Starkey, oh happy Starkey! I'm perfectly orangey tonight.

Fanny turned Grannies chime off the other night, and now the dear has the sulks and wont strike at all. We must investigate her when you come.

Your very humble and devoted servant,

Barnes.

I thought that sounded wrong somehow, of course, its "obedient". Oh well, Im both "devoted" and "obedient" to you.

♦ Barnes seized the opportunity which Uncle Arthur's revised rules allowed: he gave Molly the letter, written a year ago, in which he had tried to explain the ineptitude of his proposal made to her while walking down Finchley Road in the pouring rain – the undelivered letter which he had taken away with him on that miserable Christmas day of 1922. The letter has no envelope: it was tucked into Molly's hand when they were together on 19th January. When that letter was written, Barnes had hoped to venture into 1923 sharing with Molly his profound emotion, whatever the final outcome might be. He had been honest in stating his position then, and in making clear his resolve to demand nothing except the right to hope. Guided by his ideals of chivalry and devotion, he left on that Christmas day with his declaration in his pocket. But although sadly tried, he had certainly not been defeated; he was now free to recall the earlier letter and its contents, and to lay the beginning of 1923 beside the beginning of 1924.

Barnes: 24th Dec. 1922. New Cross
It was most <u>awfully</u> plucky of you to come out with me on Thursday [21st Dec. 1922] – in all the pouring rain you must have been simply soaked.

I fear I made a sad mess of what I had to say, but it really under the circumstances was rather difficult, and I do ask you to make some allowance for that. It is awkward to carry on <u>any</u> conversation, when the instant you bend your head a long stream of water shoots off the brim of your hat. Now I feel that perhaps there is still a little ground to be cleared, both in justice to you, and to myself. This isn't going to be a love letter, but a plain statement, or as plain as I can make it of just how we stand – at least as seen from my point of view.

In the first place then Molly, I love you. How much and how deeply I must compel myself not to say here. And I have done so ever since the first time we met. You said "Why not Baba?", and almost in the same breath accused me of lack of experience. What is it Molly that guides a man in making his – I was going to say in making his choice – but I didn't make a choice, it just came. Surely a man doesn't solemnly sit down and argue out to himself the respective merits of the various girls he meets and knows? What a dreadful thought. I can understand that to a man who does things that way experience may be useful and necessary. I have met quite a lot of girls, but I know intimately – none. And yet I realised that first Sunday that you were the one. What on earth can experience have to do with it. Would you have me try falling in love with every girl I meet? There was never any question of Baba – or anybody else – I am not making comparisons – I like and admire Baba very much indeed, but I should never have fallen in love with her, even if you had not existed, any more than I have fallen in love with any of the other girls I have had the pleasure to meet before I met you. Do not imagine that I dont know why you are the only one – I jolly well do – but that must be reserved for my first love letter – when I am allowed to write it.

In the second place then, there rose at once in my mind the question of our ages. That is a point that you should know – I was 35 in September, nearly 17 years older than you. At first I feared it was too great a difference, and tried to put you out of my heart. But it was not to be done, and a remark of Fanny's, showing that she had seen what had befallen me, made me take her wholly into my confidence. To my intense delight she told me that she and the Pater had already seen and discussed the matter, and I had the wholehearted approval of both of them. We cannot of course tell the future, in the course of nature I must probably predecease you, but I think such speculations are idle. In any case, if we love each other are we to be separated by a matter of years?

So with Father and Fanny's approval I felt I could go firmly ahead. I did not have time to see your Father as I left England in such a hurry, but Fanny arranged a meeting with him last October and told him just how things stood. I think he felt what I myself have felt, that you were rather young, but gave permission for me to write to you, only not love-letters at that time. So I did honourably try to keep the condition tho' it has been most awfully hard. Now I have fully tested out my feeling and have no doubt about the matter. Each day does but serve to deepen my love for you. And you have advanced enormously in your knowledge of men and affairs, by even your one term at college, and have had time and opportunity to make the acquaintance of other men.

And it got harder and harder, until I felt that when I reached England I must tell you the state of my mind. But about this I consulted Fanny – as I did not wish to do anything that others might think unfair, for I did feel keenly that up to the time that you went to Coll. I must have been practically the only man you knew, and it would not be fair to attempt to prejudice you in my favour until you realised that you would be able to meet many more men.

I scarcely dare to hope that I shall ever succeed in winning your love, – but I am determined to try.

Don't say "No" just yet Molly, let me have a chance, after all I am under a big handicap, being abroad. That is all I ask, – if in the end you feel that I am not the man for you – but I am not going to contemplate defeat, its unbearable.

I wont write more, you know now as much as I do. Will you let me see as much as possible of you during these holidays?

Molly: 21st Jan. 1924. Hampstead
Barnes,

The point is how am I to begin to you – am I to go on my-dear-Barnesing you or am I to begin some other way, as I would like to. No, I think I'd better carry on the same way for a time, however much I want to begin differently. You won't mind?

Barnes, you don't really want me to keep to that bargain, and not thank you, do you? You see, there is no possible way I can show my thanks for yesterday and Saturday, unless it is by saying them. It was all too wonderful for words from the time when you showed me the roses in the taxi, to the time when you opened the door and went into the house; and from when Auntie Fanny and Baba went to get ready, to when they came in and asked if there was any tea left. Never in all my life have I spent two such afternoons, – and then this morning to have to dissect out the vascular system of dogfish! And all the time I was thinking how you had said this and how you had said that, and how you had moved away that little bit of hair, and how badly I needed you to do it for me then because I couldn't touch it with my hands, and it would keep on coming down. Why does everything and everybody seem so ordinary and dull after you? Still, I'm jolly well not going to be feeble and weak about it; it just wouldn't be worthy of you. One can do anything if one is strong enough and firm enough; and I got a jolly fine dissection anyway, though I say it who shouldn't!

But you needn't be diffident Barnes. I love, love, <u>love</u> to have you put back that bit of hair, or touch my hand or my arm or my dress, or – oh well, do anything you like with me. And I don't mind how much you stare at me

– I want to at you, only I never seem to get much further than your hands, which by the way, seem to me to be a very wonderful pair – so strong and firm and trustable, and yet at the same time so gentle, and able to do such neat little things, like putting small screws into clocks, and heaps of other very much smaller, neater things too. And if I happen to smile when you are looking at me, it is only because I know you are somewhere near and therefore I feel smile-y, or because I think of something I like about you – like the crinkles, or the way you say "Molly" or "oh Child".

Do you remember you asked me if anyone else knew about it? I absolutely forgot at the moment, but Betty does. It was after that Christmas a year ago; Nancy and I changed rooms, so that I slept alone with Betty for one night (she was sleeping in the spare room then – not with Pam). Betty and I are the very greatest friends, and before, I had always told her everything. She is always very quick to notice anything about me, and she had seen that something had happened on Christmas day [1922], because you see I felt too awful for words after you had gone. So she asked "Do you like Barnes very much, Molly?" What could I do? I couldn't just say "'yes" in an ordinary sort of way, so it all came out, and she was most awfully sweet and understanding. But I have never told a single other person except Daddy and Mother and her.

I had a ripping letter from Auntie Nellie yesterday morning. She said that for you and me "to learn to know each other" is as right and true and wholesome a process, and as necessary a one to real happiness, as the training for one's calling in life. Were we "learning to know each other" on Sunday, Barnes?

I see what you mean about the letter you gave me, which I should have had two Christmases ago. One can see a difference between the two letters when one reads them together. In the first one (the one you gave me on Sunday) [written Christmas 1922] you say you are not going to contemplate defeat, it's unbearable. And in the other one [written Jan. 1924] you say you would find some consolation in the thought that you were serving me by decently and quietly going out of my life. Barnes, that was a wonderful letter; but I love the other one too, and thank you so much for giving it to me. I love them both and have read them and re-read them until I almost know them by heart. But how you could write such a beautiful, wonderful letter as the one you sent me last week, I can't think. Oh Barnes, I don't want you to go out of my life; I want you to be in it for ever and ever.

Barnes, you <u>do</u> understand about the Airships? I never meant that I should ever want you to stop. As I said, I'm as proud as – what is very proud? anyway the proudest thing you know – of you and Airships. And anyway you know I <u>never</u> meant anything about money or poorness or anything. Barnes, if anything were to happen to you, I believe – no, I know – I'd rather it happened after – I mean when I had had you even if only for a very little time; because although I should know then much more fully what I had lost, I should be able to feel that whatever happened, I belonged to you and you to me. What I dread is that something might happen to you before. But Barnes, I'm not a coward – at least I try my very hardest not to be – and I'm proud, <u>proud</u> of you, and I wouldn't have you do anything else for all the world; and you do believe that I should be the very first person to encourage you to do anything (if you needed it, (encouragement I mean) which I know you never would) even if it were dangerous; don't you Barnes?

This letter is awfully muddly but it has been written in several little bits to-day and yesterday. I don't know if I ought to tear it up or send it, anyway I'm going to do the latter; I hope it doesn't matter. Oh, Barnes I <u>do</u> want Sunday back again.

Molly.

Barnes: 22nd Jan. 1924. New Cross
Molly dear, its quite impossible to go for another whole ten days before writing to you again. At least just this once; I'll try to be gooder next time. Oh Molly, I did enjoy that time so. Its me that must thank you, ever and ever so much for coming with me.

I hope the work goes well. If its anything like mine it doesn't, because every few minutes I have to stop to think about you. I found myself yesterday wandering about my room, looking for my pipe, tightly holding it in my hand the whole time!

Molly, do you think I might come to Hampstead at the weekend? Please trust me, and say "no" if you think Uncle Arthur would not wish it or if it would interfere in the very slightest with your work. Do please be quite honest; you do know that <u>anything</u> you say or wish, I just love to do, and I shall quite understand and be happy if you think it better for me not to come.

Molly: 24th Jan. 1924. Hampstead
Barnes, will you really come over to Hampstead this Sunday? Oh, how too lovely for words. You'll come just as early and go just as late as possible won't you? I shall expect you on Sunday afternoon. Oh good, good, good.

By the way, Daddy doesn't mind if you come over to Hampstead; I asked him.

Barnes: 25th Jan. 1924. New Cross
Molly dear, I <u>do</u> hope you haven't been worried at my being so long in answering. I was so puzzled by what to say, that I spent all Wednesday evening in simply sitting and thinking, and then Thursday, Burney kept me till 7.15. My dear Girl, I can't tell you how I loved your dear letter. I dont think I <u>can</u> answer it properly. In a way it has quite, oh bother, I cant find the word I want I dont want to say "upset", what I mean is it has completely modified my outlook. You see Molly, I never dared to hope that you cared so deeply for me. I do not think that if you only entertained a passing affection for me, that you could ever have known <u>how</u> to write to me in such a wonderful way. How you must laugh at my elaborate man like precautions to preserve your independance. But it frightens me rather Molly, because if you feel nearly as bad as I do, I dread lest it will interfere with your work. Uncle Arthur would never forgive me, and I would never forgive myself if anything did that.

What I want to say, dear, is this. If you think it would be better, I am willing for your dear sake, to face seeing nothing of you, until after your exam, and even not to write also. It would be hard for both of us. Hard things are often good things, unfortunately.

Molly, I expect you will think it funny of me, but somehow I can't get over my diffidence in writing to you. A reserve so long practised is not readily broken down. Somehow I've got the feeling that I mustn't give way entirely just yet. You do understand, dear?

You queer child, I certainly learnt to know a good deal of <u>you</u> on Sunday. I don't see how else we are to set about it, except by being together, do you? I admit I am now as much in the dark as you are, having no previous experience to guide me.

Ever your humble devoted servant Barnes.

Barnes: 28th Jan. 1924. Vickers House, Westminster
Lunch time, and a brief respite to write to you. I do wish I could see you, and talk to you and explain things. Molly, I feel very wretched about my letter. I'm <u>not</u> going to be restrained. Dear Molly, if you could have the courage to write to me as you did, surely it is just for me to be equally frank. I feel now so miserable that I did not reply at once. I feel as if I had failed you Molly, in a moment of emergency. Molly dear, you <u>do</u> know that I love you, love you and love you, and oh my dear, if I seem a clumsy

backward sort of lover, it isn't because I dont feel most intensely. You see, I'm so utterly unworthy of your love, that when I got your letter, I could think of nothing, but that you loved me, by some miracle, at last. Of course I could have written that Wednesday night, but I just sat and dreamed.

Molly dear, you see I thought of myself before I thought of you, but I am trying, really I am to put you before everything. Can you make allowances for me? I just pray every night that my life may be one devoted service to you, working for you, serving you, living only for your happiness. And then, in practice, I come so very far short of what I would be. Dear One, however intensely one loves, it must take some time before one adjusts ones outlook on life so that thought for you comes automatically first. Just as in lower things, like learning to drive a car, or fly, however much a man may be devoted to his profession, it takes long hours of practice, before at the moment of danger, his brain will direct his hands without time for thought. Molly, I just want to fit myself to be your radiantly happy servant, free in your glorious service. And then, when I come face to face with you, my wits all go blundering, and I do and say everything all wrong, and feel miserable afterwards. Please Dear One be kind. I will try Molly darling. You can help me, so much, by telling me and guiding me. I sometimes wish now that I had been in love before, so that I might have greater experience of womankind, and how to think for you, and please you, and serve you, yes and how to make love to you too. Can you wonder that the revelation of your letter left me incapable of thought? I can't thank you for it Molly, I feel as if it would take all my life to repay you.

I want you to get this letter tonight, Molly, and my men are back. I haven't nearly said all. But it was all strictly against our bargain and really, I dont know what to do to try to preserve for you that feeling of freedom which for your sake I must so earnestly desire.

Barnes: 31st Jan. 1924. New Cross
I haven't half answered your letter which you wrote after you had been staying here. To begin with, I wont mind how you start your letters to me, so long as I get letters. Yes Molly, I did wish you to keep to that bargain. You dont owe me any thanks. You dont seem to see that its a great privilege and pleasure for me when you allow me to do anything for you and the thanks are all on my side. Never in all my life either have I spent two such afternoons. To me, in a way you were something like your rose.

Molly dear, of course I understood about the Airships. But dont you see, I have to put the worst side of things before you. I personally think airships one of the safest things in the world, but of course many people would not agree with me, especially after the recent series of fatal accidents. But to anyone who knows, they are all perfectly explicable. Molly I never for a single instant thought that the knowledge of danger would in any way deter you. It was an instinctive knowledge of your high courage that made me answer you so frankly. I know you would infinitely rather know <u>everything</u> than be stupidly "shielded" by me. I've no patience with men who adopt that attitude, and I dont think you would have either.

But at the same time I think it was only fair for me to offer to give them up for your sake if you would rather. Are not you trusting something far and far more precious than <u>any</u> worldly thing, your own dear future, to me? At least of course, you haven't said so yet, but if you do, I mean. And I know you never gave a thought to poorness or anything like that. But again, it was only fair to tell you facts.

Molly, its quite impossible to begin to thank you for all the wonderful things you say. You know I am yours, dear, just for the taking, soul and body. But there again, its no gift worth your having. Molly, I'm a mean and selfish wretch, ever to have <u>dared</u> to fall in love with you. What is <u>anything</u> that I can offer you, in comparison with your perfect self. Oh Molly, if in your infinite love, you do some day decide to have me, I dont know what I shall do. Do you know dear, sometimes Im absolutely <u>frightened</u> lest you <u>should</u> accept me, and I should prove so utterly unworthy of you.

Molly: 4th Feb. 1924. Hampstead
My dear dear dear Barnes,

I can't help it; it meant to begin in a respectable way, but it found it couldn't, and I hope you don't mind or think it very wrong. This is only Monday evening and here am I writing to you, when I know the letter can't be posted till Friday. It is very mad of me, but I do so want to see you and talk to you, and since I can't do that, the next best thing is to write to you.

Oh Barnes, will it all come true do you think; and will I really go with you one day to that beautiful place where you wrote a letter to me, and the bee sat on the end of your pen – how sticky it must have been! And could we really go in the summer when it wasn't too crowded, so that we could go for walks all alone – just us two. Barnes, you know no other man could do it the same as you could. Why, I shouldn't know what to say to anyone else, and I should be bored stiff after a while and so would he be I expect. Goodness knows what there is about you that makes you different, because there certainly is something which makes me want to be with you, and as close as I can to you always. Though when I say that, I don't for one moment mean that I don't know how delightful and wonderful you are and how dear in the things you say and do, so that you are quite quite different, and miles above any other person that was ever made; but I believe if you were dumb and couldn't move and weren't a bit clever and couldn't explain things, if I could just look into your eyes, it would be exactly the same – no, I don't believe it; I'm sure of it. And Barnes, if for some reason you could never go to Switzerland or abroad at all, and if you had to live always in a little house in a little street like we saw this morning, I shouldn't care a bit, and I should be just as happy. It isn't where you are that matters, but the person you are with; don't you think so?

You are quite right Barnes, I should hate to be "shielded" by you. I want you always to tell me everything, and you can rely on me to be strong and not feeble about it. I want to stand beside you and not behind you, so that if anything happened it should happen to us both.

Barnes you don't still feel miserable because you didn't answer that letter of mine at once? You certainly need never have felt for an instant that you had failed me in a moment of emergency. You see, I never expected you to answer it. I wrote it because there were some things that I simply had to tell you, but none of them ever expected an answer. And Barnes, there is no need for me to make allowances for you. Why Barnes dear, you are the most wonderful, most perfect lover that a girl ever had. You don't say everything all wrong when you see me; you say and do everything just right. And if you had been in love fifty times before, and had all the experience of all the lovers that ever were, you couldn't please or serve me or make love to me one tiny bit better than you do, and probably not half so well; and you certainly couldn't be as dear as you are.

Barnes, how I'll love to take charge of you and sew on your buttons. Does a man really like to feel that he will be looked after like that?

Oh Barnes, what gift could be so wonderful and beautiful as yourself. It would be more worth my having than anything else under the sun. And to think it's mine just for the taking! It is most awfully difficult to realise even now. I want to take it now – this very minute.

I say, I don't believe I ought to be writing to you like this. What shall I do? I know. Look here, Barnes, please will you tell me absolutely honestly if you would rather I wrote you ordinary, weather-y letters. <u>Please</u> tell me, because I wouldn't for the world do anything that would make it more difficult for you. And anyway, perhaps you think I ought to write ordinary letters. It's dreadfully muddly.

Yesterday last year was the day we started measles so all the work now is practically entirely ne to me.

Barnes: 15th Feb. 1924. New Cross
My darling Molly,

I cant thank you enough for your dear letters. I'll begin at the end my dear One. Do you seriously think I would rather you wrote me weathery letters? Then you can't realise the exquisite pleasure that your sweet letters give. No, my dear, they dont make it more difficult for me. Molly, I just love you, worship you, adore you. My dear Girl, for me you are just the beginning and end of all things. But what about <u>you</u>?

Can you go on telling a man you love him, and at the same time preserve that independant feeling about which we are made to suffer all these restrictions and separation? What is the difference my dear, between saying you love me, and saying you'll have me, as far as your own dear heart is concerned?

You know Dear, I love you far too truly ever by word or even thought, to stand in the way of your true happiness, should you find your feelings to change. As far as I am involved, commitments or no commitments you are as free as ever.

But you have the most intensely loyal nature, and I have the feeling that you would probably rather die than desert me when once you had said you loved me, even though your love had changed. Molly dear, its of your own dear true loyal heart that both Uncle Arthur and I are thinking. If I let you write such letters, all our deprivations are useless. As I see it there are only two ways. Either we must attempt to preserve the spirit as well as the letter of your father's regulations, and you must submit to being "cured" with as good a grace as possible, or else we must convince him that your attachment to me is deep and permanent, and that he is causing unnecessary suffering by standing any longer between us.

I have tried to do the latter and failed. It is his opinion that you are still capable of change.

Its all my wretched fault Molly. I suppose I ought never to have written that first letter of explanation. How could I ever expect you to remain dumb when I broke the silence? How on earth did he and I ever imagine that I could discuss intimately our relations without stirring into flame slumbering fires. But oh Molly, I've endured so much. There's a limit to everyones endurance, and surely surely we know our own minds after nearly two years? Why <u>must</u> we lose these precious hours of happiness. I dont know.

But I can see no other way. I must just write weathery letters to you, and you must just write weathery letters to me, until you have either forgotten me or are of an age to choose for yourself.

Molly dearest, I <u>am</u> so sorry. But I <u>can't</u>, honourably, see any other way out. When you calmly think it out can <u>you</u>? Tell me truly, child. For in spite of what Uncle Arthur says, I think it does indeed involve your happiness, and I will not take any decision that involves you without your consent and approval.

Molly, you haven't agreed that we shan't write love letters yet, tho' I fear you must. So I can say, I love you my dear dear Molly. Sorry, Child, it just pops out every now and then.

No a man, this man at least, doesn't care two hoots about your looking after his personal comfort. I think, however delightful it may be, its just the last thing he thinks of. When I said take charge of me, I meant that I felt all <u>wrong</u>, and so hopelessly unworthy of <u>you</u> that only your dear patience and teaching could possibly make a man of me. Oh dear, I haven't time to explain things now. Dont be hurt Child at what I've just said. I shouldn't have put it like that. Of course one would <u>love</u> your looking after things like that, but you can hardly expect me to think of the central pivot of the universe sewing on buttons can you? Oh, my darling Central Pivot its ten oclock. Madame dear I had so much to tell you, and now its all too late. And oh my dearest dearest Child, dont be hurt or unhappy, and try to forgive me. Molly dear, I cant tell you how much I love you. God bless you, dear Molly.

♦ Meetings, usually chaperoned by the families, became much more frequent for the two and, to make arrangements, circumspect telephone calls were exchanged. But to express emotions beyond "reasonableness" on Barnes's part and reticence on Molly's, letters were still the vehicle, and both were aware that they were overstepping the agreement made with Molly's father. They were writing too frequently, and were making explicit those feelings which it would be hard to quieten or, unthinkably, withdraw.

Chapter 13
A MEASURE OF NORMALITY

In her letter of 4th February, Molly reminded Barnes that one year ago the quarantine for German Measles had kept her away from College for weeks, with the result that the current part of the syllabus was entirely new to her. Barnes dragged himself out of the transports of a lover and applied himself without much joy to physics. The two chapters which followed were far briefer than those for Calculus or Trigonometry. They were titled but not numbered, and they do not carry Molly's own pagination which had run through the previous eighteen chapters.

Note on Potential, Charge and Capacity.
Let us see if we can obtain from a consideration of hydrostatics, some clear ideas on the three things above. To begin with, hydrostatics merely means a short way of saying "problems or theorems connected with water which is not in motion"; that is as opposed to water which is flowing.

What follows then is all about water at rest. I think perhaps things are easier to grasp if one talks in definite units and with definite figures. Dont you? Here are a few figures about water:-

A cubic foot of water weighs 62.35 pounds approx.

A gallon of water weighs 10 pounds almost exactly.

This is easily remembered; or perhaps more easily still by the rhyme
> "a pint of pure water
> weighs a pound and a quarter"

since there are 8 pints to the gallon.

This also gives us the very useful relationship that 1 gallon measure is $\frac{10}{6235}$ths of a cubic foot in capacity or

6.235 gallons are equal in volume to one cubic foot.

Turning to the C.G.S. (Centimetre, Gramme, Second) system we have as you know 1 c.c. weighs 1 gramme

Though even this isn't <u>quite</u> correct, because the French as you also probably know made a mess of all their measurements when they adopted the metric system in 1790 something. And of course, if we wish to be really scientific we must also specify the temperature at which our measurements are made, because like most other things water expands with heat, and consequently 1 cc of water at 20 °C does not weigh the same as 1 cc at 100 °C. The latter is a little lighter, because the water will have expanded a little, and some will have run out of the cc measure, if one did it that way, which of course one doesn't. But this is going to be a terribly unscientific explanation, so we will not bother any more about the jolly old scientists.

Now suppose we have 4 beakers, as I have sketched them on the next page, and the cross sectional area of each is as shown. We are going to "charge" each beaker with water, and we are going to put an equal "charge" into each. (As the toastmaster says, "Charge your glasses gentlemen").

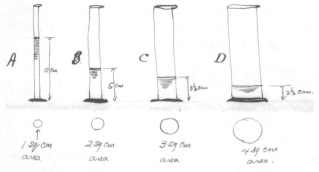

Suppose for the purpose of measuring the charge, we regard 1 c.c. as being the unit charge. We will elect to charge each beaker with 10 units of water charge (i.e. 10 c.c's of water).

10 cc's is a volume, and volume/area = height

or in other words we have only to divide the total charge by the cross sectional area of each beaker to find the height to which the given charge will fill each beaker.

The water in A will rise to a height of 10/1 = 10 cms.
    "   "   " B " " " "   "   " 10/2 = 5 cms
    "   "   " C " " " "   "   " $10/3 = 3\frac{1}{3}$ cms
    "   "   " D " " " "   "   " $10/4 = 2\frac{1}{2}$ cms

Note that the same "charge" has filled A to 4 times the height of D. Or, another way you would think to yourself, the capacity of D is 4 times as great as that of A for a given height.

Suppose for a moment we had only put 1 c.c. of water (i.e. unit charge). The water would have risen to 1 cm. in A and to $\frac{1}{4}$ cm in D.

These figures make it plain that the "capacity" of the beakers' is in <u>inverse proportion to the height to which unit charge will fill</u> <u>them</u>. That is:-

Capacity of D is to Capacity of A

$= 1/\frac{1}{4}$ cm is to $1/1$ cm

$= 4$ is to 1 (You know that 1/(the quantity) is called the "inverse ratio" or "upside down" ratio!)

It is obvious that the capacities are in <u>direct proportion</u> to the cross sectional areas, but supposing we didn't know the cross sectional areas, and we wanted to compare the capacities?

What has gone before provides us with a very ready method of <u>comparing</u> the capacities, for we have only to pour into each beaker an <u>equal</u> <u>charge</u> and then measure the <u>height</u>, and we get:-

<u>Comparative</u> Capacity = Charge/Height

If we want to define a <u>unit</u> of capacity instead of merely <u>comparing</u> capacities we may agree to say (convention) that

Capacity = charge/corresponding height.

or we might state this equation in words

"The <u>unit of capacity</u> shall be taken as the <u>charge</u> which will fill the beaker to <u>unit height</u>".

Let us see how this works out.

<u>Beaker A</u>

Capacity = 10 cc/10 cm = 1 <u>unit of capacity</u>

<u>Beaker B</u>

Capacity = 10 cc/5 cm = 2 units of capacity

<u>Beaker C</u>

Capacity = 10 cc/$3\frac{1}{3}$ cm = 3 units ...

<u>Beaker D</u>

Capacity = 10 cc/$2\frac{1}{2}$ cm = 4 units ...

You will notice that defined in this way "capacity" does <u>not</u> mean the total quantity that a beaker will hold. You may think this sounds stupid. But was it not the Caterpillar in "Alice" who paid his words double and made them mean what he liked? So, if we chose to say that when used in this particular way the word "capacity" shall mean the height to which a "charge" will fill a beaker there is nothing unreasonable about it. And we may go further, and say that the <u>unit</u> of capacity shall be the quantity of water (measured in "charge" units i.e. really in cc's) required to raise the water to unit height (i.e. in this case to a height of 1 cm).

This isn't nearly so mad as it looks, for suppose for a certain experiment you wanted a beaker that was graduated, and would measure to say 1/10 cm. Obviously beaker A would be much better for your purpose than beaker D, because 1 c.c. of water in A rises 1 cm. up the glass and so you could easily read to 1/10 cm. which equals a volume of 1/10 c.c. of water. Whereas suppose you tried to use beaker D 1 c.c. of water would only rise $\frac{1}{4}$ cm. up the glass and so it would be practically impossible to read closer than $\frac{1}{2}$ c.c.

So you would be quite justified in going to the lab. attendant and asking for a "1 unit" beaker, supposing that the convention that I have suggested were understood by you and he.

Is that clear?

Notice that the "capacity" of each beaker corresponds exactly with the cross sectional area in sq. cms. This brings home to you the fact that our conventional capacity does represent a definite property of the beakers.

The <u>height</u> of the beaker handed out to you might be anything, and would not be important when it was accuracy of measurement that you were wanting. By asking for a "1 unit" beaker you would always get a beaker of the <u>same diameter</u> (and therefore of the same cross-sectional area) and with the same size of graduations, whatever its height might be.

Now we have already seen that the pressure at any depth of water is solely dependent on the height of the water above. Perhaps I might just go over this again.

Let us take our "A" Beaker, with its 1 sq. cm. area. Fill it with unit charge (= 1 c.c.) of water. It rises to 1 cm. The weight of 1 c.c. = 1 gram. Therefore since the area of the bottom inside is 1 sq. cm. we have a weight of 1 gr. distributed over an area of 1 sq. cm. or as we more usually say, the pressure at the bottom is 1 gramme per square cm. written for short like this "1gr/cm^2".

Add another unit charge. We now have 2 grammes of water in the beaker, – the height is 2 cms., but the area of the tube remains unchanged. So that the 1 sq.cm. at the bottom has to support 2 grammes weight, – or

pressure at bottom due to height of 2 cms = 2 grammes/cm^2

Add a further unit charge, then

press. at bottom due to ht of 3 cms = 3 grammes/cm^2

Add 4 more units –

press. at bottom due to height of 7 cms = 7 grammes/cm^2

And so on.

Turn to beaker D, and fill it with 4 units charge. This will raise the height of water to 1 cm.

Now total weight of water in D = 4 grammes since we put in 4 units of charge each weighing 1 gramme.

But area of bottom inside = 4 sq.cm. Therefore

<u>Total</u> pressure on bottom of area 4 cm^2 = 4 grammes.

Therefore

Pressure per sq cm = total pressure/area over which it is distributed

= 4 grammes weight/4 sq cm

= 1 gramme/1 sq cm

or as generally
written   = 1 gramme per sq.cm or 1 gr/cm^2

I have failed to draw the conclusion – that a height of water of 1 cm. in a beaker of area 4 cm^2, produces exactly the same pressure per unit area on the bottom, as an equal height produced in a beaker of 1 cm^2 area. This can be proved for a beaker of <u>any</u> area.

∴ Pressure per unit area is solely dependent on height of water and not on area of container.

We <u>always</u> measure pressure in terms of the load borne by <u>unit area</u>. We can use any units we please thus

Pressure may be measured in   pounds/square foot
                                pounds/square inch
                                grammes/square cm.
                                kilogrammes/sq. cm.

We can even mix the two systems, and sometimes talk of pounds per sq. cm.!! Why not? All you have to do it to use unit<u>s</u> of of weight in conjunction with <u>a</u> unit of area.

Any one can be readily converted into any other:-

For instance

Pressure of the atmosphere at sea level = 15 lbs/in^2 (approx)

Now there are 144 sq ins in 1 sq ft.

$$\therefore \text{Press. of atmos. at sea level} = (15 \times 144)\text{lbs/ft}^2$$
$$= 2160 \text{ lbs/ft}^2$$

So that when you speak of 15 pounds per sq. in. it is exactly the same thing as speaking of 2160 pounds per sq. foot.

Can we turn lbs/in^2 into grammes/cm^2. Let us try.

1 kilogram = 2.2 lbs (approx.) (You have to <u>know</u> that – I mean you cant work <u>that</u> out – it just means that the French and English standards of wt. are different).

$$\therefore 1 \underline{\text{gramme}} = 2.2/1000 \text{ lbs} = \underline{.0022 \text{ lbs}}$$

Again 2.54 cms = 1 inch

With these "Conversion Constants" let us turn 15 lbs/in^2 into grammes/cm^2.

$$\text{Since} \quad 1 \text{ gr.} = .0022 \text{ lbs}$$
$$\therefore \qquad 1 \text{ lb} = 1/.0022 \text{ grs} = 455 \text{ grammes}$$
$$\therefore \qquad 15 \text{ lbs} = (455 \times 15) \text{ grammes}$$
$$= \underline{6825 \text{ grammes}}$$

and since 1 in = 2.54 cms

$$\therefore 1 \underline{\text{sq}} \text{ in} = (2.54 \times 2.54) \underline{\text{sq}} \text{ cms}$$
$$= 6.46 \text{ sq cm. (approx)}$$

So that 15 lbs per <u>one</u> sq. inch is the same thing as
$$6825 \text{ grs " } \underline{6.46} \text{ " cms}$$

To bring this to grammes per <u>one</u> sq cm. we just have to divide 6825 by 6.46 and we get

$$15 \text{ lbs/in}^2 = 1060 \text{ grammes/cm}^2$$

Which is the required pressure of atmosphere expressed in metric units. (All the figures are approximate, being worked by slide rule).

I have just put in this little digression to accustom you to the idea of intensity of pressure – that is pressure per unit area.

Before going on, I want to show you one further example illustrating how pressure per unit area is solely dependant on height of water.

Look at the two vessels on the next page. Both are filled with water to a height of shall we say 50 cms.

From what has been said before, you will probably be fairly satisfied that the pressure at base level of B is 50 grammes per sq. cm. due to height of column, whatever the cross sectional area of B may be.

But what about A, which has ll the mass of water in the expanded part above it?

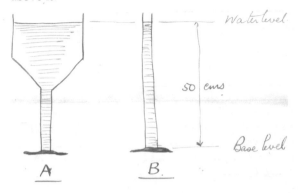

Well, let us get the glass blower to stick a little tub in the side of each at the base level, so that we can form what is virtually a U-tube out of each with very different diameters of limbs.

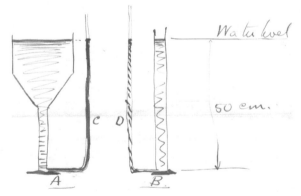

If the bore of the little tubes C and D is very small, so that the volume of water taken to fill them is insignificant, you will find that the water in all four tubes remains at <u>exactly</u> the same height.

Here we have great big vessel A, balanced by tiny little tube C, giving a good illustration of the equality of the intensity of pressure.

The water in C is pushing down with an exactly equal pressure to that in A, so neither can beat the other, and both remain balanced at the same height. You can prove this mathematically – all I want you to do is to realise the fact.

244 MATHEMATICS WITH LOVE

It has an interesting application when we come to consider the discharge between two "condensers" in electricity.

But to go back to our original purpose.

On [p. 239] we saw that if we chose, we could define the unit of capacity as the number of charge units required to raise the water to unit height in the beaker.

Now we have further established a relationship between <u>height</u> and <u>pressure</u> at bottom.

If we work in the C.G.S. system we see that the height in cms is numerically equal to the pressure in grammes per cm². e.g. 10 cms height is equivalent to 10 gr/cm² pressure and so on.

So that we might equally well define the unit of capacity as the number of charge units required to produce <u>unit intensity of pressure</u> at the bottom of the beaker.

Remember always that in this sense Capacity does not mean the total quantity the vessel will hold. It means the <u>limited</u> quantity that will fill the vessel to such a height as to produce unit intensity of pressure.

Unit intensity of pressure in C.G.S. units is of course 1 gramme per sq cm. Unit charge is 1 c.c.

And hence the "Capacity" is the no. of ccs required to fill a vessel to a height of 1 cm, thus producing on the bottom a pressure of 1 gramme per cm².

<u>If we had a beaker of infinite height and uniform bore all the way up, we could charge it up to anything we liked</u> – until in fact the pressure at the bottom became so great as to burst the glass.

But its "capacity" from our point of view would be the same as that of a beaker 1 foot high.

We saw on [p. 239] that we could easily determine the capacity, by putting into a vessel a measured charge, and then measuring the height, wh. is equivalent to measuring the pressure, and then by means of the equation

"Capacity of" vessel = Charge given/Height or pressure produced

we could derive its nominal capacity.

Let us put it into symbols.

Let C stand for "Capacity"

Q   "   " Charge given

V   "   " Intensity of pressure produced by charge Q.

Then our equation above becomes

$$C = Q/V$$

All these properties are exactly analagous to their electrical equivalents. We have only to substitute the word "potential" for intensity of pressure. It already means the same thing – only the name is different. Charge and Capacity are exactly the same – only the units of measurement are different. I may have given you the impression on that Sunday [17th Feb: Molly visited New Cross], that a conductor could not hold more than a certain charge. My examples were made up on the spur of the moment, and the one about pumping water into a container is not a good one. But I hadn't looked at the subject for 3 years, and all I know was learnt in a week or two, when cramming for Inter so you must forgive me.

An infinite charge can be theoretically crammed onto the surface of a conductor, such as a spherical ball. Electricity does not appear to occupy any space, and so you can cram the little ions or electron fellows or whatever they are, closer and closer together, the only result being that they shove about more and get their potential raised. Like when you're the last one getting into an already overfilled tube carriage and the guard simply shoves you in. Your potential goes up, and so does everyone elses!

But in actual practice, when the potential is raised too high, discharges will take place to anything that is handy, poor little electrons getting pushed off!

Molly: 20th Feb. 1924. Hampstead.
The weather continues to be most distressingly cold, doesn't it? I thought last night that it was going to turn warmer, but this morning it was trying to snow here, though there was no sign of snow at Gower Street; I wonder if you had any at New Cross. I always think a – (bother,I haven't the remotest idea what sort of a wind it is; I suppose I should have found out before; anyway, we'll say it's East) an East wind is so trying, don't you? But we can only hope it will change soon. There! don't you think that is a perfect opening for a weathery letter?

Barnes, thank you so much for sending me the notes on Potential, Charge and Capacity. It has made everything so beautifully clear; it is strange how simple it is when you think of beakers and water.

You know, I find it most awfully difficult to take a thing in when you are explaining it to me yourself – when you are saying it and not writing it, I mean – because all the time I'm listening to your voice and hearing something fresh in it which I have to think about, so that every now and then I miss two or three words. And when you are drawing or pointing to something at the same time, it makes it even more difficult because then there are your hands to be reckoned with. If you could have somebody else's voice for the occasion – thank goodness you can't. I do love to have you explain things to me, and it does help such a lot.

There is something else I don't quite understand. Supposing you have a can and you are trying to build up a charge on it (I'm not sure if that's how you say it – it sounds rather funny), you can do it by giving a very small charge – say a +ve one – to another can, and putting a sphere inside it and earthing the sphere while it is in contact with the can; then the sphere has a negative charge which you can give to the other can. And when the sphere is touching the other can you earth the sphere, when it gets a +ve charge which you can give to the first can, and so on. I'm afraid this is rather muddled, being so full of cans, but I hope you can understand it. The point is this. How does the sphere manage to get electricity of the opposite sign when you earth it? The potential then must surely be zero since the earth's potential is zero; and can you have a charge where there is no potential. And also where on earth does the other -ve charge come from? The sphere is in contact with +ve electricity.

By the way, it was Humpty Dumpty in "Through the Looking Glass" who paid his words extra and made them mean what he liked. And "glory" meant "there's a nice knock-down argument for you"; and all the words came round him of a Saturday night for to get their wages; and a word that could mean more than one thing was like a portmanteau that would open out. So if one day I say "glory" to you, you'll know what it means!

I love the part of you which starts talking about something else, or points out something which strikes its fancy, interrupting itself in the middle of its conversation. Also I love the part – I say, it doesn't matter saying that, does it? One could like bits of anyone, not that one does, but one could. Though I suppose if I said enough parts I should get to the whole of you. But also I love it when you tease me.

Barnes, it was ripping of you to wait for me at the place where the 'bus stops on Sunday. I am so sorry I was so late. I am practising walking bigger steps, but it's not easy to get just the right size, when there is nothing to measure by. Most certainly I should never <u>never</u> wear ear-rings. I don't see how people can make holes in their ears and stick things through them. I am so glad you hate them too. No man has ever kissed me either, and no man except one, ever shall.

Oh Barnes, I <u>am</u> happy – as happy as happy as can be. How could I help being, after last Sunday, and after your letter which I had last week? It was such a lovely one, and I did love it so. Barnes, why should I try to forgive you. What have you ever done? Why Barnes, you have made my life so wonderful that every morning when I wake up I think of Barnes and how jolly glad I am to be alive.

Barnes: 23rd Feb. 1924. New Cross.
My darling Molly,

I suppose thats not a proper beginning, but I really can't help it. And its only Friday, and I suppose I've no business to be starting my letter yet, but I cant help that either. If there's one thing I simply hate, its the week when, having posted my letter to you, there's nothing to be done but wait for Saturday, blessed day, to bring your letter to me.

Its a queer thing, in some ways I feel so tremendously capable of taking care of you and working for you, and yet under it all I feel as if I were one of your children. Its the most wonderful feeling in the world. I did so love your touching my hair. Will you always do that too?

Thank you again and again for your letter Molly dear. I just loved the weathery beginning. You did it awfully well. I'm jolly glad you didn't ask me to explain that induction question on the spur of the moment. Tho' I would never mind confessing that I didn't know to you. Its all very well to say you find it hard to take things in when I am saying them. Do you stop to think how fearfully difficult it is for me, in your dear presence, to collect my thoughts and plunge into scientific explanations of things more than half forgotten?

Its useless to try to single out any one bit of you Molly – its just all of you, from your dear, dear, feet to your glorious hair, oh Molly, I do so worship you. Oh hasn't it been snowing lately, and isn't the wind cold and dont you love your morning bath. Oh bother the weather.

Barnes: 26th Feb. 1924. Rudgwick.
I <u>do</u> hope this will be all clear to you? I was awfully worried for two days trying to see how it worked. Anything you dont see <u>write at once</u>, dont wait for your turn. I will always try to answer at once and I <u>do</u> love to help.

Electrostatic Induction.

This is a perfectly awful subject, which I dont properly understand myself. It always puzzles me, and having read up all my text book has to say I dont feel any better qualified to attempt to explain it to you. Anyhow, all these explanations are sort of conventions in order to help one to understand <u>what</u> happens, although they may leave you rather in the dark as to <u>how</u> and <u>why</u> it happens.

You know of course, that in order to explain all these phenomena, it is usual to assume that there are two kinds of electricity (whatever that may be) called for want of better terms positive and negative. I dont think it is necessary for your purpose that you should attempt to understand the latest real theory, even if I could explain it which I cant, but briefly its something like this:-

An atom of matter consists of a core or centre which is positively charged, and around which rotate at enormous speed a number of negatively charged "corpuscles" I believe they call them, or "electrons". Never mind the name. Lets say "electrons" it sounds nicer. Now apparently there are generally a dozen or two negative electrons too many, or at least not so firmly attracted to the centre core as are the others, and the structure of the atom does not break down if these surplus negative electrons are removed or driven off by any means.

Now the condition usually referred to as "positively" charged is held really to be that a lot of the surplus -ve electrons have been removed from a substance, while a "negative" charge is when a substance has <u>more</u> than its usual allowance of surplus -ve electrons.

This means that when you rub a plate of ebonite with fur you are rubbing or otherwise transferring surplus -ve electrons from the fur to the ebonite. The fur would then have a deficiency, although its atoms still remained as it were "fur" atoms, while the ebonite would have a surplus of -ve electrons although its atoms too would remain "ebonite" atoms.

As there is a pretty average crush of atoms inside the ebonite, all of whom are already "full up" with surplus -ve lodgers as it were, the poor old surplus have to remain on the surface, and are jolly well pushed off to find "a better 'ole" at the first opportunity. Similarly the fur has now a number of rooms to let.

Dear Mother Earth has vast reserves of both -ve lodgers <u>and</u> rooms, and if you connect the ebonite to her she promptly accommodates all the poor unwanted surplus. On the fur applying to earth to find it lodgers, she again obliges, and all is happy and contented as before.

Instead therefore of thinking of a "positively" charged body as <u>discharging</u> to earth when the connection is made, it is more correct (as I understand it) to think of a flow of negative electrons coming <u>up from</u> the earth and occupying the vacant lodgings on the so-called positively charged body. Whereas a "negatively" charged body when connected to earth does in fact get rid of the unwanted surplus.

This explains I think the phenomenon of "induction" rather better than the old theory of "two kinds".

For imagine a conductor (i.e. a body which offers no resistance to the flow or movement of surplus -ve electrons) being brought near to a body which is just in the usual or "neutral" state (i.e. just the usual number of surplus -ve electrons).

1st. – Let the Conductor have a large unwanted surplus, for whom no rooms can be found.

The happy lodgers in the neutral body see the approach of a surplus army and take fright. They are cowards, and instead of waiting to be turned out, they flee as far away as they can get, leaving their rooms vacant behind them. Here is a picture of it:-

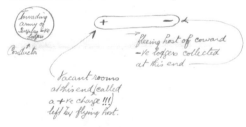

2nd. Let the approaching conductor (sounds like a busman coming to take ones fare) have a number of rooms to let (so called +ve charge).

The coward host on the neutral body (the house) say "Hullo boys, here's a perfect floating hotel, lets change our digs" and so they all crowd up towards the end where the conductor is approaching, ready to hop on board if he should touch.

You will notice from all this two things:-

1) In a "neutral" conductor the coward host of -ve electrons really hardly know their own minds; in more scientific language, they are rather "unstable". They will leave their rooms at the smallest excuse, either to avoid a <u>possible</u> incoming host, <u>or</u> to occupy new premises seen from afar.

2) Conductors with surpluses (-ve charge) or conductors whose landladies have vacant rooms (+ve charge) have <u>no uncertainty whatever</u>. The first will get rid of its surplus, and the second will make up its deficit, at the slightest opportunity.

One or two further points before we start on your problem.

Point 1. Surpluses and vacancies always lie on the <u>surface</u> of a conductor.

Can we explain or illustrate this? Lets see.

A surplus is in the 1st place not wanted and gets pushed to where there is most room. This appears to be on the surface. If the conductor is hollow (like a can) the surplus will lie on the larger surface. Clearly the outside of a can has a slightly larger surface than the inside, hence the surplus would be pushed onto the <u>outside</u> surface.

IF however a conductor with rooms to let (+ve charge) were lowered inside the can without touching it, the surplus (-ve charge) on the outside would immediately rush to the inside surface, in the hope of getting rooms, although they crowded themselves up still more by doing so. Here is a picture

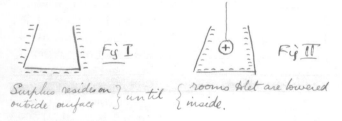

NOW if we earth the can in Fig I the surplus will bolt for the earth, BUT if we earth the can in Fig II the silly surplus will remain gazing at the rooms to let inside and will <u>not</u> go to earth. This is because besides being cowards they are also fools and the charge (or rooms) in the immediate neighbourhood is more powerful to hold them than the earth. In these circumstances the surplus is referred to as the "bound" charge, being (spell)bound by the +ve charge so that they <u>cannot</u> go to earth.

Of course exactly the same happens if the rooms to let are on the can, and the surplus is lowered into it. In this case the landladies pick up their rooms! and rush to the inside. They also are very shortsighted landladies, as on earthing the can, they <u>refuse</u> to have any lodgers from the earth, preferring to gaze on the charmers within. Here is a picture

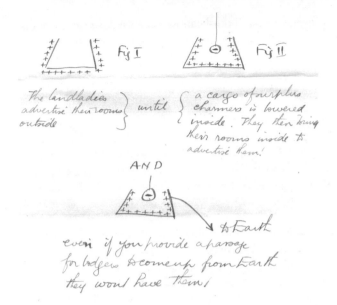

The landladies advertise their rooms outside } until { a cargo of surplus charmers is lowered inside. They then bring their rooms inside to advertise them!

AND

→ to Earth

even if you provide a passage for lodgers to come up from Earth they wont have them!

Of course if you earthed Fig I they would have up the earths surplus quick enough.

Point 2.

Supposing I start with a neutral can and lower a set of landladies (+ve charge) into it? Just as in the picture on [p. 250] the unstable lodgers leave their rooms and rush to the inside thus

Now if we earth the can?

Lodgers rush up to fill the vacant rooms of the silly ones on the inner surface and the "to let" signs (+ve charge) on the outside disappear thus

neutral

and lo! the silly ones, on <u>removal</u> of the sphere inside, find they have nowhere to go and are now a surplus. (-ve charge). Serve them right. So they dash to the outside in a fury thus

earth

If we lower a <u>surplus</u> into a neutral can we get a similar result of opposite sign thus

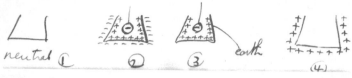

neutral ①   ②   ③   earth   (4)

Point 3.

Start with a neutral can and lower a +ve charged sphere into it thus

neutral

So far the same as point 2). Now suppose we let the sphere <u>touch</u> the can and do not earth either of them?

The lodgers after all <u>do</u> get into the promised land because for some reason or other they prefer the new quarters to the old ones and we get this result, after the sphere has been moved away again. I have marked it N for neutral.

Note carefully that the lodgers exactly fitted into the new rooms. (Induced charge = original charge).

And on removing the neutral sphere, we have a +vely charged can, just the opp. of case 1 in point 2).

The same thing happens in case 2 with opp. signs thus

neutral

and on withdrawal of sphere, can remains -vely charged.

<u>Point 4</u>

Suppose we reverse the procedure of point 3), and, starting with a charged can, lower a neutral sphere into it.

 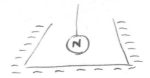

There being nothing to attract the charge it remains outside. But is there no inductive effect on the sphere? If you connect it to an electroscope the leaves will <u>not</u> diverge, and Faraday even went so far as to construct a huge tinfoil cube, into which he got together with all his electroscopes, and then closed the door, and his assistants outside charged up the cube until it was fit to bust. Nothing whatever happened to Faraday or the 'scopes.

How do we explain this apparent absence of inductive action inside a charged conductor?

Well the theory is this:-

There <u>is</u> inductive action all right, but since the induc<u>ing</u> charge is <u>all round</u> the neutral body, (referring to -vely charged can above), the +ve induc<u>ed</u> charge on the N sphere is <u>equally pulled in all directions</u>, and therefore stays where it is. Similarly the -ve induc<u>ed</u> charge is equally <u>repelled in all directions</u> and therefore also stays where it is.

Therefore to all intents and purposes the sphere remains neutral.

But suppose we earth it? (i.e. the sphere). <u>Now</u> the -ve induced charge can escape to earth and the +ve induced charge become apparent. The mutual attraction between this +ve charge on the can causes the latter charge to hop inside to say How-do like this:-

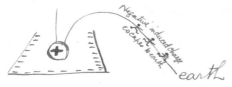

On disconnecting the earth wire the sphere remains +vely charged and we can then remove it from the can so:-

N.B. You must of course un-earth the sphere <u>before</u> removing the can. If you earthed it when <u>out-side</u> the can the +ve charge wd. be no longer bound, and would escape to earth also.

If we have a +vely charged can exactly the same will happen with reversed signs thus

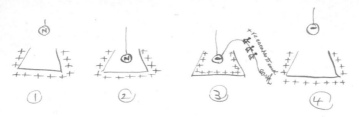

All clear?

Now we can tackle your problem. I think you have not got it quite correctly. Listen.

"... you can do it (I wish I could write as neatly as you) by giving a very small charge – say a +ve one – to another can, and putting a sphere inside it and earthing the sphere while it is in contact with the can; then the sphere has a negative charge which you can give to the other can. And when the sphere is touching the other can you earth the sphere, when it gets a +ve charge which you can give to the first can; and so on. The point is this. How does the sphere manage to get electricity of the opposite sign when you earth it? The potential then must surely be zero since the earth's potential is zero; and can you have a charge where there is no potential. And also where on earth does the other +ve charge come from? The sphere is in contact with +ve electricity".

You poor child, I'm jolly glad you <u>are</u> puzzled. It shows that you <u>do</u> understand, because what you describe couldn't happen. I suppose you took it down from quick lecture notes? Here is the real sequence – it took me the spare time of 2 days to make it out myself!

THE CELEBRATED FILM
ENTITLED
"THE CHARGE"

BY MOLLY BLOXAM AND BARNES WALLIS
FEATURING TWO CANS AND ONE SPHERE.

REEL I

Read downwards

↓

enter Can "A"
with small +ve charge

①

② enter the Sphere

who is neutral.

③

③ Neutral sphere in can
— no change.

④
 +ve to earth

④ On earthing the sphere the induced +ve charge goes to earth leaving sphere with bound −ve charge

⑤

⑤ Remove sphere after unearthing. Do not allow it to touch can.

Enter Can "B"
who is neutral
↓

⑥

N.B. his legs are insulators. That is why he is knockneed. "GLORY."

⑦

⑦ On lowering −ve sphere into "B", charges are induced on can as shown.

⑧

⑧. Allow sphere to touch inside of can. The charges on sphere & inside of can neutralise each other leaving can with external −ve charge.

⑨

Un−touch the sphere. Sphere still apparently neutral.

⑩
 −ve escape to earth

Now earth the sphere. −ve induced charge goes to earth leaving +ve charge bound as shown.

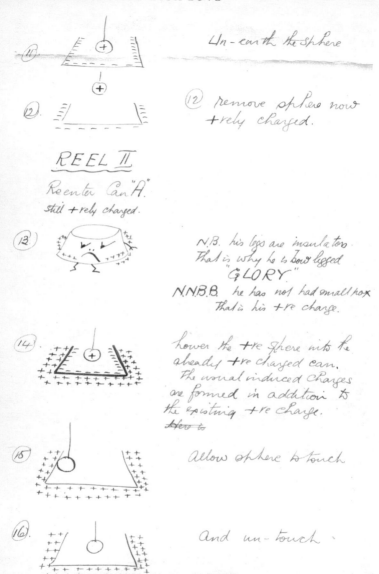

Un-earth the sphere

⑫ remove sphere now + rely charged.

REEL II

Re-enter Can "A".
still + rely charged.

N.B. his legs are insulators.
That is why he is bow legged
"GLORY."
N.N.B.B. he has not had small pox
That is his +re charge.

lower the +re sphere into the
already +re charged can.
The usual induced charges
are formed in addition to
the existing +re charge.
~~Here to~~

allow sphere to touch

And un-touch.

(17) And earth.

++ &earth

↑ sorry this ought & be <u>inside</u> the can

(18) And un-earth

(19) and remove without touching can.

Note that Sphere & Can have twice former charges.

Reel III

Reenter Can B.

(20)

(21)

lower Sphere into B.
<u>Double</u> charges are induced, making 3 negative charges on outside

(22). allow to touch inside.

(23). and Un touch.

(24) And earth.

(25) and Un earth.

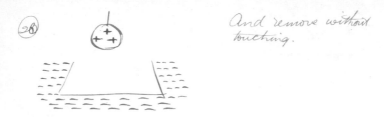

And remove without touching.

Reels IV to infinity next week.

Notice that A's next charge will be <u>5</u> times, and B's next 5 + 3 = 8 times, and so on so it increases very rapidly. I think it will go on working (theoretically) until the <u>potential</u> on the can is equal to the potential on the sphere. As the cans are so much bigger "capacity" than the sphere their potential will remain lower and I dont <u>think</u> ever catches up. But I haven't thought it out.

♦ The course ended suddenly here. Neither Barnes nor Molly wrote any further word on mathematics.

Chapter 14
LOVE TAKES THE STAGE

The Airship Guarantee Company under Commander Burney's director-ship had been formed as a subsidiary of Vickers Ltd. Barnes was now Burney's right-hand man, Chief Designer of the airship R100 to be built by the private enterprise company. Her sister ship the R101, financed by the first Socialist government under Ramsay Macdonald, was to be built at the airship works outside Bedford.

Barnes: 4th March 1924. New Cross.
My darling Molly,

I am most awfully sorry that I could not find time to write to you yester-day. I do so hope you were not disappointed when there was no letter from me? I had the most hectic day after leaving you. Soon after arrival at my office Burney sent for me, and said casually "Very busy today Wallis?" so I said "Not more than usual, Sir", so he said "All right, I want you to go down to Bristol to lecture at the Royal Colonial Institute this afternoon. I've got a cold and cannot go. Catch the 1.10 from Paddington." With that he pushed a box of lantern slides and the report of a lecture he gave 2 weeks ago to the Cambridge Aeronautical Society towards me, and turned to go on with his work. Exit me, staggering slightly. This was about 11 a.m. and I had a conference on the manufacture of gasbags till 12.30. Oh by the way, my feet were wet <u>through</u>. I <u>do</u> hope yours were not, Child dear? I felt rather guilty about you? I caught the train with about 2 min-utes to spare after having stopped to buy a pair of insoles to mop up the wet in my shoes, had lunch, and made out a list of the lantern slides in the train, and got to Bristol at 3. I drove to the place and found the Secretary busy organising the lecture room. My first enquiry was as to the title of my lecture! Burney didn't know, he's a most frightfully casual man. I shook like a leaf when I learned it was something about the Political and Eco-nomic effect of the development of the Commercial Airship on the Empire!! I rather fear that up to now I haven't cared a hoot for that side of the question, having left it to the politicians. The lecture began at 6, and I was to have tea with the Chairman at 5, and it was by this time 4 so I had an hour in which to compose an hour and a quarter's lecture. That was the time that was allowed (!) me, as I was scheduled to catch the 7.32 back to town, getting in 10.25 at Paddington.

I say Molly, it <u>was</u> an ordeal. And I wanted so badly to think of you. And by this time, what with the rush and the excitement, I had a rotten headache and beastly indigestion. And they gave me solid slabs of doughy toast with cold hot butter for tea!

I had cold feet good and proper by the time the show was to begin, and they had press reporters there too! You see I've never given a public lecture before. It was a very different matter at the school in Switzerland where they were only sort of amateur social once-a-week affairs anyhow. But do you know, I shall never be so frightened again, for I have discovered a most marvellous thing. You just give up all control of your brain <u>entirely</u>, and speak automatically. Its <u>perfectly extraordinary</u>, for in some most wonderful way, the jolly old brain carries on by itself. Effortless, rapid, sensible, out pours an even flow of words. So effortless, that I have I think two great faults. The first that I speak with monotonous continuity, never pausing, and never varying the tone, and the second, that I go far too fast. I had one moment of intense panic, when I focussed my eyes on somebodys face, and for some reason tried to <u>think</u> what to say. The minute I did that there wasn't a thought in my head! I quickly stopped thinking, and the brain got to work again. After that it was all plain sailing. And more wonderful than all; while the old subconscious is spouting away, the conscious is able to think of other things. Such as "Hullo I'm standing with my hands in my pockets that wont do", and then "I'm scratching my neck" and then "I wonder how time's going, I'll take off my wrist watch, and put it on the table so that I can see". And sure enough I did. I <u>do</u> think its all so wonderful, and absolutely nothing to do with <u>me</u>. I'm not being conceited, Molly, because its <u>not</u> me. I <u>couldn't do it if I tried.</u> Apparently it was successful, for the Secretary told me it was the best lecture they had had!! He little knew how it was done!

They had a taxi waiting, and I just caught the train and oh my poor head. And there was no restaurant on the train, so I got no dinner, but did get a few sandwiches, and some whiskey, blessed stuff (dont tell Mummum), and got home soon after 11 or thereabouts, absolutely tired out.

Oh Molly dear, I did have such a ripping week end. And you did look so <u>beautiful</u> on Saturday night. You are always beautiful, Child, but that way of doing your hair shows up the grave sweetness of your dear face as nothing else does. Oh Molly darling, I do love you so much. It gets harder and harder to go on without you.

Molly: 5th March 1924. Hampstead
Thank you very much indeed for your letter. It was most awfully interesting, and jolly exciting too. Why shouldn't it be about yourself; there is nothing in the world that can interest me more. When I got to exit you staggering slightly, I was very excited, but when I heard the title of the lecture I just held my breath and dashed on; and when I realised that you were safe and the lecture was going all right, I had time to stop and look about me. Congratulations! I'm so glad it was such a success. I <u>am</u> proud of you – or if it wasn't you, of the part of you that did it – of your subconscious part. You've got a jolly fine subconscious, though I guess there was a little Barnes in it too; I mean I don't believe just anybody, even if they knew all you did, could have done it. But wasn't it wonderful; it is so difficult to understand – in fact one can't understand it. How could you think to yourself that you mustn't put your hands in your pockets, when if you had thought about what you were saying, it would have been disasterous. It is just marvellous, and I'm so glad you went and that you told me about it, and that it was so successful, and I'm so proud of you. I do wish I could have been there to hear it. I hope one day you'll give another lecture all in a hurry like that, and I shall be able to go and listen to you and watch you. It must be a relief to Commander Burney to know he's got you to do anything if he wants it done unexpectedly or in a hurry. In fact it's jolly fine for him having you, in every way.

But you must have been tired when you got home on Monday night. Poor Barnes, with that headache, and your feet all wet, I wonder you could even think of giving a lecture. Oh dear, I do hope you haven't caught cold; it was most awfully bad for you going about in soaking wet shoes like that. My feet were a bit wet but there are hot pipes running under my bench in the zoo. lab, so I was able to dry them. It is futile that I should have to wait till Saturday week before I can know if you have a cold, or if you are quite all right. I guess I shall have to 'phone.

<u>Thursday</u>. Barnes please, I'm most awfully sorry I seemed so abrupt and hurried when I was telephoning last night. I didn't mean to be a bit – quite the contrary, I wanted to go on for ages. But I'd left Daddy sort of suspended in mid air, right in the middle of a sentence, and I had to get back to him as quickly as I could. You see we were talking about resistances connected in series and in parallel and he had taken the trouble to find a

book of his on electricity and to see what it had to say, and he was just telling me about it. It was ripping of you to 'phone; I wish you could do it every evening so that I could listen to you talking for a little while; I do so love your voice and the way you talk. Oh Barnes, you <u>do</u> read aloud beautifully; I could sit and listen to you for ever. Do you like reading aloud to people? and do you often do it? Oh, why have you got such a dear voice and such a dear face and such a dear self.

Baba had a letter from Auntie Fanny this morning. And I shall see you again in one little week – only seven days.

Goodbye for seven days, Barnes.   Molly.

Barnes: 10th March 1924. New Cross
Oh Child, its just too delightful to think of your being able to come. Of course Molly dear I didn't think you were abrupt or hurried on the 'phone. And if you had been I should have understood. And you can be, and do, just whatever you like to me. And I <u>will</u> ring you up every night if you think it will be all right. I too have often longed to, but feared to embarrass you. Please Molly tell me if I may?

I am sending a press report of the lecture. Its a bit mixed in places. Will you keep it? It was sort of dedicated to you. Molly darling, you <u>do</u> share in all I do. At least you dont share, you <u>own</u> the whole jolly lot, because I <u>am</u> yours. Oh Child dear, I do love you so much. I daren't think how lovely it will be when I really <u>do</u> belong to you, not only just pretend to myself.

Molly: 11th March 1924. Hampstead
I should simply love to be asked to New Cross for the 29th week-end; but are you sure Auntie Fanny won't be weary with having me there such a lot? Goodness knows, though, whether I should be allowed to come; I doubt it. I say, wouldn't it be glorious. I do do hope I can, but I don't expect it one little bit.

Barnes dear, you mustn't ring me up every night. Did you seriously mean it? You <u>are</u> a dear. The family'd be most surprised, and part of it would be most worried if you did. I wish you could though, all the same.

Barnes: 12th March 1924. New Cross
<u>Of course</u> Auntie Fanny would love to have you for good and all if she could. The very idea of her ever getting weary of having you! I think we will risk your not being allowed(!) to come. Why should you not? I feel rather as if I had the plague.

I am very tired, having been dashing about all yesterday and today. Yesterday I was down at some engineering works at Hammersmith testing a model of my new Airship structure – very successful; and then was up till 12 at Victor Goddard's and today went down to the National Physical Laboratory at Teddington to discuss the model of the airship itself which they are now testing in one of the wind tunnels there, back via Hammersmith to Victoria Street, where I had a long discussion with Burney.

Yes, of course I was serious about the telephone. If ever you do have me, and I am still here I <u>shall</u> family or no family.

♦ Barnes was venturing on mild rebellion. He began to plan methods for circumventing Uncle Arthur's rules, refusing to give way on the matter of visits and enlisting Auntie Fanny's connivance. Molly, over whom nobody had ever fought before, was emboldened to get permission for Barnes to accompany herself and Baba (as a brother, she assured her father) to the end of term College Ball. Barnes continued to assume an outing to the theatre would take place on the 29th, and formed his tactics.

Barnes: 18th April [March] 1924. New Cross
I am sorry I did not write this letter last night, only for some reason I was so dead tired that I went to bed at nine oclock! Also I am quite unable to get a satisfactory solution to an airship problem, and I always find it is useless to try to force ones ideas; the best thing is to get as much rest as possible, sometimes for weeks on end, and then it seems to come.

Oh Molly, it <u>was</u> so lovely having you for the weekend. I enjoyed every single moment of it. All the savour seems to have gone out of my letters somehow, and I just want <u>you</u>. I've got another fit of furious impatience, and find myself walking along composing convincing letters to Uncle Arthur.

I will get Auntie Fanny to write asking you for the 29th. Do you think in order to avoid trouble it would be wiser for her to ask Betty as well straight off? She would of course love to have her and apart from my liking Betty most awfully, she (Betty) would not interfere between us as dear Baba does, and I would like to take her to the Immortal Hour. There's a different feeling with her altogether. Do you not think so? And we could enjoy it practically as well or just as well in a slightly different way, as we could if alone. Could you let me know by return, so that I can arrange for the invitation and get the tickets?

Apparently Barbara has been "looking after us" from a sense of duty. She mentioned the matter of our being alone to Auntie Fanny on Saturday. I am most thankful the position was discussed between them, as I do not think Baba will worry us in future – she hasn't <u>wanted to</u> in the past, and I have got to that state, that if any person unwarrantably attempts to interfere between you and me, it will form a barrier which will take many years to live down. It is very questionable if <u>anyone</u> has any right after two years to separate us, but as Auntie Nellie says, I suppose I must force myself to acquire the "impossible virtue of patience".

I will come up to Hampstead to lunch [for the dance] and change there if it is in anyway possible. <u>Please</u> Molly, may I have the first two, and the one <u>before</u> and the one <u>after</u> supper, and the <u>last two</u> dances <u>at least</u> on Saturday. The others I must pray for later I suppose. I <u>am</u> so looking forward to the dance.

Molly: 19th March 1924. Hampstead
It is a jolly good idea to ask Betty as well. You are quite right; there is an entirely different feeling when Betty is there. The only trouble is that school for her doesn't end till April 4th, so I don't know if she would be allowed to come, and if she were allowed to, we should both have to come home on Sunday because she has to be at school at 9. Oh dear, if only we could say can I come alone till Monday, and if not can Betty come with me till Sunday. Because you see I'm not perfectly certain that I couldn't come alone; perhaps with lots of persuasion I could.

Barnes: 21st April [March] 1924. New Cross
Of course Auntie Fanny does not think it selfish for us to wish to be alone. She is a very understanding person. She is busy composing a beautiful invitation to you extending from Saturday till Wednesday. Its simply bristling with the most unanswerable reasons for coming, among them being:-

1. On Saturday we go to the theatre.

2. On Tuesday the two boys [his nephews] come from Canterbury and sleep the night.

3. On Tuesday I am purposing, at Fathers request, to give a lecture on Airships to the local doctors! If possible we are going to get a lantern. I shall simply hate doing it – one always feels such a perfect ass when talking to a <u>few</u> people. For some reason a hundred or two is infinitely easier. One seems so much more impersonal.

If all this does not succeed in persuading Uncle Arthur, then we must fall back on Betty. But here is a word of worldly wisdom, (which some day you may use against myself!)

If you wish to gain a point, but would be willing (faut de mieux) to accept a compromise never mention the compromise until you are quite sure you are absolutely and finally beaten on your original point.

To illustrate:-

If you go to Uncle Arthur and say "May I go alone till Wednesday (!), or may I go with Betty till Sunday" the objectives, in their order of desirability from your (and my) point of view are:-

1. To go alone   (desirable)
2. To go with Betty   (less desirable, but still desirable)
3. Not to go at all   (most undesirable!)

While the objectives from Uncle Arthur's point of view in order of desirability are:-

1. Not to go at all   (desirable)
2. To go with Betty   (undesirable but still not impossible)
3. To go alone   (most undesirable)

Now you will find that if you present all three alternatives at once – (you may not mention no: 1. above, Uncle will put that in for himself) the battle ground will inevitably shift to a discussion of 1. and 2. – in no case will you be able to hold the argument to 3.

If however you do not mention 2. it is unlikely that Uncle will put it in himself, being undesirable, and the battle will be fought on 1. and 3. You may be defeated – come again next day if you can on the same ground.

Only when quite sure of defeat on 1. and 3., do you bring in 2. Your chances of winning on 1. and 2. are now increased, as the winner of the 1. and 3. battle is probably feeling some of the generosity of victory, or the remorse of hard-heartedness whichever way he is constituted.

I hope you dont think its horrid to write like this? Living is an art, in which one is constantly having to pit one's brains against someone else's. The person who studies however carelessly, the battle ground before hand, is enjoying and exercising this art, and I cannot see anything unsporting about it. One might as well say it isn't fair to train for a rugger match.

And after all, I've been playing the game of life against Uncle Arthur for the last year or two. He has had the making of all the rules, but, while keeping to his rules I have been conducting an elaborate campaign to win you. Perhaps more elaborate than you realised. Poor child, oh Molly, darling one, I do hope you will never have cause to regret it. Anyhow, no one can now say that I have caught you unawares. You are the open spoil of my bow and arrow as it were. I dont mean that, I mean rather that I have won the great prize of your dear love in open combat, and can be very very proud of you before everyone.

Oh, I do hope you can come. All part of my wicked campaign again of course!

Have you been having a jolly good week? And can I now have <u>every</u> dance please?

```
 – – – – – Then every dance but one? – – – – – – – –
 – – – – –    "     "     "     " two?? – – – – – – –
 – – – – –    "     "     "     " three??? – – – – – –
 – – – – –    "     "     "     " four???? – – – – – –
 – – – – –    "     "     "     " five????? – – – – –
             – – – – – – – – – – – – – – – – – – – –
```

"Oh, <u>please</u> Molly" ... "Oh bother Hurley" ... "oh bother Hugh" ... "oh bother conventions" ... Oh <u>please</u> Molly" and SO ON.

Molly: 21st March 1924. Hampstead
I shall remember your advice and not ask for (2) until I am <u>quite</u> certain that (1) is no good; as a matter of fact I had decided not to, anyhow. Of course I don't think it's horrid of you to write like that. What you say is very true; and after all why shouldn't we; there's not the least harm in it, and its perfectly futile to think that it will make any difference to me in the way of being cured or not being cured – I am absolutely <u>incurable</u>; and it only means that I shall have that much less of you and enjoyment in my life, and I want every bit of you, and therefore the other, that I can have. Oh, won't it be lovely – to see the Immortal Hour with you! Will we go in the evening? And shall I really hear that lecture on airships, and see Sherard and Hugh? [the nephews]

About to-morrow evening – I s'pose I ought to be very gracious and rather doubtful – Well, Barnes, I might possibly be able to manage two, one in the first half and one in the second, but as for the supper dance and the one after – remember Hurley and Hugh – no I really don't think more than two would be possible; besides how do you know I'm not longing to dance those dances with Hurley or Hugh or – Oh my dear, of course of course I will dance those six with you, and as many more as you want and can conveniently manage, you shall have. I quite agree – bother Hurley and the others; never mind we will have as many as we can.

I have had a long talk with Daddy and Mother. And I can come next week-end till Wednesday morning! I will tell you all about it to-morrow.

♦ Barnes and Molly were in paradise at the dance. For the first time, Barnes could hold Molly closely with propriety, although as ever he was not satisfied with his own behaviour. The joy of the evening provoked Molly into a mild rebellion of her own. She asked permission to attend a dancing school in Victoria. Mr Bloxam was suspicious; it was too close to Barnes's office for his peace of mind. In spite of Molly's anger, he insisted on a class in Hampstead instead. Molly bravely demanded the reason for her father's decision, but Mr Bloxam did not give it. He took his cue from his father, Professor Bloxam, an exemplary Victorian pater-familias, loving, controlling, family-centred. Arthur was the only one of three sons to marry, and then not until the age of 36. Of the Professor's three daughters, only Fanny married, at 52. Barnes had had hints from both Fanny and Auntie Nellie about the influence of their father on Arthur's attitude. Old Professor Bloxam had kept his family close; but times had changed, and Arthur agreed to allow Barnes to ask for an answer from his daughter on her 20th birthday, September 12th of 1924.

Barnes: 23rd March 1924. New Cross
My darling Child,

Every time I see you I seem to long for you and love you more than ever. Today I feel particularly bad. I've got that awful ache-y sort of feeling. Do you know it? Quite a physical sensation in one's inside. Child dear I do love you so. And thank you ever and ever so much for asking me to the dance. I do think it was ripping of you, and I did enjoy it so. If only I had had more time to talk to you. I am so sorry you were feeling tired. I wasn't a bit sympathetic, but oh, really truly I am. I never seem to be able to behave just as I wish in your presence unless we are alone – I am getting more used to that now, but in a crowd, I always seem to make a mess of things. Its partly I think due to the difficulty of being natural. To be natural with you now means to gaze at you like a large and faithful dog gazing at his mistress. In attempting not to show this, I sometimes adopt a tone of hateful dont-care sort of flippancy, and loathe myself afterwards. And oh Child dear, I didn't mean to make you feel uncomfortable in that one-step. I said "Sorry" in a dry, unsympathetic kind of tone, just as if it had been your fault instead of my own, and just hated myself for it afterwards. Darling Molly, I'm not a nice fellow, but vain and selfish and conceited but if you will help me, and give me time, I will so try to be better for your sake. I do so want to be always gentle and tender, just living to please you, and not myself. Please Molly, be patient with me, and dont be hurt by me. If you could just remember that I can be taught, and am longing to learn from you it might make things easier for you to bear, and you wouldn't feel miserable because you had fallen in love with a hopelessly selfcentred man. Please Molly? If you will just let me see when I have been unkind or thoughtless, you will I think find me very quick to feel your thoughts, and Molly I never mind saying when I have been wrong, and coming to beg your dear forgiveness.

Oh Molly, dearest, dearest, dearest of all girls. It is a rum sensation being in love. I wouldn't change places with anyone in the whole world. And oh, little wicked One, why didn't you tell me you had done very well in all your exams, the results of all, physics excepted, being known?

Dearest Child, I have been thinking carefully of as much as you had time to tell me of Uncle Arthurs talk with you, and, I do perhaps feel that it will be best to do as he wishes, provided that, on your 20th birthday, if you wish, we may be formally engaged? We must talk it all out when we meet. I will do nothing that you do not wish for.

God bless and keep you, my own dear dear Molly, you know I love and worship you dear One with all my heart and soul.

Molly: 24th March 1924. Hampstead
I know just exactly how you feel, or how you felt on Sunday; it's the same with me. It feels as if part of you – a big part – were at New Cross, and you don't know how to carry on without it.

I haven't done at all well in the exams – quite the contrary, I have done very badly. I had 48, 58 and 68 for zoo. botany and chemistry. The zoo didn't matter so much because Professor Watson told us that the paper was a very difficult one and above Inter standard, and nobody got very high marks; but the Botany was very bad indeed. Physics I only know I haven't failed; he hasn't told us our marks and I don't suppose he will. I'd have told you the results before, only I didn't have an opportunity. Oh dear, I wish they weren't so rotten; it isn't that I haven't been working. Please Barnes, don't mind very much.

It's funny, I was most awfully tired at the beginning of the dance, but by the end I was feeling quite the contrary; and when I was in bed I felt so untired that I stayed awake for a long time. And when I lay down the clock kept on saying over and over again "Barnes, I love you, Barnes I love you" until I went to sleep. It was the clock, mind you.

You never have been unkind or thoughtless or selfish or anything to me, you are just wonderful and dear and lovable and I – I wish I were my clock. I would tell you, really Barnes, but there has never been anything to tell you yet, and I don't suppose there will.

Yes, it is a delicious sensation discovering one's feelings. I think that is the very nicest feeling, to know that you and I know each other so well that we can do things together almost as if one of us were doing them alone. Oh bother, that is the most absurd and impossible sentence; I don't know if you understand in the least what I mean, but the wretched post is going in three minutes.

Goodbye Barnes till Saturday. You are beloved.

Barnes: 24th March 1924. New Cross
Heres jolly old Monday passed away. Tuesday, Wednesday, Thursday and Friday hurry along <u>please</u>. I say Molly, do you get more pleasure or more pain out of being in love? Its a queer mixture of both I find, only even the pain is enjoyable, being for your sake.

I can't think of anything to say Molly, except how much I love you, dear, and I suppose you must be getting tired of that. For Molly, one human being, this side up with care. I feel as if I simply belonged to you – I wish I did in reality. It must be frightfully nice to be yours, just to live and work for you, to do your will and obey all your wishes. Do you feel very important at having one complete man-power at your disposal? Like having a new bicycle sort of feeling?

Heres jolly old Tuesday passed away. Wednesday, Thursday, Friday, hurry along <u>please</u>!

Wednesday. Here's jolly old Wednesday nearly gone. Thursday and Friday hurry along PLEASE. I am writing in a little restaurant in Chelsea, waiting for my dinner, as tonight I am a gallant sergeant, and am going to drill at the Chelsea Barracks.

Confound these tube men; if a settlement is not reached very soon, I shall have to fetch you in a golden chariot after all.

Thurs. Child dear, I haven't had a moment to add anything since I wrote the last.

Since the bus strike is certain not to be over and a tube strike is possible, I will definitely arrange to fetch you, as early as may be, on Saturday, or brilliant thought, why should I not fetch you on <u>Friday</u> night, when tubes will still be working?

◆ Wednesday evening was Territorial Army drill evening, and while the gallant sergeant waited for his meal he took his lady's letter, written on the previous Wednesday 19th March, from his pocket. Molly's latest letters went with him for love and for a talisman, ever there to keep her close. On the back of this envelope Barnes scribbled the drill schedule until the end of May. Letters from a loved one have accompanied many a man on active service, an experience that Barnes still regretted having missed.

March	26th	Descr^n of Gun. Gun Drill.
April	2nd	Breach. Foot Drill.
"	9th	Buffer. Rifle Drill.
"	16th	Gun Drill. Theory of Sights.
"	23rd	Sight testing. Conduct of Fire.
"	30th	Platform drill. Instruments.
May	6th	Instruments drill. Calibration.
"	14th	Illegible [Direction of Parties?]
"	21st	Revision.

Molly: 3rd April 1924. Hampstead
Yes, it has been the most glorious [weekend] and I don't believe I have even attempted to thank you for Saturday evening, I enjoyed it so very much. Only, I enjoy the times when it is just you as much as the times when it is the Immortal Hour and dinner first. Barnes, I am not down cast, at least not very much now. It is only that just at first after one has been with you one does miss you so very badly; it feels most awfully incomplete and even lonely, which seems rather absurd when you think what a lot of family there is. But I shall see you again soon, and I s'pose September isn't awfully long to wait – at least, one minute I say that I just can't wait till then, and the next I tell myself not to be stupid, and it's quite easy and very good for me to wait; and always when I have been having a discussion about it with myself, I end up properly and sensibly; but Molly is a dreadfully intractable person, and needs a good deal of keeping in order. And I am glad we are going away for the summer because it will make me broader and more educated and better for you.

Friday Barnes, you do make a ripping uncle; I just love you with the boys, and they are a goodly couple. They came upstairs and helped(?) me do my packing on Wednesday morning. I got hairpins and nail and tooth brushes and fountain pens and pencils all mixed in the most hopeless confusion; and the precious pair put in everything they could find whether it belonged to Auntie Fanny or me – all amid much laughing and excitement. Altogether it was a very uproarious time, and I enjoyed them both very much.

One does get a lot of pain out of being in love, but the pleasure far far outweighs it. I agree it is a very rum sensation being in love, but at the same time it is a very very pleasant one and also a very wonderful one. It is inexpressibly strange to feel you have "one complete man-power" at your disposal and also very beautiful and a little frightening. Oh Barnes, I <u>will</u> try and do it properly. Your feet and hands send you their love – no, I suppose I mustn't say that, I mean my feet and hands which will be yours in September – and I suppose that's just as bad – oh never mind; and the rib is quite all right, he just loved being squashed.

Barnes: 7th April 1924. New Cross
My dear One, there's no need to begin thanking me for anything. Is not all yours? And I'd heaps rather you didn't. I'm jolly glad you like the times when we are alone just as well. I like them best of all, but still its good to have little jaunts together sometimes. And its perfectly ripping of you to like the little insignificant things. But you are quite right, they do count more than anything. Its really very easy to do the big things in life, and awfully hard to do the little things.

I've got one of the horrid spasms when I feel too utterly unworthy of you for words. Its quite the biggest pain that loving you has brought me. You are so young and fresh and beautiful and I'm so much older, and ugly and experienced. Oh Molly dear, I do realise how very far below you I am. I can't <u>think</u> why you ever should care for me. I'm very sure <u>you</u> need not be frightened about not doing it properly. Its me that has to feel worried.

Molly: 16th April 1924. Hampstead
About the dancing – I am not going to Victoria. I told Daddy I was going there and he said I wasn't, but I would have gone all the same because he had no reason at all; he didn't think it was too expensive. Only afterwards Mother asked me not to go. She quite sees the unreasonableness of it but she wants me to make things as easy as possible for him. You know what his father was like, well Mother says it is most awfully hard for Daddy not to be exactly like him, and that if it makes it easier for Daddy if I have lessons in Hampstead, will I have them here and not go to Victoria. I can't think what difference it can make to Daddy if I do go there, but I shan't because of what Mum said. And you don't mind Barnes? You see she says he simply hates the thought of me going (not to the dancing lessons, to you next year I mean) and I must realise how much he is giving up – not

only of me but also of all his cherished ideas, such as that we shouldn't begin to think of doing things like this (again I mean you) until we are 24 or 25. So I am realising, and therefore I shall go to a place at Hampstead. Of course if he were too unreasonable, and if he is next year, I just couldn't help it, I should have to do what he didn't want.

Good bye, my Barnes.

♦ Easter fell on April 20th, bringing with it the anniversary of their first meeting. Barnes turned sergeant and set off for training camp. With Molly in his heart, he could relish the male company and the controlled, energetic life.

Barnes: 19th April. 1924. Black Down Camp, Sussex.
My darling Child,

Just a hurried line to catch the post this evening, to give you all my love; and let you know that we are having a jolly fine time with most glorious weather. My face is as red as a beetroot with the sun and wind. I had to leave without getting your dear letter – never mind, it is something to look forward to on getting home. I came away in such a rush I brought no sleeping things and no notepaper.

Childie dear, I think of you always and am writing a proper letter.

Goodbye my own Dear One, Ever your devoted, loving Barnes.

Barnes: 22nd April 1924. New Cross
Many happy returns of our meeting-day to you. Isn't it jolly nice having it on St. George's Day? "Was there really a St George?" I neither know nor care since I have discovered Mary.

Dear One, I sent off some of your favourites on my way home. Another jolly coincidence is it not, that Roses are St Georges day flowers too. Whatever the book says, these particular roses mean, "Molly, dearest Girl, I love you, worship you and adore you, and I will if you will let me, love you and serve you tenderly and faithfully all my life, and with my life".

Of course, dear, I understand about Victoria. I think you have done quite rightly.

I think it most likely that whatever happens, I shall now stay in London, so as to be in touch with Burney and the Admiralty, so I'm jolly pleased.

Child, why do you say "next year". When (and if) we are engaged, we shall do just exactly what we please, or rather, what you please, next September?

Molly, I <u>do</u> so love your ending. Dear One I just am yours, every bit of me, to do with whatever you will, and oh Molly how I love you dear.

God bless you, my Childie dear. Ever your loving devoted Barnes.

I <u>am</u> sorry, this isn't in the least the letter I meant to write, but I <u>cant</u> write to you in a hurry. Hurrah, I've known you two years.

◆ A second note from camp was tucked into this letter, a note which Barnes had not had time to finish, much less to post. Drilling and parading called, and amidst hurry and emotion Barnes had added a year to their joint life, writing 1921 in mistake for the date engraved on his heart – 1922. Five roses arrived to greet Molly on the blessed day.

Barnes: 19th April 1924. Continued 20th Black Down Camp.
My darling darling Molly,

It seems funny to be starting another letter to you just after finishing my last. But Molly dear, I do love you so, that I must write. I'm sitting on my bed, in shirtsleeves and since writing the other letter I've had the only apology for a bath that is possible. There are as usual nothing but taps in an open shed, and I got the Battery sergeant major to come and pour buckets of cold water over me. I was <u>pink</u> all over when he had done – you know the glorious glowing feeling one gets.

<u>Easter Sunday</u> I thought of you Child dear early this morning, and tried to be with you in spirit. There was no chance of going to church at all, as we have been parading and drilling all day. I am <u>so</u> glad that I was able to go to early service with you. Just think, we're nearly two, Molly, counting April 1921 as our birth day. It will be jolly to grow up together.

Molly: 23rd April 1924. Hampstead
Thankyou, thankyou for them, how exactly like you to send me them on our second birthday. I love all five – the one dear white one and the four dear red ones. The littlest and babiest red one is leaning over and kissing my forehead as I write. He is most distracting because every now and then I have to stop to look at him and kiss him, he is so very sweet. Also I can't see to write straight because his beautiful leaves are in the way, so you must blame him not me.

It was funny, I dreamt about you hard all last night. I dreamt I was in bed in the sky parlour and it was that Saturday night two years ago – my first night at New Cross; and I was most awfully excited because I was going to see you for the first time, only the queer part was that I knew you already quite well. Then in some inexplicable way I found myself sitting at the piano in your drawing room (you remember there was a piano there then) waiting for something – I don't know what. Then suddenly I heard your voice outside in the hall saying "Molly, roses" and I jumped up and ran to the door, but just as I was opening it I woke up, and there was no you and no roses, only my dear little Duck sitting on my table staring at me and saying good morning to me. And when I got down to breakfast this morning I saw a long box, and directly I read the word Victoria on it I guessed whom it was from, and there were my real, beautiful ones inside! Oh Barnes, it was dear of you, and it was such a ripping idea to send me flowers on our meeting-day. Yes, it <u>will</u> be jolly to grow up together; we are really getting quite old – two years is a good age for two people to be, together. It is jolly nice having it on St. George's Day; and I'm very glad that roses are the flowers. All the morning while I have been sitting here, trying to do some zoo. mine have been telling me your message over and over again until I know it by heart. You choose very sweet messengers to give a very wonderful message, Barnes.

I thought of you when I was in Church on Easter morning. On Friday at about 6 o'clock I woke up and stayed awake for about half an hour and imagined you getting up and ready to go, but I was so awfully sleepy – it was a hopelessly unsleepy night – that I turned over and slept till eight. Did you see the moon on Sunday night? She had just risen when I went to the post, and I think she must have been full, she was so huge and round and beautiful. I went into the garden afterwards, and it was full of the scent of daffodils and hyacinths, and strange weird little shadows, and mysterious little rustles; and when you knelt down and put your ear to the ground beside the roots of the oak tree, you could just feel how excited the earth was because of the spring and everything going to happen.

Barnes, I am so awfully glad about Commander Burney and the Admiralty, but I believe I'm even gladder that you are going to stay in London. Does that mean there will be no Airship works at Bedford at all, or only that you won't be there?

Chapter 15
TWO YEARS ON: THE CHANGING FUTURE

A flood of emotion had been released in Barnes, pressing him towards meriting and winning Molly's love. But his creative energy, his achievement, even his affections, had been absorbed in his work for many years; he could not now lay these purposes aside, nor did he wish to do so. He fought for his love and for the outcome of his design. Molly had one purpose, to reconcile her father to her decision; her ambition to work in the world had faded. She knew her fate; Barnes, although he prayed, hoped and battled with uncertainty, did not know his for sure.

Barnes: 27th April 1924. Rudgwick, Sussex.
How I did envy the littlest and babiest red rose who was able to lean over and kiss your forehead while you were writing. A faithful little messenger. How strange that you should dream like that Childie. "Molly, roses". I <u>am</u> so glad I sent them. I do feel so frightened that sometimes I may have, or shall, disappoint you. We can be together so much in spirit, and I think it helps you when you know that I this end am true and faithful to you, loving you truly dear One beyond all words. Wont it comfort you, one little bit, Molly, my loved one, my own dear girl? So please dear One dont be unsleepy, but rest gently and I will be strong and tender for you.

There is a queer complication about the Airships. The Air Ministry are determined to build a ship, so as not to allow the technique of the business to pass entirely out of their hands; so they are going to stick to the Airship works at Bedford as a Government works. On the other hand the Admiralty say they do not trust the Air Ministry's ability to build a ship, so they insist that Burney (that is us) shall be given a contract to build a ship too. The details of this contract are not yet fixed, but we shall have to get our own aerodrome. The only other airship sheds left standing are those at Howden, near Hull, in Yorkshire, but these (fortunately from yours and my point of view), have no workshops for making the parts. So it has been decided that we shall manufacture all the parts at a large half empty factory belonging to Vickers at Crayford in Kent, and that I shall have my headquarters still in London, and go down to Crayford two or three times a week. Then, when ready we shall erect the bits at Howden. I wish it was all really settled, we have been hanging about in uncertainty so long.

I have altered [the wireless here] slightly and it is working better than ever. I managed to hear Bournemouth, as well as London just now. Yesterday I got in a walk between tea and supper. On the way home I stopped at the little village inn to have a chat and a drink, and astonished all the assembled yokels by being able to tell them who had won the Cup tie at Wembley, having just heard it on the wireless before going out.

Oh Childie dear, I used to be able to write long letters to you, and enjoyed doing it. Now somehow it makes me impatient to write anything but how much I love you. Just ordinary news seems so stale and uninteresting. Do you find a change that way too? I dont mean your news is like that. Every minute of your day is of intense interest to me, from bath time when you give my message to dear Right Foot and dear Left Foot, to bed time when you do the same for my dear dear Hands. I mean that anything I do away from you no longer interests me.

I wonder if you love the idea of dear Right Foot and dear Left Foot, and my dear Hands as much as I do. And the little Rib who got squashed. I suppose any ordinary person would think it awfully silly. But you dont mind do you dear One? They always come in the same order, and in the same words, said in just the same way, and I do love it all so dearly. Its funny how one makes it all up as one goes along isn't it. Thank goodness they dont put things like that in books. It is so much more beautiful to discover it all for oneself. I often wonder whether other people in love do the same. One hopes not, not quite the same anyhow. No, thinking over the men I know, I dont think any of them would love you quite like I do. And you always say "Yes Barnes" in the same breathless little voice that I love and know so well.

Sometimes, Childie dear, you must give me time. You see a man like me is so entirely accustomed from childhood up, to planning and settling every action that it is a little strange to have someone else's will to bow to. I will soon get used to thinking only of you and not of myself. At present it doesn't come automatically as it were. You see being single does make a man so horribly selfish. But please Childie, be patient with me, and dont be annoyed with me, because I will try. Write to me soon, dear Molly, just a little note?

Ever your humble, loving Barnes.

Molly: 1st May 1924. Hampstead

Thank you very much for your letter, it was such a dear comfortable one. Really of course I know all the time that you love me, but it does make it easier when I can read it if I can't hear you say it.

And I'm not unhappy – quite the contrary, I'm so awfully happy that sometimes my great happiness almost frightens me; the real happiness is right down inside me all the time, and sometimes it sort of comes to the top and then I'm so excited I can't go to sleep for a while, but that's nice; and sometimes a sort of restless feeling comes, and that's the other sort of unsleepiness which isn't nice.

Oh Barnes, it <u>does</u> help most awfully to know that you are loving me so hard; I know I'd like to hear you say it for ever and ever. Every word of your dear letter made me feel happy and glad and comforted and safe. Barnes, you have <u>never</u> disappointed me; you do everything that I love.

Yes, I know just what you mean about ordinary news seeming stale and uninteresting; but with me what else can I say? the only thing I want to, I am not allowed to.

Of course I just love the idea of your hands and the others; indeed I don't think it's silly, though as you say, I suppose it would seem so to an ordinary outsider. So am I – jolly glad they don't put things like that in books. You and me – it's much nicer than any book.

Molly: 7th May 1924. Hampstead

I want to know something; it isn't really important, but I just want to know it out of interest. If you have two curves, how do you find the resultant curve? and when you have got it, is it of any use to you? Suppose you had two curves like this [an untidy diagram] and they were supposed to represent sound waves (I don't s'pose you ever could have them that shape) what is the resultant curve supposed to convey to you, if it is supposed to convey anything. But there's no hurry and it will do quite well when I see you.

Goodnight my Barnes, Molly.

Barnes: 5th May 1924. New Cross

I dont know what there is to say when I have seen you so recently, except dear One, how much I love you. It was so delightful to get so much of you to myself, tho' Daddy was evidently determined that we shouldn't overdo things, wasn't he? I do hope work goes well Child dear. I wish I could take the jolly old exam for you, tho' I should look funny in the Zoo. parts, not knowing a hyena from a glow worm.

I dont know why one gets that ache-y feeling – its a miserable feeling really – when one is feeling rather hopeless and cast down.

Friday I am so looking forward to seeing you on Sunday. I do hope we get a decent time together. I will tell you about the curves then, and anything else if there is anything. It's been a most awfully busy week, seeing people all day. We have at last bought Howden Aerodrome, so things seem to be looking fairly certain at last. I have got together the nucleus of a very fine staff.

♦ On the morning of May 15th, Molly received a registered letter at College.

Barnes: 14th May 1924. Vickers House, Westminster
I did intend to write a letter with this pen, but time is against me. If the nib is not right go down to the Swan shop in Oxford Street, ask boldly for Mrs Oliver, who is expecting you, mention my name, and she will spend endless trouble in giving you a nib to suit your hand. If it is not what you like, change the whole pen. I will settle with them for any difference in price.

All my dearest love Childie dear, and you are not to bother to write, because you know I love you, and am yours – ∴ pen is yours also. Q.E.D.

Molly: 15th May 1924. Hampstead
I am writing with it now, and the nib is just right. It is the dearest little pen. This isn't a letter and I'm not going to thank you; it is only a sort of business note to acknowledge the pen and to accept my own property from my own Barnes. Your proof wasn't half long enough.

Data A	pen
To prove that	The pen belongs to Molly
Proof	Barnes bought the pen
∴	The pen belongs to Barnes
but	Barnes belongs to Molly
∴	The pen belongs to Molly.

Q.E.D.

You can't think what a delightful and wonderful sensation it is having a Barnes belonging to you.

Barnes: 19th May 1924. Brighton
Thank you ever so much for your little business note. You should not have bothered dear One, but I am so pleased that you like the pen. I loved your little proof. Molly dear, if its a nice sensation to you to feel that I belong to

you, its a thousand times nicer to me when you acknowledge the owner-
ship, and treat me as what I really am, your servant. I feel more than usually servant-like tonight and just long to lie at your dear feet. At times like
this I can quite understand poor Pendennis longing for a silver collar with
"Ethel" on it, tho' I have often wondered whether, if that young lady had
sent him one, he would have consented to wear it? I hardly think so, for
he was a conceited and selfish young man, and at that stage I think
scarcely loved Ethel as truly and devotedly as I do you. Which rather
reads as if I would wear it, if you were to send a collar to me. I wonder if I
would? I say Molly, what a test of your authority and my obedience if you
did. Just fancy if I came down one morning early and found a little parcel
from you containing a collar complete with a little padlock and key and a
note saying that you ordered me to wear it day and night during your pleasure! Not a silver collar Child, too cold and hard and too 'spensive, a
leather dog collar would be much more comfortable. I dont believe you've
got the courage to risk my disobedience, so I jolly well dare you to! I say,
what a sell for me, if you do dare and I have to obey!

And how long would it be my Mistresses pleasure to make me wear a
collar, (if she can), – a year? But it doesn't matter, for I'm sure I shall be
independent and disobey, and anyhow she'll never take my dare! Nothing
like being able to hoot at one's Mistress every now and then. I <u>do</u> love
teasing you Child – its sure to be a Victory for me anyway. How funny, I
started by longing to be your abject servant, and I've argued myself into
feeling as bold as brass. Cheers, I shall be giving orders to <u>you</u> soon. Buck
up B.N.W. she's only a girl!

20th May. On reading what I have written, I dont feel nearly so bold as
I did, and for a moment thought I wouldn't risk sending the letter. I never
quite know Molly how strict and how severe you really could be, or how
much you really like holding me in subjection to you. I'm inclined to think
you're not half as firm as you pretend to be, because you're never very
good at making me be your servant and I dont believe you really enjoy it,
do you? Oh dear, I suppose I shall have to pull myself together and be masterful again.

Oh yes, that reminds me, I have already decided how I am going to
start! I went a most splendid walk on Sunday from Newhaven to Brighton, including climbing a bit of cliff – in all about 11 miles along the cliff
edge, and I then thought that when it becomes necessary to subdue <u>you</u>, I
shall make you put on a pair of very very high heeled shoes, with tight and
pointy toes, and then make you do that walk, including climbing the cliff!

I'm so glad I've found something that you dont want to do, like being made to wear high heels, because I can always hold it over you as a disciplinary measure!

Whereas there's nothing to correspond that you can make me do. Men dont wear uncomfortable things thank goodness. Oh well I suppose some men do, – lots of men wear corsets and have their shoulders strapped back to hold them erect. I was awfully afraid of getting round shouldered once and tried to make myself wear shoulder straps, but to begin with you can't adjust the wretched things properly by yourself, and when you have got them on they are so uncomfortable that you promptly take them off again. But I should think if you wore them for 2 or 3 hours every day it would be a very good thing, because their very uncomfortableness makes you hold yourself erect, so that after a month or two you could leave them off again. But you will have to have me <u>very</u> firmly in hand before you could make me wear things like that! Poor Pendennis, he little knows what trouble he may have landed me in. I'm a silly to send this letter really, but having once 'dared' you, even though you hadn't had the letter, I'm hanged if I'll be a coward and be frightened of you.

I say you could always ring me at my office Victoria 6900 you'll hear a little voice say "Vickers" and then you say "Extension 144 Mr Wallis please". There are sure to be boxes at Gower Street station or at any post office and Im there roughly from 10–12.30 and from 1.30–6.

Molly dear, why shouldn't you always end that way. I dont see that possessing me for this while prejudges the issue. I would infinitely rather feel that you had owned me for 6 months, even if at the end you rejected me, than not be yours at all. But I suppose its too great a privilege, indeed I <u>know</u> it is, forgive me Molly dear. You can't think how unworthy I feel of all the dear kindness you show me. Molly you are <u>utterly adorable</u>. I simply worship you, my loved One, with all my strength.

Molly, you dont think it silly or feeble for me to feel servant like, and to wish for and long for your dominion over me, and to be just subject to you and to be your servant? If you only realised how my spirit prostrates itself before you, perhaps you would understand. It <u>must</u> find expression somehow, and I simply cant bottle it up and be dignified and superior. I suppose one does look silly when one flops on ones knees, but the act to me is one full of worship and reverence. And mixed (shall I admit it?) with a little wholesome awe of a little person who, for the first time in my somewhat masterful career, I find myself for some inscrutable reason compelled and not only compelled, but <u>longing</u> to obey and serve.

Goodbye Childie dear. God bless you and keep you and help you in your work. And dont forget to play now and then.

Molly: 22nd May 1924. Hampstead
To: Barnes Wallis,

It is my pleasure that you wear this collar with my name on it round your neck as a sign that you are my really truly servant and will do whatever I wish – from taking me to the theatre, to eating chocolate blancmange (if I should ever want you to!) You will wear it always, except when you are bathing and when you are wearing tennis flannels with a shirt which is open at the neck, until next September 12th when I will tell you if you are to keep it or throw it away. And whenever you see or feel the collar you will remember that I am your mistress and you will have to do what I order you to do for ever and ever world without end.

(signed) Molly.

♦ Tucked into the letter was a velvet ribbon.

Barnes: 22nd May 1924. New Cross
Its just sacred to me, far to good for me to wear after you have worn it. Molly I simply <u>can't</u> wear something that you have worn. It makes me feel all miserably mean and unworthy. Molly dear look, – my letters were deliberately cheeky to you. Frankly they were written to cheek you. I admit I was experimenting, – I wished to see if you were of stern enough stuff to rule me as I <u>must</u> be ruled. For I am one of those wretched people who, if he isn't simply <u>trampled</u> on ends by turning out a sort of spoilt bully.

Dear One, I am not going to promise always to obey you…. My obedience I fear is not of the dull meek order. Oh Molly dear, can you understand what I mean…. It would be so dull, always to obey straight off. So I reserve my right to disobey your orders.

Ever your humble devoted but still cheeky servant
    Barnes.

Barnes: 24th May 1924. New Cross
Molly dear, I'm not playing – I really think it is necessary for you to form in me strong habits of service and obedience to your dear Self right from the beginning, and I welcome and rejoice in the actions that will impress your Mistress-ship upon me, and accustom you to treating me as your servant. Agree?

Molly dear, your hair ribbon was the <u>sweetest</u>, <u>happiest</u>, <u>dearest</u> thought that you could ever possibly have had. Will you let me wear it <u>always</u> if you have me? Its not a question of <u>ordering</u> me to wear that, you will have to order me to leave it off. <u>I'm prouder of that</u> <u>than anything else in my life</u>, because to be allowed to wear your dear badge is the greatest privilege a man could ever, ever have. You funny Child, how <u>could</u> you think you were being <u>severe</u>?

No time for more. I have read up some Sound and will explain on Sunday if you wish. <u>Do</u> let us be alone please Dear One, I have so much to tell you.

Till tomorrow, my dearest, sweetest, most adored Mistress,

Ever your humble, when-he-chooses-your-obedient, and <u>always</u>

your devoted, loving, Servant.

Molly: 27th May 1924. Hampstead
My Barnes, my very very own Barnes; I don't care what anybody says to the contrary. You mustn't come here next Sunday, and I don't want to be asked to go and stay at New Cross alone after Barbara and I have been there. Barnes don't think I'm saying that quite glibly and cheerfully; it has taken six long weary hours, half of them asleep and half awake, to decide that.

I will tell you what happened last night. I lost my temper, Barnes, really truly and completely lost it. Ever since I was the merest kid I've had a dreadful temper, and I've been trying to squash it. Oh, I do wish I hadn't got one; it must be so easy to be Baba for instance, who never has to hold herself tight to prevent her flying into a rage. Anyway last night I asked Daddy if you could come here on Sunday; and he said no. And then we had a long argument. I asked him why he had such an antipathy to you; and he said he hadn't – quite the contrary, if he didn't think you were a good, upright honourable man he would never trust me to you. He said that the reason why he didn't make such a fuss about Barbara and Carruthers as he did about you and me was because Carruthers is so much younger, and he himself couldn't think of getting married for four or five years, during which time Baba would have plenty of opportunity for changing her mind. But you know even now I can't get at his reason for not wanting you and me to see each other often. All I can gather is that he has made the rule and is going to stick to it, which is simply obstinacy.

So I went on contesting the point, and up till then it was quite all right and I wasn't in a temper a bit. And then he said something which made me simply furious and I rushed up into the spare room and sat in the dark getting angrier and angrier. But Daddy came upstairs and he was most awfully sorry and he said he didn't mean that at all, he hadn't meant what he seemed to, and would I forgive him. Then he asked me to do what he wanted because it hurt him when I didn't. I simply can't understand why it should, but Barnes he is old and pathetic – at least he was then – and we are young and strong, so I said I'd do what he wanted till next September. After that I can't help it, if it hurts him, it will have to. Barnes I'm not cruel, but after all it is our life – yours and mine – and we've got to live it, and I don't see that Daddy has any right at all to interfere. I'm not sure that he has a right to do as much as he does now, but when a person asks you to do a thing in that way, it makes you feel different from what you do when they sort of order you to do it. Anyway I'm going to do what he wants because it hurts him when I don't – but only till September. About going to stay at New Cross; of course I never actually mentioned that because I haven't been really invited yet, but from what Daddy said I'm practically sure he'd say no. Of course I could just ask him but I don't believe it would be any good.

♦ In the midst of her diatribe, Molly put her finger on one of her father's fears. Barbara did in fact change her mind long before Carruthers, the young man in question, had achieved any situation or financial stability and the courtship came to nothing.

Molly: continuing 29th May 1924.
<u>Thursday evening</u>. Barnes, it wasn't really anything very bad that Daddy said, only I was sort of worried, and it hurt coming on top of everything. He said that it was always after seeing you that I was most trouble and bother, inferring you see that you made me horrid, or at least I thought he meant that. To begin with it isn't true, because generally after you've been here I don't ask for you to come for some time; and then I <u>won't</u> have people saying things like that – it's mean and unfair and – oh bother, I'm getting heated again. I know its most awfully childish to lose one's temper, but I honestly don't believe I do it as often as I used to before there was you. Why, I'm doing my very best to be decent and nice because there's you – I've done so ever since you first began; and then when he said that after I'd seen you I was horrid, I was furious. But Barnes there's no need to bother about it now, Daddy told me he didn't really mean that and he said he was sorry and so did I and it's all over.

This is a selfish letter. Barnes don't think I'm putting Daddy before you. I s'pose I am just for a little while, but it won't be for long, and it isn't that I love Daddy anywhere near as much, but he has done a tremendous lot for me and I'll please him just for three months. The worst part of it is that it hurts you. Don't think I don't realise it; I know just how it feels to want somebody very very badly every minute of the day, until your whole being is one big want, and to feel so lonely in the night that you have to stop yourself calling out Barnes, Barnes, Barnes, aloud in case you wake up Baba.

Oh Barnes, I almost wish I hadn't written that. It isn't like it often, really truly it isn't. I know all the time that you love me, and that makes anything and everything all right. I only said it so you wouldn't think that I didn't realise what it feels like for you.

And Barnes please don't worry; I'm not unhappy; Daddy is really most awfully ripping and kind to me, and he means to be to you, though I agree it's rather a weird sort of kindness. And you do understand why I'm doing what he wants? We can manage to hang on till September.

Goodbye. You <u>are</u> my Barnes whatever Daddy says. Molly.

Barnes: 30th May 1924. New Cross
My Poor Childie, you can not think how sorry I am that you should have all this distress and bother on my account. I know so well how miserable such scenes make you feel, and how much they take out of one too. But you mustn't take it too much to heart Molly dear. On the eve of an exam everyone gets hot and bothered and irritable and cross. I dont believe you have such a temper as you think, and anyhow if you have, I just love you all the more for it. I can so well remember Molly darling, my own dear Mother coming to me after a childish outburst and telling me how if only one used it properly a strong temper was a great gift, and not the curse that most people tried to make you think it was. I cant recall all her words but the substance has remained with me and helped me all my life.

What is really meant by what is called a violent temper? In most cases you will find that what are termed "bad tempered people" are the most faithful friends, the truest servants, the hardest workers. I rule out certain types which might perhaps more properly be classed as mentally deficient. I mean ordinary folk like you and me, who at times, for one reason or another are swept by gusts of passion, which find expression in one way or another, more or less violent.

Well, Mother's idea was that it was all the result of great depth and strength of feeling, so worked upon, that in the inexperienced it led to loss of self control. That is bad, but if only you can realise that it is your <u>good</u> qualities that are the cause of it all, it gives you heart and strength to go on and find out how and when they lead you astray.

It's because you love freedom – you feel restraint <u>too much</u>; because you hate injustice – you feel the covert sneer or unwarranted insinuation; because you love truth, you most bitterly resent the slightest imputation that you have not acted quite sincerely.

All these feelings are <u>good</u>, but they must not be allowed to obscure our sense of proportion. Properly directed they will make you one of the "salt of the earth". The possession of them in all their strength and intensity is a thing for pride and joy. Only remember when trial comes that if anything, you feel too acutely and make allowance accordingly. Say "Steady Feelings dear, dont run away with me, like Barnes, you are just my <u>servants</u>". And you will find that, like me, when handled firmly they are docile faithful slaves.

I know its easy enough to preach, but very hard to do. Its just the same sort of crisis as being in a motor accident, or steering an airship. If, when things begin to happen quickly, you can keep your head and go on consciously directing your thoughts instead of letting your mind become a blank and acting in panic, all will be well. Never lose control of your thoughts, and you will never lose your temper.

Molly dear, I know it all so well – I'm a bad tempered person – its all <u>part</u> of the reason why I fell in love with you I think. I almost instantly realised, somehow by instinct I suppose, that here was a stout hearted, great souled little person who was fighting much the same battles as I fight, losing sometimes, just as I lose, coming on again for another go, just as I come.

You simple cant think, little Great Heart how much I love and adore you for it all. Why Molly, you may take a stick and beat me if you like, I'll know just how you feel (and I shall probably deserve it) and just shake myself and come and kiss you, because I shall know too that if you didn't love me you wouldn't bother to beat me.

Its a feeling I've had all along my Childie, that for some reason I can just take you by the hand and say "Childie dear, <u>I've</u> been there too, I know how it feels, mayn't I help you along the way?

And look here Molly, of course I understand. Where I am concerned it is for you to make decisions. Thats not <u>my</u> job – I'm not one who says "I must decide this, and that; I dont think you right". How ever do you think I get thro' my work if I dont delegate 90 per cent of my responsibility to others? The men <u>know</u> they have responsibility and never fail me. Now

Childie all the responsibility for this show is yours. You <u>may</u> be only nine-teen; but if you <u>know</u> that you have the power, and with it the responsibil-ity, you will find no difficulty in making wise and right decisions. Its nothing to worry over; you know I am utterly loyal. "What you say goes" as the Americans say. I have the most utter and complete confidence that whatever you decide is right – I mean in the big and serious things. Would I give my life to you if I hadn't? And just because I <u>do</u> trust myself to you, you will find <u>you</u> have the selfconfidence and the ability to act.

Molly: 1st June 1924. Hampstead
I got your letter yesterday morning, and thank you very much indeed, Barnes. How I should love to have known your Mother. It has made me feel much less despairing about being quick-tempered; and you can't think how comfortable it is to feel that you know what it is like too. It is like head aches; you feel that a person who had never had one would find it practically impossible to understand how you could feel tired or cross because you had a bad headache, while a person who had had a bad one some time would be able to understand and sympathise with you.

Barnes: continuing 2nd June 1924. New Cross
My Mistress, thank you most awfully for your very welcome little letter. I do hope the work is going well? I will write for Thursday. In the meantime dear One, just dont bother or worry. You know that whatever happens I love you with all my strength. But I am perfectly certain that you will simply walk through. Dont be discouraged by the first day. When I took Inter it was pure maths all day first day. I think I did one sum in the morn-ing and about one and a half in the afternoon, and came home feeling completely done for and wretched. But I got through. And with you, they dont only go by this exam. All your past year of good work is taken into consideration.

Post time and I must stop. I have got 4 ripping seats for June 14 – front row of dress circle, so just fix your mind on the time when all the worst will be over, and you and I are sitting together seeing Romeo and Juliet.

Barnes: 4th June 1924. New Cross
My dearest sweetest Molly, this is just a line my dear to send you all my love and duty, and all my longing for your good fortune in the exam. Childie dear, if I could do it for you, how gladly I would.

My own darling Girl, just remember that to me who love you so dearly, the exam means nothing. Just as <u>you</u> might care for any one in success or failure, so I care for you, with all my strength and heart and soul. Failure or success means nothing – its the long effort that counts to those who know. You and I know that you have striven jolly well, and have more than done your best.

Pray God you will be successful – I <u>have</u> prayed for you Dear Childie; so much the better if you are, for it is nice to be successful, but often dear One still nicer to fail when failure draws you, as nothing else can right into the hearts of those who love and worship you. Just keep calm, and try to enjoy the whole jolly show, and you will find it isn't half so bad.

Goodnight my own dear One. I will pray for you and think of you all the day.

I'm just <u>yours</u> Molly dear, when you choose to take me. – Kept in cold storage as it were – (sorry couldn't help a joke).

♦ This letter arrived in one of the well-loved, large, grey, crinkly envelopes with the darker grey lining which had brought the post from Switzerland eighteen months before. On the back of the envelope was written "Other envies all exhausted. Old ones <u>so</u> pleased and proud".

A second old envelope followed quickly, bringing sympathy and encouraging advice for a day of physics, morning and afternoon.

Barnes: 6th June 1924. In the train to Rudgwick, Sussex
My own sweetest, adorable Molly,

I suppose I've no business to say "my own". I know you are not, but it is so comforting to write it. That the rest is true anyone can see for himself, and no one can stop me adoring you to my hearts content. So I do, Molly dear, from your dainty foot to the top of your glorious head, I adore and worship every single part of you. My darling gentle Hands, with their soft pink motherlike palms; the dear firm curve of your proud neck, the little nestling curly hairs hidden where only I may look and kiss; best and most revered of all your sweet dear mind, Oh Molly how I long for you.

But I didn't mean to worry you like this – I only meant to write a little note of cheer to reach you tomorrow. I just hate this stupid exam. I can't bear that you should be worried and bothered with things like this. Please Childie dont worry about it, it will soon be over and then you can be your own dear happy self again and can set yourself to making up your mind about me, your servant.

Barnes: 6th June 1924. Rudgwick

I do hope I'm not worrying you with all my letters, Dear One? I seem to have written quite a lot lately somehow. You must just tell me if I am doing it too much.

How did things go today. Chemistry is a horrible subject, but I expect you have done all right. Besides Molly dear, you <u>are</u> most awfully brainy, although you will try to pretend you are not. But I know better. All that you suffer from is bad teaching.

When you give up coll. would you care to do some interesting maths and engineering? Tho' really you know quite enough to be able to do all the ordinary engineering things. We shall have to learn all about motor car engines and things like that, so that when (if ever) we can afford a car you will know all about running it. Dearest One I'm no earthly use at keeping up a pretence of superior knowledge about anything. I want to teach you all that you care to know, that I can, so that you will be absolutely independant of me. But Molly, <u>dont</u> stop caring for me, when you have explored the limits of my small knowledge. I'd hate you to think I was anything but what I am, a very middling sort of fellow, whose only claim to greatness is that for some inexplicable reason you have taken him for your servant.

Auntie May is most awfully sweet about you, she knows I want to think and talk of nothing else. She has thoroughly fallen in love with you herself. I <u>do</u> think I'm a clever fellow – I dont mean to see that you are beautiful and sweet and clever; anyone and everyone sees that, but to have made <u>you</u> think of me, who am so far far below you Mistress dear. (Note: this is seriously meant and only too true). Never mind, I shall beat you with a broom handle when we are married. (Note: this is cheek).

Molly: 8th June 1924. Hampstead

My dear Barnes,

I do hate that beginning, but I s'pose it will have to be that for a little while. I love the way you say very vaguely "I seem to have written quite a lot lately somehow" – only five letters in three days, Barnes! Of course you aren't worrying me with them – quite the contrary, I just love having them, the more the better.

<u>If</u> I ever marry you, I shall jolly well get in first with a poker, so there. What would I have done if you had been a masterful sort of person whom I had to look up to? Shouldn't have done it. I'd have told you I wasn't going to look up to you, and let you be my master, and do everything you

told me, and we should have quarrelled and I should have won. Yes of course I really truly like being your mistress and you being my servant, and it does feel wonderful and strange and exciting. Don't you worry, I shall get severer and severer. But don't forget that sometimes I want you to be just Barnes and nice and close and near me.

I had a perfectly ripping night on Thursday, thanks to you. I did everything you said. First I went into the kitchen and asked Nan for some hot milk. She said there wasn't any milk at all, she had just used it all up to make some cocoa, which she was going to drink. So of course I left it and said I didn't really want any, and went upstairs. I had a hot bath and sponged myself with cold water afterwards, and you are quite right, it is delightfully refreshing. Then when I got back to my room, there I found Nan with a cup of steaming hot cocoa, and while I drank it in bed she did my hair. Wasn't she a darling? Fancy giving me her cocoa. I went to sleep directly I lay down, in the middle of kissing the Duck good night, so by mistake he slept in my bed all night because I went to sleep before I had time to put him on my table. However nothing happened to him because when I woke up in the morning I was in exactly the same position as when I had gone to sleep; I simply hadn't moved the least little bit, which was funny because I generally wriggle about a good deal.

I'd love to know about motor car engines and things so that I should know about running a car, and so that I could understand as far as possible the sort of things you do. Of course I shouldn't stop caring for you if I ever did explore to the limits of your knowledge. I'd much rather you weren't the sort of person who got everything absolutely perfect first try because you'd never be able to understand how other people find things hard and puzzling, at least I should think it would be jolly difficult.

I am sending you my two Physics papers. I'm not a bit cheerful about the first one, and I don't believe they average you, I think you have to pass in both. I made a dreadful mess of most of it. Couldn't do the example in 2 [Kinetic and Potential Energy], nor the one in 3 properly [Archimedes' Principle]. Couldn't think of a single experiment to illustrate the laws of electromagnetic induction. I did 6 beautifully with the most elegant diagrams and worked out how to find the strength of the pole of the magnet, and when I got home I nearly wept because I had mixed the two up and had drawn a diagram for a magnet in the meridian and had called it at right angles to the meridian, and vice versa with the other one! Also, I did 8 last and at 2 minutes to one I discovered that I had been gaily describing the moving coil galvanometer! So you know what to expect. The morning chemistry was horrid, but the afternoon one wasn't too bad. By the way, you know you suggested writing down any formula you happened to remember when you got in, well I wrote down $R = pl/\pi r^2$ and then I discovered that I wanted it in 9 [Resistance and Specific Resistance]. Wasn't it lucky? Anyway the four worst exams are over, and there isn't such a rush now.

Barnes: 9th June 1924. Rudgwick
You would have laughed at me yesterday afternoon. Just as I finished your letter, I saw the postman flash past the cottage on his bicycle. I simply tore out, letter in one hand, envie in the other, and shouted and whistled. But he was already too far down the road to hear. As he was coming from Rudgwick village P.O. I thought the worst had happened, and dressed in Sunday clothes, and full of Sunday lunch, started to run down the road in pursuit. Alas, he went down hill on a free wheel and any chance of catching him was of course impossible. I was too blown and had to walk, but I chased on down to Bucks Green about one and a half miles away where there is another little P.O. No signs of the postman, and the box had no collection ticket on it at all so that one couldn't tell whether the letters had been collected or not. I stood wondering what to do for a few minutes and then thoroughly miserable and annoyed with myself for having failed you I slipped the letter in. Walking back up the hill I was overtaken by a heavy shower of rain, and having neither hat nor coat and my best suit, my temper wasn't improved, till just as I was reaching the house it being now nearly 4 oclock, (as I had done a 3 mile walk) who should overtake me, going <u>to</u> Rudgwick P.O., but the fleeting Postman. "Oh" says I "have you collected Bucks Green letters yet"? "No Sir" says he, "I don't collect Rudgwick till 4!!" I was too happy to be annoyed, to think that my letter would reach you today after all. But I might have had another hour to write to you. Poor Childie, are you getting tired of me?

Talking of walks, I had no end of a one on Saturday. I went into Horsham by train just before lunch, posted your letter, and they assured me it would be delivered that night, saw some friends, bought some wireless wire, had lunch, and then walked out to Housey – that is the school name for Christs Hospital. Thats about 3 miles, and I got there about 3 and didn't leave till 6. Incidentally I made friends with two of the Headmasters children Margie and Christopher aged about 10 and 6. Christopher played stokers with me, which mainly consisted of tearing about on my shoulder and thumping my head vigorously whenever we ran out of hot water. What we really ran out of was breath, which no amount of thumps on the head would help me to recover. I parted with them regretfully, but one cant be stoked for ever, and set off for the station, only to find when I got there, that there was no train for nearly an hour. Having a good map with me – never stir in the country without a map – I decided to walk as far as an intermediate station Slinfold, about 4 miles on the way to Rudgwick, by devious and very muddy footpaths. Only last Wednesday they were 3 feet under water with floods. The going was heavy and I saw my train leave Slinfold while I was yet about a quarter of a mile away. Nothing for it but to walk on to Rudgwick about another 4 miles.

This really was to wish you luck in the exam tomorrow afternoon. <u>Please</u> ring me up in the evening.

Barnes: 12th June 1924. New Cross
Of course I wont forget the "other side" Childie. I am both at once always, which is a very nice arrangement. When you are tired of me close, you can banish me to your feet. Tho' I dont believe you could ever be firm enough to keep me there for long. Being sorry <u>does</u> matter very much, but should never interfere with punishment. Do you remember how Bagheera said to old Baloo, "Sorrow never stays punishment" when Mowgli was rescued from the Bandalog? And forthwith fetched him a swipe with his paw? In fact, when you are very sorry, you actually welcome punishment, because it sort of restores your self respect by squaring the account.

Thats why I want you to be so jolly strict with me, because I <u>know</u> I'm a clumsy brute, and I dont know the first thing about a girl really, and I <u>do</u> so want to be taught and made to be decent. Only its such an awful bother for you, that I'm always afraid you wont think it worthwhile.

I haven't had time to do the physics papers, several of the questions I should have to look up. I dont think you need be at all worried about it. I daren't ring you up tonight, Dear One, tho' I'm just longing to do it. It soon becomes a habit that I feel as if I could never do without. I haven't time for more, do you think we could speak together tomorrow? Dont risk it if you think it will lead to total prohibition, but I shall be home all evening.

♦ Telephoning was not a private matter in the Bloxam household, with the telephone in the hall in clear sight and hearing of all the surrounding rooms; and Barnes was nervous lest he should overtax Mr Bloxam's patience. Molly's letters were infrequent while she sat the final exams. They did, however, meet on three weekends in succession and letters were not quite so necessary. On 14 June, as Barnes had promised, they sat together through the great celebration of young love, *Romeo and Juliet*. It was a season that made theatrical history, with a meltingly handsome twenty-year-old called John Gielgud in his first London appearance in a leading role, partnering the accomplished and beautiful Gwen Ffrangcon-Davies. They shared at least some of the next weekend; and on Friday 27 June they went to see Sybil Thorndike, widely acclaimed as Bernard Shaw's St Joan. The browned, shrivelled and almost unrecognisable remains of a red rose lies in each programme. Molly passed Inter., but whereas the outcome of the mock exams earlier in the year was detailed in her letters, this time there is no written record except for the formal pass list issued by the University, which she kept.

Barnes: 29th June 1924. Box House, Minchinhampton, Stroud
My darling darling Childie,

I dont know how you felt yesterday, but I felt utterly deserted. Its at night time that its worst – I behaved like a perfect baby last night. You cant think how precious you have become to me Dear One. I suppose its good to be parted, certainly it has the effect of making one realise more acutely than ever before how dear and how sweet you are and how unbearable life seems without your dear presence. But I suppose I mustn't write such things, and I will try not to in future. I wish these months were over. I know one thing, that if when you come back [from Switzerland] you are of the same mind, I'll never part with you again, not for long anyway, unless you wish it.

Dear One, I <u>did</u> enjoy Friday night so much, thank you again and again for letting me take you.

30/6/24. At home again.

Childie dear, I haven't much time to write now. I'm going to write to you as often as ever I can. Do you mind?

Oh, I <u>am</u> so <u>very very very proud</u> of you for passing Inter. Its just <u>splendid</u> and I do think you have every reason to be very proud too. Auntie Fanny says it speaks volumes for your self control and hard work during a very trying and difficult period. She is most pleased too, and so is Uncle Charlie.

I told Auntie Fanny how sweetly pretty you looked on Friday. Oh I <u>did</u> enjoy it so, but I mustn't start again, for it makes my heart ache. Childie, Childie, Childie my own dear Mistress, I couldn't bear parting from you

♦ This one brief congratulation was the last mention of the subject of Inter:, and indeed of any further serious consideration of a College career.

Chapter 16
MR BLOXAM'S "CURE"

The women of the family, still working against Mr Bloxam, suggested that Molly might do well to take a course on domestic rather than medical skills when the new term started. Her expertise on such matters was negligible, limited to a modest ability to dust, tidy and mend. The only cookery she had encountered at close quarters was that of Auntie Fanny and of Barnes himself; and her mother's capacity for household administration was not such as to inspire confidence. With the prevailing expectations surrounding married life in her class and time, running a house for a busy and preoccupied husband would be a problem. But Barnes could not forget the possibility that it might all be a dream, and that Molly might prefer to learn independence while returning to College and a career. Nor could he honourably overlook the word they had both given to Molly's father, that she should not commit herself in any way to a future with him. Constantly threatened by these thoughts he began counting down the days until the fateful date.

Since April Molly's father had been planning his final bid to defeat Barnes. Away from the family, her familiar surroundings, and from Barnes, Molly was to be introduced to wider horizons and the attractions of other men. A trusted friend of his wife's was to chaperone Molly and Barbara on a month's visit to Switzerland, a prospect about which, in spite of everything, Molly was excited. But she was in no doubt about the likely failure of any 'cure'. Barnes was to spend two weeks' holiday again in Borth, where the activities and atmosphere would be the same as the last two years.

Molly: 2nd July 1924. Hampstead
You shall write just as often as you have time, because even if I do have letters from you, it doesn't prevent my looking about and seeing likely cures. Stupid old cures, how I hate them.

Barnes: 3rd July 1924. 70 days New Cross
Thank you more than I can say, for letting me write to you, dear One.

I have consulted Auntie Fanny and Father about the [domestic science] course at the Hampstead place, and they think I am quite justified in suggesting to you that you delay entering your name <u>anywhere</u>, until your return from abroad.

For the following reasons:-

1. Such an action prejudices your decision and is there contrary to your Fathers avowed wishes, and stultifies all the pain and separation that we are now undergoing at his desire.

2. If you reject me, you will wish to return to Coll., which you can easily do if you are free.

3. If you accept me, it is most desirable that you should go to some place in Central London, accessible from both my home as well as yours.

To your father, you had better only advance 1 and possibly 2, or base your objection upon a general reluctance to take any premature step.

If we become engaged, Auntie Fanny will see Uncle Arthur on my behalf with a view to securing your greater independance from the childish discipline that you are still under. That you may fully develope your sense of judgement and learn to act and think for yourself, I think it is absolutely necessary that you should leave home for rather longer periods than you have been accustomed to doing. You can only do that here. Also you will see me under ordinary workaday conditions at close quarters. That is why I want you to come to a central place.

I dont want a wife who will treat me as if I were "in loco parentis!" but an independant little comrade, who is as used to controlling her own movements as I am.

No time for more, – I am sorry that neither Auntie Fanny nor I can fight the immediate battle for you. We have, at present no reason to interfere and would probably be sent about our business and do more harm than good.

But I cannot see any possible objection to your wishing to postpone your entry. It isn't difficult to get into places like that, even at a moments notice.

All my dearest love,

Ever your loving humble devoted servant, Barnes.

Barnes: 6th July 1924. 68 days Borth, N.Wales
The last day or two was an awful rush for me, and Thursday evening was occupied by altering my room to make it better for visitors. That kept me busy till half past ten, so that I had no time to pack, and had to leave it till Friday morning. And of course lots of things turned up at the last moment

at the office, so that I had some difficulty in getting away to catch the 2.10 at Paddington.

I got to Borth about twenty to nine, after a very wearisome journey, and found all my good friends waiting to meet me. There are ten of us in the house, Dr and Mrs Boyd, and Mrs Dennis Boyd, wife of one of the brothers, who is in the Navy, and Norman Boyd, who is a parson and Doc's twin brother, and Doc's two children Leslie, and the new baby Jolette, and Mrs Dennis' two children Arthur and Ann aged 8 and 6 respectively, and a very nice Nan, and me makes 10. Not to mention our dear Miss Davies who is beginning to teach me Welsh, and her Mother aged 80 something who hops round the house like a good'un.

Three hefty children came and sat on my tummy in a row first thing Saturday morning and then although it was vile weather made me bathe before breakfast, tho' the bathe of the shameless Ann consists of nothing worse than a glorified paddle. Nevertheless she came and splashed her Uncle Woggins to her hearts content. Sort of thing a girl would do.

<u>Monday 67 days</u>. We were home too late yesterday afternoon to catch the post, which goes out about half past five here. Sunday was a glorious day, and we bathed before lunch and played golf in the afternoon. I didn't go to early service as I thought you wouldn't, but we did go in the evening, to church. Today its simply awful again, and we bathed before breakfast in pouring rain and a gale of wind. The best of this place is that one can undress in ones bedroom and just walk across the road. Doc's got a new car this year, an all weather Hillman, and we adults went in to Aberystwyth about 9 miles drive on Saturday night to go to the funny little theatre. We always enjoy that, because even if the plays are bad, the actors are jolly funny.

Dearest One, I will send this to Hampstead, as these posts are so jolly erratic that one never knows when letters are going to reach town. I do hope you haven't minded my not writing for so long. I always feel rather miserable about it, for I should hate you to feel I was neglecting you, or not thinking of you; because really you are in my thoughts day and night, Mistress dear.

Do you go to New Cross on Friday? I have put my watch in the left hand corner of the left hand small drawer of my chest of drawers for you, and you will find the chocs, our chocs, there too. Do have one.

Molly: 6th July 1924. Princes Risborough [Mary's home]
When I get home I will tell Daddy, though I don't think he will be content unless I give him the third reason as well. No, I don't see why I should give him that reason; I won't, I'll only tell him 1 and 2.

<u>Wednesday</u>

When I got home yesterday afternoon I found your letter waiting for me. Thank you so very much for it. Barnes look here, I really and truly don't expect you to write to me when you are away, at least I only expect very little – about two letters. I know from experience what it is like. Why, even in the short time I have been away with Mary, you can see how much of a letter I was able to write; and there there are no children, so I can quite understand what it is like with Arthur and Ann and Leslie and Jolette (though I don't suppose you have much to do with her yet, she is so little) besides all the grown up ones. And I guess they all want as much of you as they can get, and you simply can't go shutting yourself up to write letters. I know I am in your thoughts all the time because always, whatever I am doing, you are in mine; and I know how sometimes you wish to goodness you could get a minute just to write even a little bit of a letter, at least that's what it's like with me; but really truly I shall understand if I don't get many letters till after the 21st (is it then you come back?) and again when you go to camp, when I s'pose you will have still less time. So it's all right. See Barnes?

Oh, and Mary has been trying to teach me to drive the car. It isn't so very difficult, but the gears are so muddly, and the trying to change gear without making a sort of raspy noise. I wish I could have stayed a bit longer, and then I might have got on more. And can you drive a car, I mean practise driving it – accelerate and change gears and put on the brake and so on – while it is still and the engine isn't working, without spoiling it? Because if you can, it's a jolly good way to practise driving if you are alone.

I am going to New Cross on Friday and coming back on Wednesday evening. This is supposed to be a very cousinly and weathery letter, and it is, and they will all be. Every morning when I wake up I think 66 days, 65 days, and so on. I'm glad you put them at the top of your letters.

Barnes: 8th July 1924. 65 days Borth, Wales
My darling darling Childie,

I cant help starting another letter to you, although there is little to tell except how very dearly I love you. You cant think how much I look forward to getting even the smallest of notes from you. Every little detail of your day is of interest to me.

Padre has been praying hard for an anticyclone and he must have made great medecine for today has been really quite good, no sun, but that awful gale has subsided. My golf has got steadily worse. Like tennis it is often the way that when you haven't played for a long time, you play quite well for the first day, and after that simply go to pieces.

<u>Wednesday. 64 days.</u>

Today has been simply glorious. A cool northwesterly breeze, and not a cloud in the sky. The early morning bathe was the best we have had, as the sea for the first time was flat and calm. After breakfast I had a lesson from the professional at the golf club and he showed me what I was doing wrong. Its a fearfully tricky game. I shall have to have several more lessons before I become any good again. One so easily gets into bad habits.

In the afternoon we went for a picnic at Tal-y-bont. The mill stream was very full and rapid after all the recent rains, and we had a ripping bathe. Part of the pool must be over 8 feet deep, and one can dive from the rocks. After tea we went down below one of the weirs and played making salmon jump the weir, a game I invented. You stand on a rocky spur, and make a flat stone do ducks and drakes so hard that it leaps over the weir, and goes on ducking and draking in the pool above. The Padre was most awfully good at it.

I am beginning to feel a little happier. I was rather miserable the first day or two. You see I think of practically nothing but you, and how much you would enjoy it all, all day long. Of course the others know nothing about it, so I suppose it makes me feel rather lonelier than I used to.

<u>Thursday. 63 days.</u>

My dear One, I do hope you are being happy, and enjoying your dear self? I've given up trying to be exuberant, I simply cant, Dear One, I do love you so much, love you and worship and adore you. I shall love to think of you in my room, sleep well Childie dear.

I expect you are full of your foreign tour. I do hope you didnt mind my last week's letter Dear? I know I've no right to interfere Childie, I love you, love you, love you.

Molly: 11th July 1924. New Cross
Barnes dear, I have just arrived, and I have found your letter and the most beautiful flowers awaiting me. You can't think how glorious the flowers are. There are red carnations, very wet and slender, and white roses, tall and stately, and some littler, dearest of all red roses. I shall have them all beside me when I am in bed, but in the day time only the red ones will be on the table, and my family shall sit beside them and enjoy them; the white ones and the carnations will be on the chest of drawers. Barnes dear

dear, thank you and thank you for them; every time I see them in the day-time or smell them at night, I shall think of my Barnes; and their sweet-ness and beauty will remind me of the sweetness and beauty of your love. Oh Barnes, Barnes how I long to see you or to hear you or to feel you; how I love the flowers.

Molly.

Molly: 12th July 1924. New Cross

I have just been woken up by the sun kissing my eyelids open, and the first thing I saw was sunshine streaming through red rose petals and into my eyes. It is just six o'clock and I can't waste any more of the beautiful day sleeping. I have had a splendid night – slept solidly from 12.30 to 6. There is something restful about the atmosphere of this room which sends you to sleep when you are ready and keeps you asleep; I think it is because everything reminds one so of you. You are in everything – flowers, books, pictures, airship, animals on the mantlepiece and in your untidy cupboard whose door won't shut prop-erly. And you were certainly standing by my bed when I went to sleep and when I woke up this morning, so that the last and first things I said were good night and good morning to you. You know sometimes, very occasionally, you are there when I am in bed at home – there's a queer sort of tingly feeling and you're there right close to me for about 30 seconds and then you are gone. It looks as if I were quite mad; maybe I am.

I can't help this writing sloping and being a bit funny; yours would if you were doing it on your knees in bed with nothing behind you but one small pillow which isn't much support, and knobby bed rails. The sun has moved round a bit now, and he is kissing my ten toes with such warm warm lips.

Your letter that I had yesterday is the sort of letter I want when you are away. From it I can tell just exactly what you have been doing and I can imagine you bathing and playing golf and "making salmon jump the weir". You surely must have had a fine day yesterday. Here is was glorious – not a cloud in the sky all day, and the night was so hot that I slept with nothing over me and only the flimsiest, laciest of nightdresses.

<u>Saturday morning</u>. Auntie Fanny and I have just been out shopping and we have come back more dead than alive. You've no idea how hot it is. We are about to have the last meal we shall in peace without the thought of clearing away and washing up looming ahead, for Harriet goes away this afternoon. Not, mind you, that I don't like doing it; I enjoy every bit of it.

<u>Saturday night</u>. It is still as hot as ever; your thermometer says 79.5. All afternoon Auntie Fanny and I sat on the verandah and did needlework and Uncle rested in my room. Then we got tea ready in the garden, and I cut the bread and butter, and Uncle ate 7 pieces of white and 1 brown, and, quite uncalled for, he remarked that the bread and butter was very good.

And after supper we cleared away and saved the washing up till tomorrow; and Auntie Fanny and I looked at Uncle from the garden, and thought how ripping he looked, sitting in his consulting room with his white hair and his white coat and the light shining down on him, reading a book. And then I took teapots and clocks and things upstairs, and locked up the dining room, and wound up Granny and the cuckoo clock; and then we sat in the garden till about a quarter to eleven. And now I am half lying down the wrong way round on my bed, ready to go to sleep, writing to you – no easy task considering the position I am in. Is that a detailed enough account of how I have spent my day? I must stop; it's twenty to twelve and I'm wasting the Doctor's gas; what would he say. But perhaps you'll have to pay for it and it won't matter. Goodnight you dear dear Barnes (That's quite cousinly really; I might be rather a fond one).

<u>Sunday morning</u>. I like your room like this, it seems to look larger; and I love your great big Airship. I have just been standing on a chair to examine it closely. Did you draw it and paint it? One day I want you to tell me what every single part is. I do like that [photo] of you downstairs in the drawing room. I don't think I realised it before, but whatever I am doing, you are always looking at me. If I am dusting the mantlepiece or looking at the wireless or winding Granny, you always stare solemnly and seriously at me. And sometimes when I smile at you, you begin to smile too, but you never go very far.

I have your watch, and thank you very much for it. Also I have examined the chocs. and if I can find a not too juicy one, I think I know what I shall do.

Barnes: 14th July 1924. 59 days Borth
My darling One, this is not to thank you for your wonderful letter, but just to say in case I dont get time to write before you leave New Cross, – Go to my untidy cupboard; fish out the big wooden box on the floor, – the padlock isn't locked, open it, and inside you will find a large folder tied with a little tape thing, crammed with loose photographs. In that you will find a photo of me. If you care to have it, take it by all means, Dear One.

Yes, I am by you sometimes – I try to be anyhow, and call out to you.

Barnes: 16th July 1924. 57 days Borth

Thank you again and again for your lovely letter. I <u>do</u> think it was ripping of you to find so much time to write to me. I can so well imagine you half lying on the bed, in that hot room, and I do so love to hear of all the <u>little</u> things you do. My dearest One, how very dearly I do love you.

I am so glad you had a good time with Mary. It was jolly learning to drive the car. One can do very little really, if the car is not running, except familiarise oneself with what the various levers and pedals are for. In any case you can not possibly injure the car in any way. You will find that occasionally the gears will not slip in to mesh, owing to the wheel teeth not being properly together.

I am so glad you liked the flowers Childie dear, and slept well in my room. Molly, I believe sometimes I can come to you, when it happens to coincide that you are in that half dreamy state when they say our sub conscious minds are most active and receptive, and I am thinking intensely of you, and perhaps calling to you at the same time. Some day we must try an experiment to see if it really does so happen.

Lucky lucky sun, what right has he to kiss my dear toes, and I so far away.

Yes, I drew that airship for my lecture when I was in Switzerland. Its really out of date now, as I've invented so many things since then. I had no proper things to draw it with and no board, I just had to pin it on the wall of my room, and made it up out of my head. I shall never forget wandering round Montreux trying to buy "un pistolet pour le dessin" which the Swiss master told me was the French for what we call a French curve, a piece of wood cut in to several curved shapes. Of course I will tell you every part again and again, until you know it all as well as I do. Then I can talk over difficulties with you, and you can help me.

I wonder if you went to my box and took the photo? I do hope so. I know what you have done to the not too juicy choc; bitten him in half, and left the dearest little bitten bit for me? Thank you ever so much, what a ripping thought.

Please, please dear Childie tell me exactly where you are going with the dates and addresses, so that I can write in time to meet you at each place. Oh how I wish it was me that had the privilege of taking you. I would so love to tend you and serve you and look after you, my dear dear Mistress. (Though I would let you do all the French conversation).

We climbed Cader yesterday, just us 3 men. We had a most glorious time, and came down a different way, down almost a precipice leading to a little lake Llyn Can, about 1500 feet below the summit. Part of the descent was so steep on grass, that we sat down on our little behinds and tobogganed about 200 feet down. At one time I thought the speed was really dangerous, as it is difficult to steer. Needless to say our trousers were ruined, and at the bottom I ran over a concealed stone which bruised the base of my spine rather. When we got to the lake we had a glorious bathe, although we had no towels or anything and just dried in the sun. We had only had a few sandwiches that we brought with us at 12 oclock, before starting the climb, and it was after 6 when we regained the inn, so you can imagine how famished we were, as we had had quite a hard time. We ordered 12 eggs and eat 4 each without stopping to think.

Oh Molly dear one, how I should love to show you all these beautiful places, just you and I together. I always think of you, and often when I see some quaint little flower or shell or stone I pick it up and put it in my pocket rather pathetically, for you.

Molly dear I love and worship you and adore you, dearest sweetest Mistress. I am yours absolutely, oh my dear Loved One. Goodbye dear dear Molly, my dearest Childie.

Molly: 17th July 1924. Hampstead
I'm so sorry I forgot to tell you – I did speak to Daddy, and he has agreed to let it wait till September. Of course he hates it, and I guess he will oppose the plan pretty strongly.

How do you know what I was going to do with the not too juicy choc? There is a note for you with your watch. I half wish I hadn't left it there, but had torn it up and thrown it away. But I didn't have a chance to, because Elf [Barnes's cousin] came and had my room yesterday. Barnes I <u>do</u> love your Father; he was so awfully dear when he kissed me goodbye.

Oh dear, here am I with boxes and cases to be ready in three hours and I haven't started doing a thing, and I may not have a button to any of my clothes as far as I know. I must stop.

Here is a list of the places we are going to. When do you go to camp?

Barnes: 17th July 1924. 56 days Borth
Just a brief note to tell you again how much I love you, my dear Mistress, and to wish you the best and happiest holiday in all the world. Oh Molly darling, <u>do</u> enjoy it ever and ever so much, because you have earned every minute of it by your faithful hard work in the past.

Childie, my own sweetest dearest Childie, all my heart and soul go with you, where ever you may be. I am there loving, worshipping, adoring you, your own loving willing servant, who only wants to live for you and serve you and obey you, and most of all to bring you love and happiness. If you feel after all you are changing, my Childie, do not fear or worry; I am always yours, and your welfare and happiness are my happiness – I just live for you, and if it be your will, will die for you too – or go out of your life, which is much the same thing. So enjoy yourself to the full and do not think that you have any duty or obligation to me because of our friendship in the past.

Darling One, if you were to dismiss me at once, I should still be for ever your debtor and your lover, for the dear dear kindness and sweet love you have shown to me. Oh Childie, these things I'll treasure always and always in my heart as the happiest things in my life, never to be taken from me whatever the outcome may be.

Molly, dear dear Girl, I love you and love you and love you and love you.

Ever and ever your humble loving adoring servant Barnes.

God bless and keep you my Childie dear.

♦ The next day Molly left England with Barbara and the chaperone Miss Erskine Scott, a seasoned traveller and one full of propriety. Known affectionately as Sersky by Molly and her sisters and by the next generation of Bloxam offspring, Miss Erskine Scott was a life-long spinster. Whether she had loved and lost neither Molly's generation nor the next ever knew; but she had a romantic heart and was no more in sympathy with Mr Bloxam's aims than were his wife and sisters. Her practical good nature persuaded her to carry out her duties as she saw fit, rather than fulfilling Mr Bloxam's purposes to the letter. He had no idea that the upright, upper-class lady with the gentle manner and well-modulated Scottish accent would effectively undermine his plan.

Barnes was left in Borth trying to be sociable. He had Molly's itinerary and he immediately besieged her with post in spite of the impatient taunts of his uncomprehending friends. Two dreary sepia postcards of the Borth coast line and three more letters hastened from Cardiganshire to Switzerland.

Barnes: 18th July 1924. 55 days Borth
This is just a little note to welcome you at Lucerne. It wont meet you, of course, but it may come in time to cheer you up if you are feeling the least bit home sick. Tho' I dont really see why you should, as I expect you will

be much too excited and interested to think of anything else. Darling darling Childie, it seems like the last stage of this endless waiting to me. Just think, when you return it will be only 23 days, a short 3 weeks; and by the time I return from camp, it will be only 11 days.

I am most awfully glad that you have got the domestic course postponed. When once we are engaged, I think Uncle Arthur will see that as my future wife, we are the people who are most concerned in any arrangements that are made.

I had a private letter from the secretary of the Airship Company a few days ago, saying that the Directors had voted me a salary of £1500 a year, to date from June 1st last, and to become payable directly the legal contract for this ship is concluded. So your Father cannot possibly object to me on financial grounds, which is a great relief as this salary is really a big one, for a man of my age.

I must stop my dear One. I am so looking forward to getting home, and finding your little note. Why did you wish to tear it up? Was I right about the choc?

Goodbye Childie dear, – everyone is looking at me and making fun of me, so I simply cant write a really truly love letter to you. But you know that I love you with all my heart and soul, always and always. Ever your loving devoted adoring servant Barnes.

Barnes: 19th July 1924. 54 days Borth
I seem to get more and more impatient as the days go on, and now regard 54 days with as much disgust as I did the 70 when I started my holiday. I feel as if I never, never, never could bear to be parted from you again. People say husband and wife should often go away separately, but I dont think I would ever like to do that with you, tho' it shall be just as you wish.

Are you tired of letters from me Sweet One? I do hope I'm not being a nuisance, but whenever I can find a moment, I feel as if I must write to you, although I have nothing new to say.

In a way I shall be jolly glad to get back home again, as the time seems to go quicker when I am at work, than it does when I have all day to think about your dear self. Mrs Audrey Boyd and her two children Arthur and Ann are coming with me [by train]. I shall get your dearest of all notes about 9 oclock on Monday. Molly, why did you wish you had torn it up. It isn't to say you dont care, Childie, but I'm sure not, only I'm in such a miserable state, I've had just about as much suspense as I can bear I think.

God bless you, my own dearest most loved Mistress. All my love to you, whom I worship and serve.

Barnes: 21st July 1924. 52 days In the train
My own darling Child,

Please if you dont want me to bother you so with letters, while you are away, will you frankly tell me? I suppose its rather cheek beginning as I do, because you are not my own at all really, although I am yours, as long as you want me, that is. But anyhow you are my Mistress, because you've said so so I can begin "My own adored Mistress" which is just as satisfactory as the other I think, tho' I never can tell which way of beginning I like the best. Which do you prefer. Its funny that although I begin in several different ways, my ending up always seems much the same. Your going abroad reminds me so much of the time that I left to go to Switzerland, and had the most ripping letter from you on the morning that I left England. I wrote to you in the train going to Dover, and I think that was the first letter I ended "Your humble obedient servant". You cant think of the thrill it gave me, for it was utterly true, even then, and I am afraid I used Jane Austen only as an excuse for writing what I really felt. I often wonder if you understood that, or whether you thought I was a bit mad, and just did it for fun? Do tell me. Oh dear, I do wish this time was over.

I've talked all the time as if I were going to join the party again next year. Of course they dont know anything about you. Why, if you have me, we may be married by then! And in any case if we are only engaged, we shall go away together. And if you dont have me, I shan't care where I go or what I do so it wont matter.

Oh Childie I do love you so very very dearly. I tried most awfully hard to be with you last night (Sunday) about a quarter or half past ten or thereabouts, after I had said my prayers. And then I tried to go to you – I didn't feel as if anything happened, except that I could see your dear Face in my mind so clearly. I've never tried actually consciously to join you before. Wouldn't it be delightful if we could do it at will.

Now I must leave this letter for while and play with Anne and Arthur. Goodbye for a moment my adored Mistress.

Home again. 9 p.m.

My darling darling loved Molly, I have just read your little note. Oh my Dear, my Dear, you couldn't have written me any word more sweet. Oh my darling, do you think I could have any eyes to see the moon, if you were beside me?

I have kissed the chocolates, but I mustn't eat them until I go to bed or the time will be wrong.

I <u>must</u> go, and talk to Fanny and Father, or they will wonder whatever has become of me as I only kissed them in the hall and then came straight up here. Goodnight, Molly dear, if only I could hold you close to me and tell you how much I love you. I will stand where you stood, all ready for bed, and pretend hard that you are there.

I will write tomorrow.

♦ Barnes found this note in the drawer by his watch:

Molly: Tuesday night [15th July 1924.]
You will open our chocolate box, and in it you will find the halves of two chocolates. I have eaten half of each, and you will eat the other halves. You will do this on Monday evening when you go to bed so that I at Lucerne will be able to think of you while you are doing it.

And Barnes you can pretend that I am there, and have just bitten them and am giving them to you, even as I have just been pretending that you are here beside me now. Dear, I wish you were. It almost feels as if you were, and as if in another minute you would put your hands on my shoulders and your head close to mine to see what I am writing as I stand here against the chest of drawers with the candle beside me. I'm all ready for bed and my feet are getting cold with standing on the oil cloth, but I can't get into bed because if I do I shall have to turn round and then you won't be there. Dear, it's a lovely night with a great beautiful moon, but half her beauty is missing because I want your eyes to see her with mine. My flowers are very sweet, but I do so want to show you one particular little darling red rose. Barnes, I'm going to bed; your watch says 20 past 12; I've been standing here for ages doing nothing. I turned round, and you're not there; Barnes dear, it is so <u>very</u> lonely without you. I want to go on writing to you all night; it makes it feel better. Barnes, I'm <u>not</u> unhappy, really truly, and I've enjoyed myself very very much; and I'm going to enjoy myself and to have the loveliest of holidays. Thank you so very much for your watch. Goodnight dear dear my Barnes.

Molly: 21st July 1924. Hotel Belvedere, Lucerne
We went to Boulogne instead of Calais on Friday, and I was as nearly sick as I could be when we were crossing. I'm afraid I'll make a very bad sailor. We had a horrid night and were scarcely able to sleep at all because there were five very lively people in our carriage who would talk and laugh all

the time, besides there's nothing to prop your head against, though Baba and Miss Erskine Scott let me have the corner because I was the youngest and also I was still feeling a bit sickish and also I was a little bit bruisy because I tumbled over at Boulogne station. However I spent most of the night in the corridor watching a French night, and I enjoyed it. We arrived at Basle at 6.45 and there we had coffee and rolls of which we were very glad. I'm learning to drink coffee without sugar and I eat the sugar afterwards by itself because it is such lovely stuff to suck, quite different from ours.

All the same I <u>was</u> homesick at first, and your note was very welcome. As a matter of fact I was more Barnes-sick than homesick, so you can imagine how glad I was to get your letters. I got the first letter, posted on Friday, at breakfast time, and the second one and the two postcards at lunch.

Miss Erskine Scott knows all about you and she has been talking to me and giving me valuable advice. She says that she knew a girl who went to the Domestic Training College at Hampstead, and she learnt nothing at all. It was probably the fault of the girl, but I shall certainly tell Daddy! She also says there is a very good training college place at Victoria – Buckingham Palace Road I think – which would be much better than the one at Hampstead. Is that anywhere near Vickers?

We went for a glorious walk this afternoon. We started on a funicular railway and arrived at the top of the hill where we had tea. Then we walked back to Lucerne through woods and fields filled with cow parsley; the woods were pine woods and smelt delicious specially where there were lots of cut logs. When we were just out of the woods a huge great thunderstorm came up; we had seen large black clouds settling lower and lower on the hills at the back of Lucerne, but we thought we should get home in time. But apparently storms come on very suddenly here, and we were thoroughly caught in this one. I enjoyed it to the full, but I'm afraid the others hated it. You can't think what a gorgeous sight it was to see the great black clouds rolling down the valley and then the lightening flashing from one end of the lake to as far as you could see towards the other end. And the rain! Every garment I had on was soaked, and I could wring the water out of my coat and the front of my dress. I could have enjoyed that most awfully if only I had had someone who enjoyed it too, because once you were thoroughly wet, you couldn't get any wetter, and you couldn't help enjoying it then.

It is a quarter past nine; and I s'pose you know now if you were right about the choc., and also why I almost wish I'd torn up the note. Barnes

I'm most awfully sorry; I wouldn't have said anything about it if I had thought it would worry you. Of course of course I'm not tired of letters from you; write as often as you can – the more the better.

I say Barnes, how splendid about that salary of £1500; it seems the most colossal sum to me. Barnes you know you can always tell me anything and I should never never repeat it to a soul, not even Betty. I expect you know that, but there might be something, like this for instance, that you wanted to tell me, but you might think "Oh, she'll tell someone because she's a woman and they all prattle", but really truly I wouldn't; you needn't even bother to ask me not to because I shouldn't without your permission. See?

It's nearly half past ten; I wonder if you've eaten your half chocs. yet. I've been thinking of nothing but you since half past nine; and I shall think of nothing but you till I go to sleep.

♦ Before the anxious Barnes received this, he had news of Molly through a letter to Auntie Fanny. Not content any more with a simple counting of the days he began to calculate separately the days in July, August and September. Molly could sleep peacefully in the assurance, many times reiterated, of his devotion while he could not find peace. The thought of Molly bruised and lying on an alien pavement distressed and worried him; but on the back of the envelope he drew a delightful set of matchstick people illustrating the accident. Two women equipped with bags stand by a rail track and a third, with bag dropped, lies on the ground. A train is coming towards her, and a man is rushing to the scene. On the edge of the envelope stands another man, arms outstretched, at the end of a winding road and on the edge of a hill or cliff – Barnes himself, helpless and far off. He had a charming way with matchstick men.

Barnes: 22nd July 1924. 9 + 31 + 11 = 51 days New Cross
Did you feel me near you when I eat the chocolates last night. I tried hard to feel you, and to be with you, but I can never feel anything. It was one of those nights when one simply cannot sleep. Strange that you, because of its association with me, should have found this room so restful, while I, returning to it, because of its association with dear You, should find it impossible to rest. I thought of you all night, tho' after 2 oclock, I did get to sleep in snatches.

Your lovely letter to Auntie Fanny arrived at 9 oclock last night, just as I was finishing my letter to you, and I read it afterwards. My dear One, you <u>must</u> be careful, I do hope you haven't really hurt yourself seriously? It must have given you an awful shock and a fright. Please, please be careful, and always remember that the stupid traffic runs on the opposite side of the road to what ours does, so that you have to look to your <u>left</u> when stepping off a pavement, instead of to the right. And the law in France is, that if you are run over, it is <u>your</u> fault for being in the road, so all motors go at a reckless speed, and do not slow up for you, because they expect <u>you</u> to get out of <u>their</u> way. Paris is simply awful.

Please Childie dear, tell me if you were badly hurt, because I know you would make light of it when writing to Auntie Fanny, but I must know the truth, or you will leave me miserably anxious.

Both Auntie Fanny and Uncle Charlie were delighted with your letter. He said it was no good my going on trying to win you, because he had fallen in love with you himself, and by virtue of his age would have precedence over me!

Have you met any nice Cures yet?

I'm very very thankful you <u>didn't</u> tear that note up, Childie dear. Why should you have? Would you rather I didn't write love letters to you, my own Mistress? Why can't I say "I love you" as well as any potential Cure? Because I <u>do</u> love you, and love you and love you, my own Dear One, and simply worship you.

Ever your loving adoring humble servant Barnes.

Is Sersky shocked at your getting so many letters from me, and does she try to convert you?

Molly: 24th July 1924. Lucerne
My dear Barnes,

Of course I am all right – as fit as can be, really truly I am. I was only a little bit bruisy at first, but I was quite all right next day and on Sunday we went for a long walk and I didn't even feel stiff. So please please don't worry. Isn't it a bother about the traffic; though really it doesn't matter much here, as there are singularly few things in the road. I am just beginning to get used to it.

I got your letter to-day at lunch time. Thank you so much for it and for the two others. You can't think how I enjoy having them. Of course I wouldn't rather you didn't write love letters to me. I love them and love them; they always go to bed with me. The letter stays close beside me all night, and whenever I wake up in the middle of the night I grope around and find it and I – well, you know what I do with it, and then I go to sleep again. They used to be kept sedately under my pillow but now I need them closer to me. So you won't stop writing them to me?

Oh Barnes, do you know you spend an unnecessary halfpenny on every letter? It only costs $2\frac{1}{2}$d not 3^d.

Sersky knows that I am s'posed to be away to get new experiences and see new people because of you; but she isn't the least bit shocked by me getting so many letters from you. Now she more or less knows your writing and when she sees a letter from you on the table, she discreetly turns to Baba and says "Well, Baba, we must just study the menu", though I never never read your letter at the table but always save it up till afterwards. She is a funny soul – she asked me the other day what sort of a place I would like to go to for my honeymoon! Of course I wasn't going to tell her anything about it, so I just said I hadn't decided yet, which was quite true, or more or less so. But it shows that she takes it all for granted and considers it quite settled. Poor Daddy, that's just what he didn't want; he would be worried if he knew.

No, I didn't feel you the same way as I do sometimes, on Monday night. I imagined myself in your room giving you the chocolates. I am so sorry you couldn't sleep; I didn't go to sleep for a long time, but I had two dear letters, and two discreet postcards under the pillow. On Sunday night when I was half asleep at about half past ten (we went to bed very early) I most distinctly did have the sort of feeling I get when you are beside me. But it only lasted a few seconds because the next thing I knew, you and I were scrambling down Cader together and you threw a huge stone into a lake down below and then you informed me that it would be quite easy for you to do the same to me. I said you couldn't, so you lifted me up, but instead of throwing me into the lake you carried me down to the bottom and deposited me beside it. It was the most delightful way of going over the rough stony road for me (it wasn't any real way down Cader because it was just a rough road), but how you managed it I couldn't think.

Uncle Charlie's is only cupboard love, because I can cut him nice bread and butter, so you needn't worry.

I remember that letter you wrote to me in the train going to Dover; you wrote it in pencil and it was the first time you ended up "your humble and obedient servant" and you said afterwards you liked Jane Austen-y endings. You know it is most awfully hard to tell whether I guessed then or not. I know it gave me a thrill to read it; but it seems now as if I have known you and that you loved me ever since I first began, so it is difficult to tell what I really did think then. I think in my very heart of hearts I guessed that you really meant it – only of course I never dreamt of acknowledging it to myself – because really truly I knew then that I –

bother. Anyway I knew it was true after the 3rd or 4th time you had used it when you had said you didn't like variable endings and wouldn't use the French ones, but said "Jane Austen endings for me", though it wasn't till after Christmas when you stopped using the ending and became my affectionate cousin, that I fully realised all you had meant by it, and could tell myself so.

Barnes: 25th July 1924. 7 + 31 + 11 = 49 days New Cross
Thank you ever and ever so much for your very dearest letter. How much I wish that I had been with you to look after you Dear One. Oh Molly, I do long to <u>do</u> something – anything, for you so much. You cant think what it is like to sit inactive here while you investigate cures abroad.

Didn't you get pillows for the night journey? You can get them, nice clean ones for 1 or 2 fr each, and it adds greatly to the comfort of the journey. I'm <u>so</u> sorry you had such an uncomfortable time. You jolly well wont when you go abroad with me. We shall have a jolly good dinner in Paris, and sleeping berths.

Try to imagine Burney and I and the interpreter that we took with us at Basle at about 5 am only we being MEN ordered bacon and eggs.

I <u>am</u> so glad about the Hampstead Domestic T.C. Good old Sersky. Yes Buckingham Palace Road is quite close to Vickers and we know a girl who went there and speaks most highly of it. Gently urge Sersky to give evidence to Daddy also.

My dearest One, I am so sorry I couldn't write to you yesterday. I've done nothing but work and drill ever since I returned and didn't get home till ten last night and nine tonight, so this is only a scribbled note. I had so wanted to write to you every single day, but it seems to be impossible, tho' my thoughts are always with you.

Just think, Childie, only 7 weeks today I come to you for my answer. The time <u>does</u> go quicker when I am at work, but oh how slowly even then.

Barnes: 27th July 1924. 4 + 31 + 11 = 46 days Wilts
I came down here to stop with Air-Commodore and Mrs Masterman yesterday, and I am <u>so</u> sorry that I didn't get time to write to you.

My good resolution that I would write to you every day is getting sadly broken.

I am getting very little time altogether, for I have been to drill every evening since I have been back, except Tuesday, which means getting home late and rather tired. You see I have a good many arrears of drill to make up before camp on Aug 17th – we travel down on the Sunday, the day you arrive in Paris. If only August were gone!

Waterloo station yesterday was almost impassable, but I had a fairly comfortable journey. The Commodore has got a car now and they took me straight away out to a picnic among the hills, where he said there was a fine natural harbour for building Airships in the open. We did a long walk, inspecting the various valleys, and did not get back till nearly 8 oclock.

Then this morning they drove me to see Stonehenge – a most wonderful place but rather spoilt by the proximity of the road and the number of cars and sightseers. You feel as if there is only one person in the world who you wish to be with, when you see a thing like that for the first time. I console myself by storing up in my mind all the beautiful places to which some day we shall go together, if you so decree. Nowhere used to seem particularly beautiful to me, until I began to associate everywhere with you, and now there are heaps and heaps of places that I long for you to see. I am so glad you love the country.

♦ While Barnes was marvelling at Stonehenge in the afternoon light of an English sun, Molly was gazing at the unbroken blue of the sky over the Swiss–Italian border. The trio of ladies had moved on from Lucerne to Lugano and Molly was delighted with the adventure, practising her French and her charm.

Molly: 26th July 1924. Paradiso-Lugano
Just after I arrived here this afternoon came the films from Montreux. Your Mistress thanks her servant very very much for his present; it was a splendid idea having them sent from Montreux. Barnes, it was ripping of you to think of it. There was a very polite note in French with the films saying "par ordre de Mr B.N.Wallis je vous envoie ci inclus 12 roul. films V.P.K." and he also said "si vous désirez que vos photos soient développées soigneusement vous pourriez les envoyer chez nous pour les développer et imprimer sur papier". So I've written back a polite note in French! to acknowledge the films, and I said I'd think about sending them there to be developed, but it wouldn't really be worth it, would it. It was fun; I just loved having the parcel and the note in French.

The only reason for my saying that about not telling people what you told me was because when I was at New Cross Uncle had a letter from you in which you told him about your salary, and he told us, and then he said to me "but don't tell anyone because Barnes might not want everyone to know" and I thought to myself "as if I should", and I didn't say anything about it to you because I knew you would tell me when you wanted to. And I thought I'd make sure with you, though I knew really and truly it was all right. Then I'll never ask your permission, no I certainly won't; and I'd like to see you trying to give me orders. It shall always be <u>I</u> who give or do not give permission, and <u>you</u> shall be my obedient servant and do as I tell you always, not only sometimes.

It was good of you to write to me when you came back tired from drill; thank you so much. Barnes I do say definitely that I want you to write love letters to me. I don't believe I could manage 46 more days if you didn't. Oh yes I could I suppose, but your letters do help a great deal. I will tell you exactly what letters I have had from you – one on Friday morning before I left Hampstead, the important one which I loved and loved and have read and re-read till I almost know it by heart and which I can't answer properly except by saying that when you do ask me, I shall give you a true and honest answer, either yes or no; a letter and two cards at breakfast on Monday; another letter at lunch on Monday; and a letter on Tuesday, Wednesday, Thursday and Friday but not one to-day. If by any chance you wrote to me at Lucerne I shouldn't have got it because we left before the post came in. That makes 7 letters and 2 cards; I am lucky.

We had the most beautiful journey here. Before you go into the St Gotthard tunnel it is all very Swiss, and when you come out it is much more Italianified. Instead of grassy hills and pine woods, all the hillsides are covered with trees – strange chestnuts and accacias and beautiful flowering trees; and there are ancient ivy-covered chateaux or castellos as my Italian youth called them, and it is heaps warmer. Of course there are still snow-capped mountains and great waterfalls and rushing streams. That Italian youth would have been a very good cure, only he lives in the South of Italy. We stood in the corridor all the way and he showed me places of interest and gave me chocolates. He couldn't speak English so we had to talk in French, and I found I simply didn't know a thing. However we managed to understand each other fairly well. Sersky was rather doubtful about letting a strange young man give me chocolates, at first, but afterwards she decided it was all right, that was when I had given her some for they were truly delicious. You see first of all he just offered me one and I took it and said merci and smiled very sweetly because I couldn't think of anything else French to say, and then he promptly gave me the whole box and when I sort of expostulated he said I must have them because "vous êtes une demoiselle".

Sersky was most awfully thrilled when the films arrived. She takes a great interest in you and your doings and Airships. It's ripping for me because she never seems to get tired of hearing me talk about you.

Sunday. It is a most glorious day with the bluest of skies, the sort of sky you expect to get in Italy. Though Lugano is in Switzerland it is much more Italian than Swiss, and everybody talks Italian, though they can speak German as well. I believe I'm going to like Lugano better than Lucerne though of course I loved Lucerne. We went out after dinner at about 9 o'clock last night, and everything that could be, was lighted up so that the edges of the lake seemed to be one mass of twinkling dancing lights.

We have to go to Church. Sersky is most awfully particular about Church going.

♦ While Sersky guarded her wards, never letting them out of her sight, Auntie Fanny did her best for Barnes; when he was not at home to pick up the morning post as it arrived, she forwarded the letters immediately to his office. They arrived at mid-day, re-addressed in her bold, black writing.

Barnes: 28th July 1924. 3 + 31 + 11 = 45 days In the train on the way to town
I do so wonder how you are getting on, and whether you are enjoying yourself. I expect Lugano is very beautiful, but also probably unbearably hot at this time of year. I suppose you go about in a body with all the other people – sort of conducted tours business? I am afraid I should simply hate it. If ever we go abroad, we shall go on our own.

Harriet is still away, so Betty was coming over to be with Auntie Fanny. Uncle Arthur wouldn't let Nancy come, because she had been at her Godfathers the weekend before.

Poor Betty says she has failed for Matric. I am so sorry, but with her maths as they are, I dont think she could ever hope to pass. Why is it that you have all been taught maths so hopelessly badly? I haven't heard what she failed in, but I should think it was that. I must write to cheer her up.

I wonder if there is a letter from you for me I do look forward to them so. Are you beginning to feel cured yet? Are there any nice young men in your party, and do you dance and mix with people a good deal? Do tell me. I think to be strictly fair, Uncle Arthur ought to provide me with bevies of beautiful young ladies. Yes, thats a good idea. Only I can see you perfectly well, tossing your proud little head and putting the most delicious little nose in the air and saying "I dont care if he does, you can go to your beautiful young ladies".

Home again. Monday evening.

Molly, it isn't only words. Its all true, that I am yours, soul and body. I could never resist or disobey you in real life, tho' I may often do so in order to experience the delicious feeling of compulsion by you. Why then will you never grant me that unless I first ask you? I dont believe you have ever given me a single order on your own initiative, without my having suggested it or dared you first. My only dread is that it all displeases or bores you, but oh Molly, if you knew how I worshipped you, I do think you would at least be able to sympathise with my longing to give it all some outward visible expression, however foolish it may seem. Oh Molly, my Mistress, my Childie, my All, say you understand?

I found your first dear dear letter, awaiting me this morning at Vickers House, where Auntie Fanny had readdressed it. I've left no time to answer it properly, but oh my Dear, it is the most satisfactory letter you ever wrote. No, Childie, please God, I will never never never cease to write to tell you how deeply truly I love you, my dear One, my loved One.

And now the post has just brought your letter, finished only yesterday (Sunday) at Lugano. Two letters from you in one day – I dont know what to do with myself. Thank you very very humbly and deeply Mistress dear. You cant realise quite what it means to me. Thank you ever so for telling me to write. Yours, yours, yours, how I do long for it to be really true. Goodbye. Ever your loving humble worshipper and servant

Barnes.

Chapter 17
PROBLEMS AND REWARDS

The problem of maintaining the spirit of romantic courtship throughout married life, if that situation should bless him, was really troubling Barnes, and he felt compelled in the last weeks of his suspense to explain his thoughts. Four letters chased each other across the Channel before another from Molly made the journey.

Barnes: 29th July 1924. 2 + 31 + 11 = 44 days New Cross
I didn't thank you half properly for your letters that I got yesterday. Please Molly thank you ever ever so much. I just loved them both, and you dont know how they have cheered me up. I am so awfully glad that you have got over your tumble at Boulogne, – do please be careful of yourself, you dont know how precious you are to me, dear One.

Yes, I can guess what you do with the letters, Wicked One, but I have an advantage over you, being a MAN, because men's jama-jamas have a pocket, just over their hearts, specially designed I suppose by a sentimental outfitter for the benefit of love sick sleepers. So I can carry a whole packet of 5 or 6, and as I generally sleep with my arms crossed, I can very conveniently hold them in my right hand at the same time. How funny – your letters used to be kept sedately under my pillow too, until I found out the real use of a jama pocket. Childie dear, of course Ill never stop writing love letters to you, now that you have said you do not mind. And I hope Ill never never stop in all my life. If you teach me properly Molly, so that reverence and obedience become from the start my natural attitude towards you I do not see why we should ever cease to be lovers. That is partly why I so love to be your servant, I mean really in fact treated as your inferior and servant. Just now, my love consists so largely of utter worship and longing for you, but I know that if you yield yourself without reserve to me these wonderful, delightful feelings are bound to lose their freshness. So I want us to use that art which is the secret of all true enjoyment, the art of restraint. To me you are on a height, wonderful, unattainable, and I dont see why you should not always be so when you wish. I dont mean Molly that I make the mistake so many men do of thinking you are a sort of angel who never gets cross. I know perfectly well that, like me, you are only human, delightfully human, and I should just hate it if you were not. When I say that to me you are on a height, I mean that I <u>feel</u> your servant, utterly unworthy and being a man, an inferior creation. I think most men when in love at first feel that. I have felt it ever since I first saw you. From my observation however, after engagement and marriage most men lose that feeling, because I am sure, their wives spoil them, and gradually, however much they may try not to, they take their wives for granted as it were, and seem to feel as if they no longer had to pay the homage that they did when first they met.

I think I have these feelings of service and homage more strongly than most. I <u>enjoy</u> being ordered and being your servant, but also, rather curiously, when once I have the upper hand, I tend to be rather dominating. Molly, sooner than have all our wonderful romance spoilt in later life by your indulgence, I would rather you refused me altogether. You must be prepared to make, and keep me, your servant. You will not spoil my character, there is nothing of the cringing henpecked husband about me, rather you will make a man of me. Please Molly, can we be the same? Or would you simply hate it?

I must stop to catch the post, only just back from drill. I will go on answering your letters tomorrow.

Barnes: 30th July 1924. 1 + 31 + 11 = 43 days New Cross
Not much time for a letter tonight – actually I am in the train at Victoria Station, 9.5 p.m. just got away from drill, and my hands are so sore I can hardly hold a pen. We've been getting one of the guns out of a very difficult and muddy corner of the Chelsea barracks, ready to go up to camp and as it weighs about 7 tons and has to be man-handled you can imagine its some job. Ive been working on the ropes and placing blocks under the wheels to prevent their sinking into the ground, and hot isn't the word.

I say Molly, they've asked me for the second time if I will take a commission, that is, become an officer in the battery, and as I couldn't very well refuse I said I would. I hadn't time to ask you if you wished it, because I only had a day to decide. Of course its a very considerable honour, so I'm most awfully pleased really, only I'm never so happy holding a commission, you have to be dignified, and dont get such a jolly time because there's a good deal of ceremonial connected with the mess and so on. Still one is very comfortable in camp, has a servant, late dinner and so on.

Are you pleased, Mistress darling? Thank heaven, the day after tomorrow I can say "Next Month". My Mistress you are always in my thoughts.

Home again. You haven't minded my last few letters? Sometimes I do so dread that I shall be too strong for you to manage. Molly, I'm quite sure that having me under your orders so strictly wont be bad for you either, because wives dont get lazy and spoilt; because their lives are one long devotion to home and children, and however much you make me wait upon you and serve you, its really nothing compared to all that you will do for me and others. And the sort of things that you can make me do are very little use I am afraid, from the point of view of being really helpful to you. After all, you can probably change your shoes for yourself in half the time that it will take to call me and make me do it for you; and when you make me wait upon you, everything is really within reach. So that keeping me as your servant really puts you to even more trouble. Molly, I am so sorry, but please please dont do things for yourself when you can make me do them for you.

And again, I am very certain that just as now, when I try my hardest to win your love and approbation, so you will be able to keep me in that state all our lives, and I shall never drop the habit of trying to please you, with loving offerings as I do now. Dont you agree? I would love you to write and discuss it, and tell me if you dont, Dear One.

Barnes: 1st August 1924. 30 + 11 = 41 days. Rudgwick.
Thank you so much for telling me about the $2\frac{1}{2}$d stamps. Do you remember when I tried to make you believe it was only $1\frac{1}{2}$d?...

Did you really know Childie, that you loved me (bother!) when I went to Switzerland? Oh, Molly, I do love you so, and just long for you. How I've endured these two years I dont know.

Barnes: 2nd August 1924. 29 + 11 = 40 days Rudgwick
Mistress dear, you've had a letter more or less every day for nearly a month now, haven't you? Anyhow I've ceased to note in my business diary "Wrote to Molly" as I have done for 2 years now. It looks simply absurd when one has to put it down every day and sometimes twice.

♦ In his diary for 1923, Barnes recorded letters sent to Molly and received from her. In the diary for 1924, after Uncle Arthur's relaxation of the rules, he recorded an ever increasing number of letters interlaced with brief telephone calls; but for the past month he entered only the arrival of letters from Molly. Unlike her earlier records on scraps of paper, Molly made no note of letters received or sent nor, when it came, of the final outcome.

Molly: 31st July 1924. Paradiso-Lugano
All last night, or at least most of it, I was thinking about what you had said. I believe I do understand how you feel, though you must remember that it is difficult because I don't feel like that in the least – quite the contrary, I feel that it is glorious and wonderful to possess somebody so completely (though even now I don't believe I fully realise the extent of my possession; I suppose I can't yet) and to have him sometimes a close and equal lover and sometimes a servant adoring at a distance, but never never do I feel the least bit servant-like myself, only either mistress-like or motherly or – oh hang it, why should I have to write it all. If only I could see you and talk to you. Six weeks is longer than all eternity.

Why I haven't ordered you to do things unless you first suggested it was simply because it is all so new and strange, and scarcely believable even yet. Don't worry; I will.

You needn't ever fear that it all displeases or bores me, but you do take some getting used to and I can't do it in a hurry. Probably you think I am very slow; maybe I am. You must remember also that I have known Molly for nearly twenty years – no not all that time I suppose – fourteen or twelve perhaps, no not even as long as that, in fact I don't believe I know her now; anyway I've been acquainted with her for fourteen or fifteen years – and to me she has always been one of the Bloxam family with four sisters and one small brother, with nothing special about her – less imaginative than Betty, more imaginative than Baba; more strong-willed or obstinate than the others and with a quicker temper; not so good at music as Baba, at painting as Nancy, at essay-writing as Betty; enjoying school with other girls, and College with other women and with men; loving the people she does love with all her energy; thinking she knew lots about everything but finding she knows nothing about anything; still childish enough to like playing games and if necessary getting grubby and untidy with Pam and George; and yet woman enough to be loved by you and to be capable of loving as a woman; and with the undeserved good fortune of possessing the very best friends in all the world. There you have an ordinary common-or-garden Molly, as I believe I remarked once before, who is only just beginning to realise the limitless power she has over you, her lover and servant. The full realisation will come soon; I do understand a little of how you feel and I am growing up quickly.

There are lots of things we should have to do if you and I went abroad together. For one thing I should have a cold bath every morning if it wasn't a place you could bathe in, and not a hot one once a week and a sort of bath-in-your-basin business in the morning.

That is just exactly the way I feel about beautiful places. Because a place is so lovely I just long to have you there with me to see its beauty and to love it with me. But I felt like that two and a half years ago on May 1st 1922, the Monday after the Saturday we came back from New Cross, when we went for that ramble. The woods were all green and fresh and the little leaves were just beginning and I heard the cuckoo for the first time that year; only then of course I hadn't the remotest idea why I wanted you to be there to see and hear and smell it all. Here it is the little places at the side of a path or a little way in a wood or up a bank beside the road that I love. Suddenly as I am going along I hear the running of water so I turn aside while the others go on; Sersky is getting fairly used to me now. Once when we were on a walk at Lucerne, I came across the dearest little place – a little blue, blue pool with a baby waterfall only about four feet high, all surrounded by rocks and flat mossy stones just made to sit on; ferns in all the crevices, harebells nodding at the reflections in the water, scabious and wild cyclamen and love-in-a-mist and forget-me-nots all round, and the whole place shaded by great tall fir trees. I'm always dying to take off my shoes and stockings and paddle, but as it is I always have to dash to catch the others up, and there'd never be time.

No, we haven't been on any of the tours that the other people have done. In fact I'm afraid we have done none of the show things that everyone is supposed to do – oh except the Rhone glacier to which we went in a char-a-banc, and that was wonderful. We just go for walks and see quaint little villages and old churches within walking distance. Yes, we have had some beautiful walks and I love Lugano; but I can quite imagine that if you weren't lucky enough to have a slight wind it would be unbearably hot.

As to cures, I have only seen two that could come anywhere near being cures. One was my Italian youth in the train from Lucerne to Lugano and the other was a German youth at the hotel in Lucerne. At least he wasn't a German because he was much too dark and lively, but he could only speak German. It was a pity because he looked most awfully nice and we had got as far as me saying "gute morgen" (haven't an idea how you spell it, but that's how it sounds) and him saying "goot morning", and I know he was just longing to talk to me, but he didn't know a word of English and I know no German, so there you are.

The people of our party are all scotch and middle-aged or more, and they sit at a table in the middle. This pension is only a little place more like a boarding house with only people of our party in it. If you really think you are being treated unfairly I can easily write to Daddy and ask him to supply a bevy of beautiful young ladies; I shouldn't care in the least what you and the beautiful young ladies did; shouldn't trouble my head to think about you at all. So there.

It is after dinner and the wisteria is smelling delicious and the grasshoppers are chirrupping madly and the lights in the town are twinkling and dancing in the water and it is so dark that I can't see, and must go in.

Barnes, I think – no, I <u>know</u> – I shall be able to manage you and make you obey me and do what I want. Yes Barnes, we can and we shall be the same. I should simply hate it if you were any different from what you are, and I jolly well <u>will</u> keep you in order. I will, really truly, Barnes, provided of course that you always submit to being ordered; if you didn't I couldn't make you because you are much much stronger than I. But I know you promised you would, so it will be all right. It will be a trouble of course – no I mean it will be difficult, but I don't think I've ever avoided trying my hardest to do something even though it was difficult. And after all life is much more interesting that way.

Barnes: 3rd August 1924. 28 + 11 = 39 days Rudgwick
I do so long for a word from you. Perhaps there is already a letter for me on its way. I often wonder if you realise just what it is like, and has meant to me all this time, to go on in faith telling you of my love, yet unable to hear of yours in return. I suppose if you love me, it has been no less hard for you to hear and to be forbidden to respond. No one can doubt that we have each had our time of trial. And I daresay that in a way your part has been the harder. My part has involved long waiting; but a waiting tempered and made endurable by the thought of coming action and during the last eight months, I have had the privilege of wooing you openly. Whereas poor you has just had to be a more or less passive spectator of your own courtship. Somehow the picture of you actually returning my love, caressing, touching me of your own free will wont come to me. Poor Childie, I wonder what you are going to tell me in September. Are you an active person? I think so far we neither of us realise that, while you must have a fair idea of what sort of a man I am, yet I have very little more knowledge of what sort of a woman you are than I had by instinct two years ago.

Molly, supposing I had been a man of less determination and your Father had succeeded in keeping me from you, what would you have done? Would you have taken any active steps to retain me? Perhaps you would have just consoled yourself with the thought that if my love was not sufficiently strong to overcome him and wait, then I was not worth having. I do not think that would have been altogether just, because your Father has always pointed out that my love is not only selfish, but unfair to you, and that you would be better off if I took myself away. All of which seems so terribly true that when I look back I dont know how I have had the courage to go on. Especially when you remember that I was completely in the dark as regards your own feelings. The only argument I had for over a year was "Surely Molly would not, could not, go on writing to me if she did not care for me". That was in the affectionate cousin stage.

I think the real answer lies in the fact that I have undoubtedly passed thro' stages; not stages in intensity of love, but stages in the quality of my love. When first I met you, certainly I was selfish. I wanted you; you seemed the only girl in the world for me. Your point of view didn't seem to strike me forcibly until some time later. I dont wonder you didn't accept me that Christmas. I must have been an awful pig.

By the time that I had begun to think more of you than of myself, you had grown to be so much a part of my life that the restrictions between us were almost welcome as giving me a chance to show true unselfish devotion, and as giving you the safeguard that my selfishness and haste had robbed you of. So I was able to go on then with more of a clear conscience, until now I haven't the slightest hesitation in writing you the loviest of love letters that circumstances permit.

Oh dear, I do wish I were by you and could hold you to me. How often I think of our last dance together. Molly, somehow you seemed sweeter to me then than ever before. Was it that you were more yielding, or that I was bolder; it was most wonderful to hold you in my arms, and look down into all the beauty of your upturned face. Dear, dear eyes, so true so gentle, dear firm lips that I do so long to kiss. Darling, this is a really truly love letter I fear. But you see, we have never properly been lovers as yet, for there has always been the restraint between us, that I do not know for certain that you are mine. Tho' I think you were almost mine that night. If I had given way, and kissed you what would you have done? After all I knew it was the last opportunity. Why shouldn't I have done it. Thousands of men would I'm sure. Do you think I'm slow or cold Molly? Is Mummum's fear that I'm undemonstrative really true. Nearly all the lovers one reads about seem to lose their self-control, or "His passion overwhelms him, and he seizes her in his arms and covers her face with kisses".

I wish things like that would happen to me. But there's always a cold little voice in my brain that says, "No you mustn't, you will be breaking your word to Molly", and so I shut up and let you go. Molly, do you think, or have you ever felt that I have been cold? I suppose it would have been far more romantic if I had written that Christmas, and said let us ignore your Father, I <u>must</u> see you, and write to you, and we must be secretly engaged. What would you have done? Would you have been pleased at my ardour?

Molly: 4th August 1924. Murren
I had two letters this morning, one on Saturday and one on Friday. Rather, I remember last time when you wanted to post me a letter from Rudgwick, and I still have a vivid picture in my mind of you dashing down the hill full of Sunday dinner! Yes, of course I still want letters even though I've had them nearly every day for a month; but I still mean what I said about you not writing when you come home late after drill, or are very busy; or if you want to too badly you can just scribble about half a page so that it won't take very long. I shall understand. I should think I do remember how you tried to make me believe it was only $1\frac{1}{2}$d to Switzerland, and I did for a while, though all the time I thought it was rather queer.

I'm glad you've accepted the commission, Barnes. I know what you mean, though. It's the same sort of idea as being voted form captain at school; you have more or less to keep order and make people shut up and you don't have such a jolly, irresponsible time. I know what it's like. Will you be an officer for next camp? If you had had time to ask me, I should have said yes. Please give me your camp title and address.

No, I shouldn't think Sersky had actually undertaken to cure me; she is a cautious person, and anyway I believe she thinks it rather ridiculous to try and do it. I can't say anything about a honeymoon now, and you know I can't say anything about how I felt when you went to Switzerland.

Though Murren is such a tiny place, this hotel is huge, at least it is to me. There are, I believe 250 bedrooms and there seem to be endless lounges and reading rooms and smoking rooms, and crowds of people, and a lovely dancing room. There is quite a good ping pong table and Baba and I have been playing, but of course we, specially she, can easily beat the other people, male and female. There are lots of nice people here and, I should think, some quite respectable cures.

Yes, Barnes, it was very wonderful my feeling you by me on that Sunday night. I hadn't meant to, or tried to in the least, and I s'pose that's the reason why I did. I was practically asleep, and you didn't wake me up a bit, and you were only there for a very few seconds because I went to sleep directly after. That's the worst of it; I never can keep you.

I don't like Italian men so much as French, they do stare so dreadfully. When you go out in the evening after dinner (with Sersky of course; she'd never dream of letting us go alone) every man you pass stares as though you were a sort of curiosity, and the only thing to do is to look as haughtily as you can right past them. Baba said to me the other evening "You notice Molly, they always stare at you and not at me", but that jolly well isn't true, and I told you so that you won't believe it if she ever said it to you. I'm quite sure they don't, why should they; and I'm not a horrid sort of person like that.

When we were at Lucerne we went to an organ recital, and the organist played something which he had composed himself. I would have loved above all things to have had you there to hear it. It was too wonderful for words, and yet I have tried to put it into words because I hate you to miss something that I know you would have loved as much as I did. I sent it to Betty too. Of course it is only the briefest outline and I couldn't attempt to say a quarter of what the organ said, nor can I say it in beautiful language, but it will at least give you an idea of what it was like; and you can imagine one of the best organs in Europe telling you in the beautifullest grandest way about a storm in the Alps. You could hear everything – the peaceful evening, the yodel and its echo, the cattle with their bells, the shepherd saying his prayer, the thunder getting louder and louder – that was truly awful – the lightening and the rain, and in between, the frightened shepherd trying to pray, the gradual ceasing of the storm, the setting of the sun, the moon, the shepherd's thanksgiving, and the light of his lantern as he went home. Of course it might all have seemed quite different to you if you had heard it, but I think it must have been something like that.

Barnes: 4th August 1924. 27 + 11 = 38 days Rudgwick
Alas, the one and only wretched post has come, and still no letter. I try not to feel depressed and miserable but with very little success. These posts are simply awful, for suppose a letter from you had reached New Cross on Saturday evening, it could not get here till Tuesday morning. And a good many – at least I've only had three, and two out of the 3 have been delivered at 9 in the evening, so that if Fanny did repost it, it wouldn't get here

yet. And now the only post out goes at 11.50 this morning so that I haven't much time to write.

Molly, dont think I'm complaining, or really expect to hear of you more often – I know I've no right to get any letters at all, and you mustn't take any notice. Only its natural when you are the one being in the whole world on whom all my thoughts and longings are fixed, that I should live simply for a word from you. I often wonder if you ever stop to think of how much time I spend in writing to you. I suppose ever since I started the maths in Switzerland, it has rarely been less that 10 or 12 hours a week. Those articles used to take me ages to do. I dont really write very quickly, one stops to dream of you so much. Yesterdays (Sunday) letter took about $2\frac{1}{2}$ hours, Saturdays an hour or more Fridays $1\frac{1}{2}$ – the whole train journey from London to Horsham, and the train was late; so it would soon add up, even in these busy days.

I wish it were tomorrow morning. Yesterday it poured with rain most of the day but after I had finished and posted my letter to you I felt I must go out, so I went for a long walk. I got simply soaked, so wet that my shoes got filled with water, from running down my legs; and in addition in trying to find a new footpath I lost my way, so that I did not get home till half past eight nearly an hour late for supper. Stupidly I only brought one pair of outdoor shoes, so today I cannot get out till they are dry, and that is a long business. I'm going for about 12 or 14 miles when they are, as I simply cant stay indoors just to mope about and think of you. Work is impossible. I hate holidays, and wish I were not going to camp. When ones mind is not occupied the time simply drags. I used to marvel at how dear Mother used to write me a letter every single day; nobody is more surprised than I at being able to do it myself. But I know the secret now, and I suppose you will have your time some day when you write your daily letter to a regardless son, from whom you will with difficulty extract a brief and grudging reply perhaps once a week if you are lucky. I can tell you exactly what he will say:-

Dear Mother,

Thanks for your letters. Isn't the weather (or whether, or wheather, or wether) awful. Jones broke his leg yesterday playing rugger. There is no news.

Your loving son

———

Please send me some clean collers.

Leaving you in an agony of apprehension as to

1) how your darling really is,

2) whether he is likely to break <u>his</u> leg – you dont care two hoots about said Jones.

3) whether Daddy (whoever the lucky man is – lets hope he's patient, and in his time has written many letters) hadn't better write to the Headmaster saying that darling – – – – – – was not to play that dreadful rough football, and couldn't he be taught to knit instead?

4) how your darling really <u>is</u>

and 5) how your darling really IS

and 6) HOW YOUR DARLING REALLY IS? ? ? ? ?

Certainly the little wretch wont know the pain he is inflicting, and if you try to tell him he will simply say "Well, you know I'm all <u>right</u>, so what is there to make a fuss about" or words to that effect. Ah well, thinking of your future troubles has almost restored me to good humour.

Barnes: 5th August 1924. 26 + 11 = 37 days In the train on the way home This is a most app(stop)allingly jiggley train, being an express to London Bridge, and its really almost impossible to write at all. However, I may not get another opportunity, as I shall probably be very late home tonight, but I am not sure. No letter for me at Rudgwick, but perhaps one will come tonight, or perhaps Fanny has redirected it to Vickers. I go on hoping. Childie dear I know you haven't any time to write to me, but I cant help being most awfully anxious about you. You see there are so many things might happen to you, you may be ill, or have got run over, or tumbled on a glacier or something. I just <u>hate</u> your being abroad. You can't realise what a time of anxiety it is to me, because you are young and perhaps dont understand quite how my heart yearns for you. I know its selfishness really on my part, and I hate to be a bother to you. Do you think you could spare the time just to send me a post card, say twice a week? You see Childie dear, even if you are not going to have me, and want to stop writing too many letters to me, I <u>care</u> just the same, and it <u>is</u> such a relief to know that you are well and happy.

<u>Home again</u>. My Darling, your wonderful wonderful letter [31st July] was waiting for me at Vickers house. Oh my dear Mistress, thank you and thank you again and again for all your great kindness and goodness to me. Oh Molly Darling, you cant think of the relief – I almost cried when I got it. Please forgive me for being impatient – I have left what I have written above. You must know me just as I am, and I see no reason for destroying what I wrote when anxious and afraid.

Molly, most of my letters have had a sort of purpose behind them – to show you as fully as might be in the short time that is left for you to chose, something – as much as possible – of what I am like. The question of our relationship naturally comes first, and I have shown you how I should like to be treated by you. The reason <u>why</u>, I have never fully asked myself, or tried to analyse; for to be your servant seemed so utterly natural; but on thinking it over, it seems to be based on fairly logical grounds, and it may help you to understand if I tell you of them.

In the first place, for all "real life" serious things the question of Mistress and servant would never arise between us. Love as deep, and as tender and (as I pray to be made) as unselfish as I feel for you, settles all real differences between us. I <u>never never</u> could do anything that I <u>knew</u> would cause you a moments real grief or pain (I might have to be told – many men offend or hurt in ignorance).

But in all the minor details of our daily life I have, for you and for you alone in all the world, the servant instinct. – I wish to be ruled by you, my sensible reasons I have told you – I dont wish to become spoilt – my (perhaps to you) unsensible ones are the pure joy of being made to serve you.

Now here, you might imagine my love would again step in, and compel me to say meekly "Yes Molly", "No Molly", "Certainly Molly" to every order you gave. But not a bit of it. The lap dog type of obedience I simply cannot <u>bear</u>. If I am to be made to obey, I must be made to obey by one who is prepared to assert and maintain her authority absolutely, despotically. Can you see the difference between a crawling lap dog, and a high spirited horse – I mean the mental difference. The lap dog comes crawling and whining at the mere stir of your foot. The horse must go thro' a long period of breaking in with bit and bridle, with bearing rein and whip. To break him in is a mans job, and only a ruler of men can do it. When it is done, he is your devoted, utter servant, not mean and cringing, but still manly and independant, and still inclined when feeling gay to take the bit between his teeth and go off, carrying you along too.

Which is the better servant, which the better companion? I know which you would prefer.

Well, I feel just like a mettled horse. If you do not wish to do this, I shall try to fall back on the lap dog type, but I <u>know</u> it will lead to surliness and domineering on my part.

Somehow, Molly, rightly or wrongly it is necessary for me, to be <u>com-pletely</u> confident and happy in my service to you, to feel that you are a ruler fit to rule; not only to feel it but to experience it.

In the process, I shall deliberately disobey you again and again, until you learn to control me. Then for a time I shall be quiet and docile, but with a twinkle in my eye all the time.

But if you are not prepared to enjoy Mistress-hood it will not be much fun for you. That is now the only thing that worries me. Molly, I would never do it if I knew you were worried or tired, or out of sorts. I'd just long to tend and serve you then, with all my love and gentle care. But if you were well and strong, dont you think you too could learn to enjoy the battle, your will against mine?

I must fly for the post – I loved every word of your dear letter.

Chapter 18
DOUBTS AND FEARS

Barnes: 6th August 1924. 25 + 11 = 36 days New Cross
Only a brief note tonight to tell you of my constant love for you. And to say this – that after a nights reflection I am quite certain that one thing I said last night is not true, namely that my complete happiness would depend on whether you could rule me or not. Childie darling, I'm quite quite sure that it doesn't depend on that at all – it only depends on love. In all the big, serious things you already rule me by love and I really know and feel that so complete already is your rule and dominion over me, that if you chose, you shall rule me in the little things by love also. Childie dear, I <u>do</u> love you so. I read and reread your dear letter. I <u>am</u> so glad that you love the little places. To lie down and poke my nose into a little ferny mossy nook makes me feel all pathetic and lonely, whereas old Mont Blanc leaves me unmoved.

Childie, would you mind remembering (that even if I were not in love with you), you are a girl in ten thousand – else why did I immediately fall in love; you are <u>not</u> at all an ordinary person. Your estimation of your comparative qualities may be a fair one, but you miss out the two really outstanding qualities that you possess – character, and a powerful intellect.

You may think me queer, but do you know I cant bear making even in thought, a sort of catalogue of you. I just like to think I love you, and loved you, from the first, because you're <u>Molly</u>, <u>you</u>. Whether fair or dark, short or tall, cross or sweet you're Molly, Molly, Molly.

Dont think Dear One, that I dont appreciate and realise all and every beautiful trait and sweetness that you possess; that I dont see and love your beauty; I <u>do</u>. But that wasn't why I fell in love with you. Even a man in love must admit that the world contains other beautiful and sweet girls. But there's only one YOU in all this great world. And if you were cross as two sticks, and lame and ugly – which heaven forbid – I would love you just the same, as long as you remained you. YOU is not dependant on cleverness or beauty, tho' it happens that you Molly possess both. I'm frightfully <u>proud</u> of you because of that, but it isn't my fundamental reason for loving you. I'm not good at explaining this sort of thing, but I can love all right.

No, I dont think you are slow – I suppose I <u>am</u> something of a freak to get used to. Did you really feel like that when you left New Cross? [in April 1922]. Do you know, I used to try not to look too much at you, and not to attract your attention that week, because it seemed so unfair. I feel as if I had lost a week of life and joy when I look back.

Barnes: 7th August 1924. 24 + 11 = 35 days New Cross, in the train on the way home from drill.

Thank you very very much for your letter [4th August] which arrived this morning. I've only just had time to read it in the train coming up, but oh Molly, I <u>do</u> think you're clever. I love the peasant and got quite excited at the thunderstorm. I will read it again and again tonight in bed.

Barnes: 8th August 1924. 23 + 11 = 34 days. New Cross, in the train on the way home from drill.

As you see, I am putting in every possible night at drill (train starts) that I can, because when it is so near camp everyone is expected to do a good deal. I went tonight expecting to have a lecture all to myself from one of the regular sergeant majors, but found a few men there to whom I had to give a lecture on breach mechanism myself. Still, it wasn't wasted time, because one is very apt to forget little details, as it is a complicated piece of mechanism; and it is just as well to make sure of one's own knowledge before going into camp.

Auntie Nelly is here, she came from Hampstead on Wednesday and sends a message to say she is going to write to you. She thinks poor Irwin Carruthers [Barbara's friend] is awfully worried about Baba, and really cares for her most awfully but has been sent away by Uncle Arthur. I fear your dear Father is responsible like his Father before him, for a good deal of unhappiness – its all right if things come right in the end – one can look back on one's suffering as something to be very proud of – but if he parts people for ever who could have been happy, it seems an awful responsibility. But perhaps A.Nelly is mistaken, she has the Bloxam imagination.

♦ Barnes whiled away the August Bank Holiday in Sussex, with Auntie May and his two frail cousins Elf and May.

Barnes: 9th August 1924. 22 + 11 = 33 days In the train, on the way to Rudgwick

Mistress dear, please forgive me for the letters I wrote last week. You know perfectly well that if you never wrote to me again, yet I would always adore you and faithfully love you. But I cant help being anxious while you are away. And please Molly, I never meant to suggest <u>for one moment</u> that I didn't get enough letters that would be the most <u>awful</u> cheek.

No – I shall not go to camp this time as an officer the papers wont get thro' in time and I'm very thankful. We travel on Sunday 17th.

Molly darling, why shouldn't people stare at you? Dont you yet realise your sweet beauty? Why, I wish to do nothing else but gaze at your dear face for ever and ever. I could watch every movement every graceful action, your dear walk – oh Molly how I wish I could see and touch you now. My Childie, how long the days seem – and yet they <u>do</u> go, but oh how slowly. Will September never come?

Barnes: 10th 1924. August 21 + 11 = 32 days Rudgwick

I had time yesterday to dash across to the up platform at Horsham and post your letter myself, instead of leaving it to the mercy of a porter as I had to do last week. You see if I waited to post in Rudgwick, I should miss the post.

This week end May is home from school, where she is a house mistress. Thats Elf's sister you know; who owing to finding your photo in my room at home knows about you. She's not at all strong, poor girl, but very plucky.

[Auntie May] and May have gone to Church and I am sitting in a deck chair in the garden. Childie dear, do you mind my letters being untidy when I write to you? It isn't meant to be careless or disrespectful, but one cant write nicely in trains, and it is so nice to sit out and write whenever I can.

What do you think will happen about Carruthers and Baba? I'm most frightfully interested, because I can so sympathise with the miserable fellow. But I cant understand why he didn't explain things to Baba, unless like myself, he was forbidden ever to mention the subject again. I do wish now that I had not obeyed Uncle Arthur so literally that Christmas, but had written at least one letter telling you of his action, and explaining that I was going to wait. It is too cruel to separate two people without allowing them a single word together, and I blame myself very much for giving in. If Carruthers has received the same treatment, as Auntie Nellie believes he has, the poor fellow has been unable to tell Baba that he still cares. The bitterness of seeing you without being able to speak was almost too much to face. Tho' I always ended up by coming whenever I got the chance, as of the two unhappinesses, that of stopping away from your dear sweet Presence seemed the greater.

I sometimes wonder if it isn't more comfortable not to love quite so much – one wouldn't get so intensely happy it is true, but on the other hand one wouldn't get so profoundly wretched. Dont you sometimes feel quite pathetically sad over some small thing? I do over nearly everything connected in the very least with you, when I am separated from you. Anyhow, its no good talking about it, because I can't alter the depth of my love for you, and I jolly well wouldn't if I could, after all. I always think I have got to the limit of intensity of feeling for you, and yet each day I want you more and more.

Oh Molly, many people say all women like to be mastered. In their heart of hearts they like the great strong cave man, who lifts them up and carries them off, whether they want it or not. Molly, dont ask me to be masterful; out in the world I s'pose I am, and I think thats just why I long to serve you so greatly; why the relation of Mistress and servant appeals to me so much.

Childie, my letters seem to be about nothing but my love. I cant seem to think or write of anything else. If I cant be in your presence, the next best thing is to write of it and think of it. And to dream of meeting you again, and once more holding you in my arms. Molly, if you refuse me, will you let us go on as before? May I kiss you even then, as far as your shoulder, which ends where your dear glorious hair begins? Or shall I be forbidden for ever to approach you?

I dont picture what will happen if you refuse me, but if you accept me, the programme – subject to your approval is that after a while we leap into a taxi – with thick blinds to it if possible, and a taximan who has the sense not to turn round – and drive to a jewellers in town, where you shall chose an engagement ring. By that time we shall be hungry, and it will be too late to go back to Hampstead for lunch. So we shall go to the Royal Air Force Club for our first lunch – you needn't meet anybody, there is a room for members entertaining ladies specially – coffee all to ourselves in the drawing room. Then, we shall leap into a train and go and get Auntie Fanny and Uncle Charlie's blessing at New Cross, and have our first tea with them, and then we shall look at my bedroom, and then we shall leap into another taxi because I shall have packed a bag at New Cross, this taxi must also have curtains and a deaf mute and blind driver, and we shall drive as slow as may be to Hampstead, where we shall have our second tea with all your family. And I shall stay the night (Molly, its not cheek – its all subject to your approval).

And on Saturday you also will pack a bag, and we shall leap into a taxi, (+ blinds and blind driver) and we shall put our things into the cloak room at Victoria, and I shall take you into St James' Park and leave you (miserable hour) for a little while, while I dash to my office, and see if there is anything very urgent.

Then I think we will lunch at the club, or anywhere, and then we'll go for a swim together at the Chiswick open air baths. Then we'll leap into a train and go to New Cross after having got our things from the Cloak room. At New Cross we shall have our 3rd tea. And you will stay until Tuesday, or Wednesday or Thursday or Friday or Saturday, JUST AS YOU PLEASE. On Sunday at New Cross, we wont go to Church (perhaps we will if you wish), and we shall do everything that we please, how we please, and when we please. And on Monday, I shall have to go to work, but we will lunch in town, together. And on Tuesday and Wednesday and Thursday, (when we might also dine and dance or, better still have a family dinner in town Mummum and Daddy and Auntie Fanny and Uncle Charlie and YOU and me). And on Friday, p'raps I'll sleep at Hampstead, and on Saturday (20th) we shall both pack our bags and leap into a windowless automatic driverless taxi, and go to Victoria, and you will look at shops, while I go to work, and then we shall lunch together, and then we shall leap into a train for Rudgwick, and you will come here. And on Saturday after tea, we shall go for a walk. And on Sunday we shall go for a walk. And on Monday you will get up awfully early, and we shall go back to town. And I hope all that week you will live at New Cross, and go to the place in Buckingham Palace Road, and we shall breakfast together, and lunch together, and dine together, and on Saturday – oh Friday's my birthday. We shall do something special, but I cant say anything about these weekdays, because by that time you will have grown up completely, and taken charge.

Molly dear, its post time and I must wake up and come back to a lonely earth again. And now I must say goodbye, but always and ever you know dearest Girl, I love, love, love you.

Barnes: 11th August 1924. 20 + 11 = 31 days. In the train to London.
I just managed to catch the post yesterday afternoon without disarranging my Sunday dinner. I wasn't a bit energetic yesterday and lazed in the sun

until tea time, after posting your letter. And then after tea we were all four going for a stroll, but May asked me to help her with her income tax, and it took so long going back over past years that it was supper time before we had finished.

Its extraordinary Childie how frightfully incapable a woman seems when she has never had to manage her affairs properly. None of the three dears has ever been shown how to do things and now that taxation has become so important they feel very helpless. Thats why I want you to become managing director from the very start Dear One. Of course I will help you when you wish and until you get to know about money and business but I think its very important that you should get accustomed to carrying on without constantly leaning on someone else – you may make a few mistakes, no one can run a show on their own without; but it does render you capable of doing without me as a matter of course, should anything happen. Agreed?

It was a glorious day yesterday, and the country is looking very wonderful, with all the corn in stooks on the fields. I didn't hear a threshing machine, but dont you love the drone of one at a distance, with the whining scream as each fresh bundle victim is thrown into its devouring jaws?

The new little maid is a poor hand at pumping water, so last night I filled my bath overnight with buckets from the well. Seven large bucket full's is as much as I dare put in, otherwise the bath might overflow when I got in. Its so narrow, I have to lie down in it sideways!

All the downs were covered with a gentle summer mist this morning when I left to catch my train soon after 8 oclock, and it all looked and smelt so sweet and dear.

Home again. 9.30 p.m.

I found your postcard waiting for me here, thank you very much indeed for it. But oh dear, what am I to do, for at least 4 letters have gone already, Thursday's, Friday's, Saturday's and Sunday's, the two latter posted at Rudgwick. And nowhere to write. Molly I think its shameful, that they should book your rooms, and then I suppose just because some horrid wealthy American tourists have promised the wretched German hotel keeper untold dollars, he has let them stay.

♦ In the gathering dusk of a hot August evening at 23 Pepys Road Barnes broke off his letter. One from Molly had arrived, a letter begun in the cool brightness of a mountain dawn.

Molly: 7th August 1924. Murren

It is a quarter to five in the morning and I have just been looking at the beginnings of a most glorious sunrise. The sun rises behind the mountains just opposite our window, so that I can watch him – rather the effect of his rays on the snows because he isn't visible yet – from my bed. I can't go to sleep again because the morning is too lovely and I can't get up because I shall disturb Baba, so I am writing to you in bed. The last two days have been absolutely perfect, and this one looks as if it would be as good. You know those pictures you see of snowy mountains in the evening with their snow all crimson and gold, well I always thought it was most awfully exaggerated but last night I saw tht it jolly well isn't. There was a sunset to match this sunrise. And the night, too, was just perfect with the little moon and the stars shining on the snow. This is a beautiful place.

Friday morning. And I never got any further. I spent the rest of the time till 7.30 reading and re-reading your letter, thinking what to say in answer to it. Watching the sun, and dreaming – not asleep dreaming. And now there are two more letters [4th and 5th August] which came yesterday.

Barnes dear, I am most awfully sorry you were worrying. I have been trying to write twice a week, but somehow my letters always seem to collect after the week end. Look here, Barnes, writing to you isn't a bother or a trouble; I like writing to you as much as you like writing to me, I guess, and I have never thought "Bother, there's Barnes to be written to" or anything like that. You did make me laugh over your letter [3rd August]; I'm sure I should never get so agitated to know "how the darling really is" – I don't know, p'raps I should. Anyway, I know how you feel about me; I felt it myself once; only it is awfully hard to realise that somebody is anxious about you when you are there with yourself and know you are all right.

How could I be annoyed with you because you were worried and told me so. I'm glad you're worried – at least I don't quite mean that, and you won't have to be again; what I mean is I should hate it if you didn't care whether I wrote or not, and if you didn't get a little impatient to hear from me. And I quite agree that a fellow may be excused a little anxiety. Oh dear, it is a muddle.

When I come to think of it you certainly do spend a tremendous lot of time writing to me. It is jolly ripping for me and I don't know what I'd do without your letters.

Yesterday we went for a dear walk, the very nicest I've been anywhere yet. Through hayfields and meadows filled with flowers, by streams and waterfalls and through many pine woods. We went on Tuesday and we loved it so we decided to go yesterday and take our lunch and walk back; before, we had come home by funicular from Lauterbrunnen. I found heaps of wild strawberries and raspberries and bilberries, and I walked behind Sersky and Baba so I could think. Wherever the road was wide enough for two I left plenty of space on my left and pretended you were there, otherwise I went on in front. And we looked at all sorts of things together – things which the other two would probably think I was crazy to bother about – strange soft little mosses, huge ants carrying things almost as big as themselves, lizards sunning themselves on stones, baby water-falls, sweet scented flowers and leaves and hundreds of other things; and also we discussed your letter, the one you wrote on Sunday [3rd August]. But it is difficult to carry on a discussion when your companion is only a pretence one.

Barnes, I cannot tell you now what I should have done if Daddy had succeeded in keeping you from me – not because I don't know; I am quite certain, because even then the possibility that you might get weary of Daddy always putting restrictions on you, often came to me; and I was quite decided what I would do. Only how can I tell you now. Ask me later and I will tell you if you want to know. And Barnes I did appreciate even then something of the unselfishness of your love for me during that time; still more I am learning to appreciate it now. It is hard not to be able to do or say anything, but to have to be so absolutely passive, or at least it has been hard, and it is difficult now to write. All one could do last year was to write pretending letters and tear them up.

There were two of me at that dance. Do you remember next morning you asked me if I knew how much you wanted to kiss me then? And I said I didn't know you did particularly just then. Well I did know all the time, only Saturday I hadn't realised it – I hadn't realised that there had been another me. I don't s'pose you'd understand; but one part of me – the hot eager part, I think all the pulses I have in my body – wrists and temples and fingers and everywhere in me – kept on saying "Barnes, kiss me, kiss me; if you'll only kiss me just once, I shan't care what happens to me for ever and ever", and that part knew it would be awfully easy for you to do it, and it didn't want to make it difficult. But the real me was very very glad you didn't, and she didn't realise it till some time after that there had been another me. It's quite clear to me of course, but I don't know if you will be able to understand. I wouldn't have told you if you hadn't asked me, because I suppose one shouldn't feel like that, but I can't help it – I always have and I always do. Barnes you have never never been slow or cold to me. It doesn't matter what anyone else says or thinks – I say that you have been and are a very perfect lover, the dearest that was ever fashioned. I think very much more of the love of a man if it is true enough and deep enough to enable him to control himself and restrain himself. Surely that love is far greater than the sort of passion which selfishly overwhelms everything; and the passion is there all the time – it is only controlled by a love which is greater than it. Of course I don't know if that is all so; what I do know is that I'm very glad you did not kiss me then; it makes it much more beautiful for afterwards.

I suppose if you had written and said that to me two Christmases ago, I should have been excited and pleased at first, but afterwards I should have been disappointed. I believe it would have spoilt it to have had to be engaged secretly and to be sort of ashamed of it; and I don't care what anybody says, I don't think it would have been straightforward, but rather mean. But mind you, there would be a limit; I mean if your father went on saying wait, indefinitely, then you should certainly do something. I think a year is about the most anyone could bear. If you hadn't spoken to Daddy then, I should have.

Yes Barnes, I <u>shall</u> enjoy it. I much much prefer you as a wild horse that has to be broken in than as a stupid, hopeless, helpless, feeble lap-dog sort of a thing. I should just hate you like that. At first I was doubtful – I mean right at the very beginning when you first suggested it – and I didn't understand what you meant by it all; but I really truly do understand now. I know you much better now than I did then. I know you would become dominating and would want me to subject myself to and be sort of swallowed up in you (I know it would be unintentional and it would be my fault) and I'm glad of it; I'm glad you are that sort of man – but mind you, I shouldn't dream of allowing myself to be dominated by you.

Barnes: 11th August 1924 continued.
Your dear letter has just come by the post, and I have delayed ages reading it.

Darling Molly, I've just hated myself for writing those whining letters. Of <u>course</u> I know how busy and fully occupied you are. Molly, it wasn't letters, I didn't ask for that, only the briefest of cards to know that you are safe and well. I am a selfish hateful pig. Please, please Childie dear, I am the very <u>least</u> of all the people to whom you have to write. I <u>do</u> know that.

<u>12.8.24</u> 19 + 11 = 30 days

Molly dear, <u>not</u> to be written to is perhaps the greatest privilege of all from you because it means that you can so trust my love and devotion.

Barnes: 12th August 1924. 19 + 11 = 30 days New Cross
I can't do more than write a very few lines tonight, in part answer to your very dear letter that came last night.

What lots and lots we shall have to talk over together if we are engaged. No, I suppose you mustn't tell me things now. I am glad you have told me about the dance. I can understand perfectly about the two you's, because you see there are two me's, just the same. But I've always been most painfully conscious of the second me. I dont see why ever one shouldn't feel the second part – its only natural to do so. But my second me gets most awfully tempted sometimes, and it was the knowledge of him, and the wish that if possible he shouldn't stir the second you, and so cloud your judgement, that led me to suggest, on our first afternoon alone together. that we shouldn't kiss till we were engaged. For it seems a most frightfully desirable thing to do.

I too am glad now that we have endured to the end. It has been hard, and has left some deep marks, but we both know that should you have me at the last, our love is not a thing of a moment, but has those enduring qualities which will enable us to go thro' life hand in hand, lovers to the end.

I think its my first me's dread of my second me – the eager, unruly, dominant passionate me, – getting the upper hand, when once you have surrendered yourself to me, that makes me so insistent on your ruling me. Hitherto the first me has been dominant, because your happiness and life depended on our remaining cool and restrained. But when that necessity has disappeared and I get slack, I do so dread the second me taking charge and stifling the first me, and dominating you; and so almost instinctively I have urged you to the subjection of him.

I didn't think of the two me's when I wrote that in all the big serious things, our loves were so great and strong as to make us safe. But it all fits in most wonderfully, and it is the second me who must be ruled. Dont you agree?

Molly: 13th August 1924. Kandersteg
All your letters have been faithfully forwarded from Wengen – four of them – and thank you very much.

We have had such a pricelessly funny evening, dancing. A band (?) came from the village, and all the fat Germans and their wives danced (they are nearly all Germans here) and solemn German youths danced solemnly and seriously with one, not bothering how often they bumped you into other people. It was more fun when I danced with Baba and had to steer her among the fat ones. But the room got most awfully hot; for as often as you open a window, some German comes and shuts it again. Then afterwards, Baba and I put on our coats and out-of-door shoes, and put up our umbrellas (it is pelting and pelting) and we were lighted up to our chalet by the boots with a lantern, for it is pitch dark without a moon, and you might miss the path. And now I am in bed in my dear wooden room.

Oh Barnes, I have wanted you here more than anywhere else, specially last night. There was a moon and lots of clouds, and it was far more wonderful than when there are no clouds, because you get great silvery cloud wisps caught in the mountains, and there are huge black clouds with silver edges lying on top of the mountains. And the mornings too are so wonderful – always so new and fresh and beautiful; and the wild strawberries you find seem much nicer when you eat them all dewy before breakfast.

I am too sleepy to write any more to-night and it's ten to twelve. Goodnight.

Good morning. It is raining harder than ever, and all the mountains are covered with an entirely new coat of snow. But I still love this place, misty or rainy, cold or hot. It is cold; I should think it might snow down here.

I was so sleepy that I could hardly see to write straight on the lines. Now don't think I didn't want to write to you; I want to write to you when I am sleepy because then I have you well in my mind immediately before I go to sleep, and it is such a cosy comfortable feeling. Only I hope it isn't too muddled.

I s'pose life would be a lot easier if one could manage not to do things so hard – love so intensely, hate horrid mean things so much, feel so dreadfully sad sometimes and so almost frightenly glad at others – but after all it is much much better than being just sort of middling about everything. Oh I am so glad you are as you are. I'm so glad glad glad there's a you. What do you think I'd like you to write about? You and your love for me are the most interesting things in the world and I could never tire of hearing of them. If I refuse you, I shouldn't let you go on as before; if I saw you at all, it would be as an affectionate cousin sort of person.

Oh Barnes, you quite took away my breath, so many times did we leap into taxis and so private did the taxis become, ending up with the windowless, driverless, automatic one! So you too dream dreams about what you and I are going to do together after my birthday. Oh well, 4 weeks more; but four weeks seems every bit as long as six did a fortnight ago; I never believed time could go so slowly when you were enjoying yourself in such beautiful places.

I must stop because I haven't an idea when the post goes; the box says 10.10, but I don't believe it for a minute, and there is only one post a day and I think that one goes when it feels like it. I want you to get this before you go to camp. I hope you have a jolly good time at camp. I _did_ love your letter. Goodbye my Barnes.

Chapter 19
THE ANSWER

Barnes: 13th August 1924. 18 + 11 = 29 days New Cross
Thank you so very much for your postcard which arrived at 9 oclock tonight. I <u>am</u> pleased that Kandersteg is a good place. I am just back from drill, and after snatching a cold bath and a mouthful of supper I am writing to you.

I seem to be very busy, and yet get little done, and seem very tired these days. The truth is I am feeling this last part of our long wait worse than any that has gone before, and it is upsetting me a good deal. I'm simply on edge, cant settle to anything and my work is simply ruined. You come into everything, and its quite uncontrollable. I suppose I've held out so long that now I'm thoroughly done up – no, I shall be all right tomorrow, you mustn't take any notice. After all Molly two years of <u>uncertainty</u> is a long time. People are sometimes engaged for longer, but I never heard of anyone being kept in suspense for so long. I've kept my patience fairly well up till now, but now, when I'm tired Dear One, its just about the limit. If you feel anything like the same, I'm most fearfully sorry; and I do think its a splendid thing for you that you are seeing so much that is new and distracting. I'm looking forward to camp most awfully, as it will occupy me pretty well all the time, and being so much with other men, one cannot think too much.

Childie Dear, this <u>is</u> a mouldy letter – dont heed it, I feel better already for writing to you and thinking of you, and all you are undergoing and feeling. I've no right, and its just selfish, to unload my worries on to poor you, who probably have heaps of your own, to cope with. I'll jolly well go to bed directly its post time, and sleep and sleep. I wish I could dream of you. I never do, tho' thats not odd, as I hardly ever dream at all. And Childie dear, you know I'll jolly well wait for ever for you – I'm only getting worked up because its so near the time. If you were to tell me you couldn't answer for another year I should probably settle down again.

new ones to go away? Do you often wear your pretty silver shoes that you got for our last dance? Molly, they <u>were</u> so dainty. I have never – (yes I have you <u>always</u> look beautiful), but I was going to say I have never seen you look so sweetly pretty as you did that night. But you looked radiant, and so <u>so</u> sweet. I <u>am</u> so pleased that you can dress so rippingly, not that it makes a difference to my love, but I do so like to feel that others admire you, it makes me very very proud and happy.

Molly, its all just part of the wonderfulness of you and me. I daresay most men dont care to worship so "actually" (bodily, would perhaps be better if you understand the word used so), as I do. They may be content with the mere form of words as one certainly reads in books. But I simply must, day after day, give actual bodily expression to my worship. It perhaps does help you, although perhaps you may not enjoy it, to realise my whole attitude towards you, and something of how intensely I love you? If I were proud and condescending and <u>entirely</u> selfish (I know I <u>am</u> selfish) I dont think I would <u>want</u> to do it, <u>long</u> to do it do you?

And so I do feel that although when I fell in love with you, I neither knew nor thought of any of these things, it is so wonderful that we so exactly seem to fit each other. I never <u>dreamt</u> that you were the very one woman who does not want to be mastered, but to master. And yet for us to be happy I am sure you <u>must</u> master me – a real fight and victory of your deep force of character against mine – no less deep I think in its way, but willing to be ruled by one who is strong and determined enough to do so.

But how in the world did we know it all, when at first glance I fell in love? I give it up. But I <u>am</u> so glad you are as you are. I'm, oh Molly, so <u>more</u> than glad, so unutterably, devoutly, thankful that there's a <u>you</u>.

But my Childie, you say "if you refuse me" I may not go on as before, but if you saw me at all, it would be as an affectionate cousin sort of person. Childie, how could I go on trying to win you if I might not see you – I wont <u>bother</u> you but you <u>dont</u> mean you wont give me another chance? Molly darling, I've told you I wont <u>worry</u> you, but equally until you marry someone else I <u>cant</u> give you up. Of course I dream dreams of what we shall do if we are engaged – all day long and half the night too.

I will finish this now as I am Battery Orderly Sergeant tomorrow, and may not have a moment. Would you rather I did not write so frequently now you are at home?

Molly: 20th August 1924. In the train from Paris to Calais
My dear sergeant,

I'm <u>so</u> excited. The train is saying "I'm going home. I'm going home" also it is wobbling most terribly, also the rain is coming in from the window, and the smuts are coming in and I'm smearing them with the rain drops over the paper. So you must excuse anything funny in the look of this letter.

I do not like Paris the least little bit, though I s'pose it's hardly fair to say that considering the very little I've seen of it. The journey to Paris was hateful. I felt most awfully sick, I don't know why, I never do on trains as a rule, and we had had no lunch except a roll and butter. Practically all the way from Berne to Paris I lay on the seat (luckily the compartment wasn't full) feeling wretcheder and wretcheder. We arrived at Paris at 12.10 and spent 40 minutes or so trying to find our luggage, which hadn't arrived. Then we drove to the hotel and spent some time filling in forms and things. Also I found your two letters and they were like an arm chair in the middle of a ploughed field over which one has been walking for days; or like it would feel to have your arms round me if I were very very tired. We were in bed soon after two, but it was practically impossible to sleep, firstly because of feeling sick, secondly because it was so dreadfully hot and stuffy and noisy after dear quiet cool Kandersteg. Sunday we spent wandering round Paris. I should have loved to have stayed in bed and not have bothered to eat or walk about, but then Sersky and Baba couldn't have gone and lunched out – at least they could have, but Sersky wouldn't. On Monday and Tuesday we went to Fontainbleau and Versailles with a crowd of people and a guide. Never, as long as I live, would anything induce me to go anywhere like that again. I have had the most delightful and lovely holiday, and I have enjoyed myself most tremendously. I am very glad I have been because it does sort of enlarge one's views and ideas, and incidentally I have learnt a great deal about you that I don't believe I should have learnt so quickly if it hadn't been for all your letters. Barnes, thank you ever and ever so much for them all; I have read and re-read and loved every separate one – 29 altogether. It would have been a desolate sort of a holiday without them.

I can't attempt to answer the last 6 now because I have to have them all spread out before me and Sersky would have a fit, but I will directly I get to Paignton if the family will give me the chance.

<u>In the train from Dover to London</u> I have come to the conclusion that England is the dearest most delightful country in the world. I just wanted to kiss the first Englishman who spoke to me – a dear grubby window-

cleaner man who came to clean the windows of the carriage in which I was keeping seats while Baba and Sersky had some tea (I was feeling too boatish).

The train is saying "I am in England, I am in England", and it is so beautiful that I don't believe I shall ever want to leave it again. Oh yes, I expect I shall one day. It is worth going abroad just to come back to England again. Poor poor French people. I don't feel so sorry for Swiss people; they are so nice, and Switzerland is such a glorious place, but fancy having to be French and live in France, or in Paris instead of London. Sersky says that to stay in Paris properly you have to have a man with you and then you can go out in the evenings, a thing which she would not let us do. She says "You must get Mr Wallis to take you one day Molly". There are scores of things she had said I must get Mr Wallis to take me to.

♦ Sersky handed back the two girls, twenty-one and nearly twenty, to their solicitous father. But, as Mr Bloxam must have realised from Molly's eager snatch at the post and her absorption in writing, she was not cured.

Molly: 22nd August 1924. Paignton
<u>The evening</u>. This morning I was seized upon by Pam and George to be shown something. After breakfast I was taken to see the town, and then we went out for a whole day walk taking sandwiches and getting tea at a little cottage. I did so want to sleep with Betty for the whole fortnight, but the others were so awfully disappointed that I have had to promise to let Pam sleep with me the first three days, Nancy the second four and Betty the last week. Pam is a darling; the bed is a double one, very small and squashy, so it gives her an excellent excuse to snuggle up to me and cuddle me all night.

Do you remember that letter in which you said you were feeling impatient and tired? I'm jolly glad you did send it. I'm glad you are just ordinary with me, and it isn't as if I didn't understand how you felt. You feel all on edge and most dreadfully bearish when anyone talks to you, and you try to do your work but it needs a lot of hard thinking – not like actual manual labour – and all the time your brain is going round and round in a circle and you are just wanting and longing and longing and wanting, and forgetting entirely that you ever had any work to do. And I do know so well what it is like to have to wait post after post for a letter and the awful feeling when there isn't one for you and you wonder if all sorts of impossible and ridiculous things could possibly have happened; but that has only happened to me once. I <u>do</u> hope the great invention is a success. I 'spect it's easier for you now you are at camp – the waiting I mean – because it is all actual energetic work, isn't it?

Another stoppage. Betty is just dying to talk to me – I can tell by the was she keeps on looking at me and solicitously asking me how my letter is getting on.

<u>9.50</u> Betty has gone to bed. Now to your letters. Only it is so difficult to answer them because I keep on wanting to read each one over and over again as I come to it.

Yes, I do agree with you that a woman should be capable of managing business affairs. It is ever so much more sensible than being a helpless sort of person who can manage nothing for herself.

You see I don't think I had fully realised until after the dance that there ever had been a second me, but now, looking back, I see that all the time you had slowly been waking her up. Before Treasure Island weekend she was absolutely asleep. All the time since then, I have wanted you to kiss me, as I said, but I hardly knew even at the time that I did because my mind was so taken up with the things you were saying and the bits of me you were kissing. No, of course there's nothing wrong in there being a second you and me; I wasn't thinking before.

It has been Saturday for a long time, since the middle of last page. Pam and George have been playing patience on the already rather shaky table. And now George is kneeling on the back of my chair and shouting in my ear that the sun is shining and I must come down to the beach and bathe bathe bathe; I must come down to the beach and bathe; the sun is shining ---- etc. ad infinitum, so for the sake of my ear drum I'm going. This letter will be finished some time.

We didn't get anything new to go away with, Sersky particularly said not. Of course I don't mind you praising bits of me – quite the contrary, I love it. If I had ugly deformed feet because I had put them into too small high heeled shoes, you couldn't possibly worship me as you do now, simply because it wouldn't be the same me; there would be something horrid in me that let me do a thing like that. Besides I should be ashamed of my feet and I should want to hide them. But I <u>do</u> enjoy it, and it does help me to realise your attitude towards me and the quality and intensity of your love. Of course I enjoy it, and love it – everything you do.

Oh, I suppose I should let you go on seeing me and talking to me if I refused you. What is the good of talking about it now. I do wish these three weeks would go a little quicker. Three weeks is an awful long time.

I don't suppose you will have time to write very often while you are at camp, but I want you to write whenever you have time. The week after we come home (on Sept 4th) I'd rather you didn't write to me very much – not because of Daddy or the family or anything, but simply because I'd rather you didn't.

Barnes: 23rd August 1924. 8 + 11 = 19 days Whitehall Camp, Norfolk
Molly, I do think you are the most wonderful, wonderful girl in all the world, because you do seem to understand me, and realise that discipline at your loved hands means life and love to me.

I am so very sorry that you had such a rotten journey to Paris and it was perfectly ripping of you to write to me on the Paris–Calais journey, because I daresay you weren't feeling any too happy then. Those French trains are simply awful, but I should think your sickness was due more to insufficient food, and the odd things you eat; why on earth didn't Sersky give you proper meals on the trains? You can get ripping breakfasts, lunches and dinners. But I was afraid you would have a rotten time arriving in Paris. You poor poor Childie how I wish I could have been there to look after you.

I'm sitting in my tent, and its too dark to write any more.

Later. I have just been down to the N.A.A.F.I. (called "Naffy") (Navy, Army and Air Force Institute) a big marquee in the camp, to get some supper, and am now lying on my blankets on the floor in my tent, writing by means of a candle stuck on a mess tin. Today I am Brigade and Camp Orderly Sergeant, and have to go round all the guards between 11 and 12 tonight with the Orderly Officer. Our day is arranged like this -

6.30   Reveille (always called "Revally", I wonder why they give things French names)
6.45   First Parade and Roll Call and physical training.
8.0   Breakfast
9.0   Second (and ceremonial) parade followed by Gun Drill etc.
1.0   dinner
2.0   Third Parade, more gun drill etc.
5.0   Tea

Tea consisting of bread and margarine and jam, and perhaps some potted meat, is the last official meal of the day!

Sunday. 24th August 7 + 11 = 18 days.

Sunday is an off day, when nothing except the necessary fatigues are done. But alas its simply pouring with rain, and has been doing so ever since we got up. I've got a most awful cold, and so have lots of other people, from exposure and wet, but I feel all right otherwise. I'm awfully sorry Childie dear that I haven't had time to write since last Tuesday – We've been working very hard, and owing to the awful weather haven't been able to do nearly so much firing as we ought. And in consequence we have remained on the gun station, about 2 miles away, much later than we ought, in the hopes that clouds would go, so that we could see the 'plane flying overhead to fire at. So that we haven't been getting back till 7 or 8 for the evening meal; and then all the sergeants have to have a meeting to arrange about duties for the next day – often we are not free till after 9, when I'm just too tired to do anything except make my bed and turn in.

Its queer that you should have learnt more of me from letters than you would if we had been able to talk. For I find it much harder to write things than to say them. But of course, you are able to read things again and think over them, so that perhaps a letter makes more impression than talking does.

<u>In bed, Sunday night</u> Just a line, my own dearest Childie, to wish you goodnight. I am in bed, with the guttering candle, and the rain pattering on the tent again. Its been a perfectly awful day, raining practically without intermission, from early morning till now, tho' there was a short interval about 7 oclock.

I went to help the Sergeant Major Instructor do accounts after tea, and then went for a walk in the fine interval with him, and then we had some bacon and eggs at a little cottage for supper – the first time I have been outside the camp.

This is a dull letter – somehow there seems nothing worth telling you, and all I ever want to write about is you, and how much I love you, and how I long and long to be with you. Somehow I always feel so lonely now – there's only one person in the world with whom I wish to be and that is you, Molly. If only I could be for ever with you, Childie dear. Goodnight my own dear Mistress, I do hope you are happy and having a good holiday – I will finish this tomorrow. Oh, I am going up flying in the target aeroplane tomorrow, while the other battery in camp here are firing at it. Quite exciting. Goodnight my Mistress.

<u>Monday 25th 6 + 11 = 17 days</u>

Your darling letter [22nd August] was waiting for me when I returned from flying this morning. Thank you ever and ever so much for it. Yes, I expect you were quite overwhelmed by your family at first.

Do you remember my starting a letter while you were abroad in which I said that I had asked Aunty May if it mattered my utterly displaying myself and my love, without reserve, to you? I dont think I ever finished the remark but she said it didn't – I could give you every single atom of me to do with as you pleased – not quite in those words of course. But I rather gathered from her that whereas I could love you with complete abandonment without a single reserve, you were better only to allow me to approach you as your equal as a privilege – I think a woman <u>does</u> keep something in reserve, however much she loves a man, so that he may always feel he must <u>work</u> for her love, just as he did when he was trying to win her. Dont you think so?

I must stop, Childie dear, as I have to go to the Mess, as they are entertaining the Officers tonight, and tomorrow we are firing very early, and I shall have no time. I'm simply longing to be back, it will mean the beginning of the last stages. Certainly, I will not bother you that last week.

Barnes: 27th August 1924. 4 + 11 = 15 days Whitehall Camp, Norfolk.
I do hope you will have received a letter which I finished yesterday, and gave to a gunner to post, when I was in a hurry. As before I cannot now remember his face – I do wish I had a better memory for names and faces for it is a real gift.

My cold is heaps better, tho' the weather has not improved very much; but we are all a good deal cheered by having done quite a lot of firing yesterday morning – not at all bad either. One is kept very much on the strain, because we are not told which shoots count for the cup competition, and there is a perfect army of people observing ones every movement, and taking marks. We were to have done night shooting as well yesterday, and paraded at half past 8, but the weather was not suitable, great heavy low clouds and rain. We are firing again this afternoon, if the weather permits.

Childie dear, you are ripping not to have minded that rotten letter. Thank you ever so much. The time <u>does</u> go somehow, tho' I never dreamed that it could drag like this. There's one – rather mathematical – consolation however. Seventy days ago, when a day passed, it was only 1/70th of the time; whereas now when a day goes it is 1/15th of the time gone. Dear dy/dx is getting bigger and bigger. But on the other hand I think each day seems about four times as long. I'm sorry Molly, I didn't mean to make you talk about what would happen if you refused me, but I can't help worrying, can I? You see, I dont know one bit what your answer is going to be.

<u>28th 3 + 11 = 14 days</u> I never, even in my most inmost consciousness think or feel that you have yet made up your mind. I suppose that you, in

whose hands everything rests, do not realise that poor me is still in actual suspense – I really truly do not know whether on your birthday I am going to be accepted by you or not. This past six or 8 weeks have been a great strain to me – no one could tell at all how your feelings would change, and your outlook develope when completely separated from me, and surrounded by novelty and fresh experience on every hand. The mere fact that you wish to have the final week free from all interruption, so that you may come to a calm and free decision only serves to emphasise that my fate still hangs in the balance. No one, especially myself, would wish it otherwise dear Molly, for your sweet sake. As Aunt Nellie said two years ago, it has been a scorching trial for me, but if you are going to marry a man so infinitely unworthy of you, it is only fair that he should be made to prove his devotion, and show that his love was no mere affair of the moment. But the fact that I have had to wait infers, – as I have so often tried to impress on you – no obligation on you whatever; you are as free as ever you were to say "no" or "wait", if you dont yet feel able to say "yes".

And whether you say "no" or "wait" will make no difference to me – I wont take "no", but shall go on waiting in either case, even if you forbid me to see you and write to you.

Put perhaps in the worst possible way, we can tabulate the matter thus –

Dont accept me because:-

1. You have kept me waiting.

2. You may feel as if you had given me encouragement by writing and seeing me.

3. You may hesitate to keep me waiting any longer. I will be true and faithful, and cheerful however long you may keep me, – you have my faithful word for this, so do not hesitate to do with me just as you will.

I know I have said all this before; I have never pleaded with you at all. My view is this, that you know I love you, the sole question is whether you love me; and being plaintive will not help you to come to a decision, but only clouds the issue.

29/8/24 2 + 11 = 13 days.

An awful day, – it has poured and poured ever since early morning, and it is now 12 oclock. We are busy beginning to pack stores etc. ready for the move to town on Sunday. All the tent boards near the opening are soaked and muddy, and the tent is leaking badly where people push in and out.

Molly: 29th August 1924. Paignton
Yes I remember in a letter a long time ago you began saying that about Aunty May, and never finished it. I meant to ask you what she had said, but I forgot. I quite agree with her, and I had thought that – not quite in those words – a long time ago, I mean that about a woman keeping something in reserve.

Do you know there was a column in The Times this morning about your camp at Holme-next-the-sea; I expect Uncle has told you about it. It is about flying in an aeroplane while the search-light is trying to find it. You are lucky being able to go in one and have the people down below trying to fire at you. But do they use real ammunition sort of stuff, because what happens if they hit you? And isn't it wonderful to think you can have huge mechanical ears which will hear the aeroplane. Do you enjoy flying as the man who wrote the article in The Times does? What does T.A. stand for after the R.G.A.? R.G.A. is Royal Garrison Artillery isn't it?

Barnes: 30th August 1924. 1 + 11 = 12 days Whitehall Camp, Norfolk.
Just at the moment your poor Sergeant is jolly damp, for it has rained and rained, and my clothes are so wet right thro' that when I get up, they simply stick to me, and I have to pull them away! On the whole, its been a wretched camp, and I shall not be sorry to leave tomorrow. We get to Liverpool Street Station about half past three on Sunday afternoon. But I'm awfully fit, and my cold quite gone.

We are having camp sports at Hunstanton this afternoon, and after that a dinner at one of the hotels, for the officers and senior sergeants which I shall enjoy, I think.

Dont you think its funny, that I who have to do nothing but give orders to grown men, who must obey me, should be myself subject to a girl! Oh, I <u>do</u> so hope you are not going to prove feeble.

Barnes: 1st September 1924. 11 – 1 = 10 days. New Cross.
Yes, I saw the article in the Times. When they are firing at an aeroplane, they use a reduced charge of cordite in the gun, and also use special deflection dials, so that the shell bursts in the air, about 2 or 3 thousand feet below the plane, but in a straight line, so that you can see whether, with proper charge and proper dials the shot would have told.

Tomorrow night I am dining with Burney, and may not have time to write, and on Wednesday I will send my letter to Hampstead to welcome you home.

I will not write after that, till I write for your birthday.

T.A. stands for Territorial Army. As I expect you know the English regular (or permanent) army is quite a small one, and in order to compete with the vast Continental conscript armies, they had first the Volunteers, which later were transformed by a clever scheme of Lord Haldane's into the Territorial Army, enabling the army to be enormously expanded in time of war, by men who normally carry on with their civilian jobs and so cost the country nothing – or practically nothing; as they are not paid for all the training they do in the evenings, but only when they go into camp.

No, I dont enjoy flying much. When first I started I did, because then one is excited, but when one has done a good deal its just the most utterly boring and monotonous job you can imagine.

Good night, my Childie dear. All my dearest love is with you always, my own sweet Molly.

Barnes: 2nd September 1924. 11 – 2 = 9 days Vickers House, Westminster.
Childie Darling, just a moment snatched to write you a line, before I go off to dinner with Burney, to tell you what I cannot imagine why you dont get tired of hearing, that I love you, Dear Little One with all my heart and soul. Actually, it is not easy to write it every day of the week, however strongly one feels it, because I cant help remembering that I said exactly the same things yesterday and the day before and the day before that, and so on.

I s'pose if I were better educated, and a poetical sort of fellow, I might find all sorts of beautiful ways of saying I love you. But as it is, I cant seem to think of anything else to say; I mean, to feel my natural and genuine self, for if I made up different things it wouldn't be my natural thoughts and way of expressing myself.

Interrupted by Burney – no time for more, Childie, I love you and love you and love you, dear One.

♦ There were two more letters before the week of silence. Molly struggled to maintain the 'weathery' cheer which had been hers for so long, but at the very end was reduced to a strangled silence. Barnes wrote a letter poetic in his own way, and with a humble nobility. They had both kept faith to the end, whatever the end was to be.

Barnes: 3rd September 1924. 11 – 3 = 8 days New Cross
My darling Childie,
    What am I to say to you in this, the last letter I shall write in this period of trial and waiting? Another week may mark the beginning of

unbelievable happiness for me, or if you so decide the beginning of yet another time of patient and determined waiting. Whichever it may be Dear One, may God give me grace to live and work for your happiness and good, and not for my own. By nature Molly, I am I fear a selfish and selfcentred man. I know that during the past two years I have not done all I might to help you to come to a perfectly unbiassed decision. But what lover worth the name could entirely resist the temptation to show his love, when on his Beloved the whole world seems to depend? I suppose it has been selfish of me to let you see how much I loved you; and to glory in the exquisite joy of trying to serve you and please you.

Strong to a certain point I can be, but strong enough to stand calmly by and allow you to be taken out of my life I was not.

You only can be my judge; if the time were all to be lived over again, I would still fight every inch of the way to try to win you, dear Loved One. So I suppose I'm still selfish. Anyhow if being selfish means winning you, then I'm jolly thankful I <u>was</u> born selfish.

My own Childie, you know that with all my heart and soul I love and worship you. In this quiet week I will daily pray, as I so often have in the past that you may be guided to a wise and right decision.

Fear nothing, least of all me. If you feel you dont want me, I shall understand. When my heart yearns so for you, could I wish you any ill?

Good night, and God bless you Dear One.

Your loving devoted

Barnes.

Molly: 3rd September 1924. Paignton
My dear Barnes,

This is the last letter I shall write you. I have had five letters and a card since my last letter to you, and thank you very much for them. I understand now why the 'planes aren't hurt when you shoot for practice. Who got the cup this year?

This evening after tea, which we had on the beach, Pam, George and I had a great time, making lakes and canals and waterfalls, and afterwards we made a great huge castle with a wall to keep out the tide, only the tide wouldn't be kept out and in the end we had to stand on the castle and be surrounded. Baba is too grown-up to do things like that, and Betty had had her foot stung by a jelly fish or something during the last bathe so she had to keep it up, and Nancy is bored by sand digging, so we three were left to enjoy ourselves.

354 MATHEMATICS WITH LOVE

Yesterday evening Nan and Baba and Betty and I went onto the beach when it was quite dark and they sat by our tent and I took off my shoes and stockings and let the waves come up and kiss my feet. It is more or less at the end of the beach where it is very dark and quiet and where there are no people. You didn't know a wave was coming until you saw a line of white appearing along the edge of the sea, and felt the water coming over your feet. Some waves were feeble and didn't dare come anywhere near my feet, some were a little braver and just kissed my toes and the underneath part, and the big bold ones enveloped them in a great rough hug. I love the sea at night time; it is so dark and mysterious and yet at the same time you think of it as very kind and loving because it is so big and untiring.

I loved your little bit of a letter. Just think, s'posing you were "a poetical sort of a fellow", how hateful it would be. No, I'd rather have it said just exactly the same over and over again. Though you might think it'd be monotonous, it never is in the least – quite the contrary.

Bedtime. I shan't get a chance of writing in the train to-morrow – much too big a scrum. I will finish this after we get home.

4th September 1924. Hampstead.
Barnes dear, thank you for your letter.

I had meant to write heaps more, but though there is no lack of time, I just can't say anything else. Goodbye for a week.

Molly.

Barnes: 11th September 1924. 11 – 11 = 0 days R.A.F.Club, Piccadilly.
9.15 pm = 12 + $1\frac{3}{4}$ = $13\frac{3}{4}$ hours.
My dearest Molly,

Just a brief line to greet you on your birthday, and to wish you many very, very happy returns of the day. I had meant to send you some flowers, but didn't leave my office till 8 p.m. and then my shop was closed. I'm so sorry Childie dear.

I've had an awful rush – funny thing how everything seems to come at once, and I have been trying to clear things up a little so that I could be with you tomorrow. And then do you know, after all these months, Burney came dashing in this afternoon and said "Wallis, at <u>11</u> oclock tomorrow I want you to come with me (and two other directors) to Crayford, to arrange about our accommodation there".

Year, month, day and hour, – you can imagine my feelings – anyhow I told him I had a most important private appointment of long standing, and arranged for Teed to go instead.

And then I was so tired and hungry, I came here for dinner.

I will call on you as near 11 oclock as I can get, as I shall have to go to work first. You <u>can</u> arrange for me to see you alone straight away, cant you? because I simply cannot face the family until I know. (I'm all shaky now).

Molly, I <u>couldn't</u> not write you this note – I know its disobeying you. Forgive me, Dear One. Ever your loving cousin, Barnes.

♦ On September 12th, her 20th Birthday, Molly gave Barnes her answer, and the two were formally engaged. Later, in a little leather-bound notebook, Barnes commemorated the highlights of their history to that time.

Molly and I met – St George's Day, Sunday April 23rd 1922.
I proposed to her Thursday December 21st 1922.
Molly accepted me Friday September 12th 1924.

And later he added:

And we were married on St George's Day Thursday April 23rd 1925, at St Lukes Church, Hampstead.

And above these entries is a motto, cut out and pasted in:

IT'S DOGGED AS DOES IT.

Dogged determination had won for the knight his lady, and for the lady the right to declare her love. Auntie Nellie, who cherished both the young people as her own children, had written to Barnes early in the year in praise of "doggedness"; her advice had inspired him over the months.

Now put your longing in your pocket, and remember no abstaining is too hard for her sake. And if Arthur is a wet blanket – <u>you</u> know and <u>we</u> know, that neither many wet blankets nor many waters can quench love.

As I told you long ago – doggedness does it. True, valiant, patient – I'm sure you are made of all three – but the last is the almost impossible virtue, and only the nobility in love have it. It means the "possession of one's soul" – what a jewel of a thing one's soul is to possess!

Keep your jewel well cut and polished to put on her fair white neck, after the ordeal is over. Won't she wear it proudly? and, I truly believe, worthily.

> Goodbye, you poor, roughly-treated piece of gold – keep your bright-
> ness, for her, and through her, for the world.

◆ No one in either family was surprised at the news that broke on September 12th and all, with the exception of Molly's father, were delighted. A
celebration dinner was held at the R.A.F. Club four days later, where both
sets of parents, Barbara and Barnes's brother John, and Molly and Barnes
themselves, signed the menu. Barnes's old friends, who had regarded him
as the archetypal "cave-dwelling bachelor", never to be caught by any
woman, rejoiced with amazement. Dr and Mrs Boyd, who had observed
his every moment when sharing holidays, wrote in the forceful language
of brisk comradeship, as was their wont.

> If ever there existed a real dirty dog in this world I think its name was
> Wog. Just fancy you sly old devil, keeping this so infernally quiet and
> here have I been working my fingers to the bone in order to find you a
> wife and all the time – really Wog its too bad and I shall not forgive you
> until Ive made "Mollies" acquaintance. Anyway I think I can safely
> claim that amongst all your old pals there is not one more glad than I
> am.

And in more decorous female tones:

> You have absolutely taken our breaths away!! Just fancy putting all
> your voluminous correspondence down to business! Probably the book
> you were writing last year at Borth was "business" too! I am most
> awfully glad to hear about it, Woggins, and from the bottom of my
> heart, wish you both every happiness.

Other rewards followed in due time. Mr Bloxam soon came to appreciate
Barnes's qualities, and to accord him respect and affection, and he and his
daughter were never estranged. The resentment against his authoritari-
anism which both Barnes and Molly had felt was soon ended, while his
standards were incorporated with their own. Barnes and Molly continued
to write long and loving letters whenever they were apart. Four children
blessed their union in Prayer Book style; the religious principles in which
Barnes and Molly had been brought up surrounded the family, together
with the Wallis sharp and imaginative humour and the Bloxam literary
bent and lively games. The airship R100 sailed across the sky from her
native Yorkshire to Canada and back, and Barnes's interests turned to

other designs. The Second World War brought almost unbearable strains. Barnes and Molly mourned the loved and lost, and protected their own brood. As with Molly's own family, marriage was the intrusion which took the young from the nest, and began to fill the vacant spaces with the next generation. But the greatest reward to both of their steadfastness, loyalty and good faith was the united and truly loving marriage of more than fifty years which supported them to the end of their days.

INDEX